ENVIRONMENTAL ENGINEERING

Fourth Edition

ENVIRONMENTAL ENGINEERING
Fourth Edition

Ruth F. Weiner
Department of Nuclear Engineering and Radiation Sciences
University of Michigan
Ann Arbor, MI

and

Robin A. Matthews
Huxley College of Environmental Studies
Western Washington University
Bellingham, WA

An Imprint of Elsevier
www.bh.com

Amsterdam Boston London New York Oxford Paris San Diego
San Francisco Singapore Sydney Tokyo

Butterworth-Heinemann is an imprint of Elsevier

Copyright © 2003, Elsevier. All rights reserved.

Permissions may be sought directly from Elsevier's Science & Technology Rights Department in Oxford, UK: phone: (+44) 1865 843830, fax: (+44) 1865 853333, e-mail: permissions@elsevier.com.uk. You may also complete your request on-line via the Elsevier Science homepage (http://elsevier.com), by selecting "Customer Support" and then "Obtaining Permissions."

∞ Recognizing the importance of preserving what has been written, Elsevier-Science prints its books on acid-free paper whenever possible.

Library of Congress Cataloging-in-Publication Data
A catalogue record for this book is available from the Library of Congress.
International Standard Book Number: 0750672943

British Library Cataloguing-in-Publication Data
A catalogue record for this book is available from the British Library.

The publisher offers special discounts on bulk orders of this book.
For information, please contact:

Manager of Special Sales
Elsevier
200 Wheeler Road
Burlington, MA 01803
Tel: 781-313-4700
Fax: 781-313-4882

For information on all Butterworth-Heinemann publications available, contact our World Wide Web home page at: http://www.bh.com

04 05 06 07 08 / 9 8 7 6 5 4 3 2

Printed in the United States of America

*To Hubert Joy, Geoffrey Matthews,
and Natalie Weiner*

Contents

	Preface	*xiii*
1	**Environmental Engineering**	**1**
	Civil Engineering	1
	Public Health	4
	Ecology	5
	Ethics	7
	Environmental Engineering as a Profession	10
	Organization of This Text	10
2	**Assessing Environmental Impact**	**13**
	Environmental Impact	13
	Use of Risk Analysis in Environmental Assessment	23
	Socioeconomic Impact Assessment	24
	Conclusion	29
	Problems	30
3	**Risk Analysis**	**33**
	Risk	33
	Assessment of Risk	34
	Probability	35
	Dose–Response Evaluation	38
	Population Responses	40
	Exposure and Latency	40
	Expression of Risk	41
	Risk Perception	46
	Ecosystem Risk Assessment	47
	Conclusion	47
	Problems	47
4	**Water Pollution**	**51**
	Sources of Water Pollution	51
	Elements of Aquatic Ecology	54
	Biodegradation	57

	Aerobic and Anaerobic Decomposition	58
	Effect of Pollution on Streams	60
	Effect of Pollution on Lakes	70
	Effect of Pollution on Groundwater	73
	Effect of Pollution on Oceans	75
	Heavy Metals and Toxic Substances	76
	Conclusion	76
	Problems	77
5	**Measurement of Water Quality**	**81**
	Sampling	81
	Dissolved Oxygen	82
	Biochemical Oxygen Demand	84
	Chemical Oxygen Demand	91
	Total Organic Carbon	91
	Turbidity	92
	Color, Taste, and Odor	92
	pH	92
	Alkalinity	94
	Solids	94
	Nitrogen and Phosphorus	97
	Pathogens	99
	Heavy Metals	102
	Other Organic Compounds	103
	Conclusion	104
	Problems	104
6	**Water Supply**	**107**
	The Hydrologic Cycle and Water Availability	107
	Groundwater Supplies	108
	Surface Water Supplies	115
	Water Transmission	119
	Conclusion	132
	Problems	134
7	**Water Treatment**	**135**
	Coagulation and Flocculation	135
	Settling	140
	Filtration	141
	Disinfection	150
	Conclusion	151
	Problems	151

Contents ix

8	**Collection of Wastewater**	**153**
	Estimating Wastewater Quantities	153
	System Layout	154
	Sewer Hydraulics	157
	Conclusion	164
	Problems	165

9	**Wastewater Treatment**	**167**
	Wastewater Characteristics	167
	On-site Wastewater Treatment	169
	Central Wastewater Treatment	171
	Primary Treatment	172
	Secondary Treatment	182
	Tertiary Treatment	195
	Conclusion	200
	Problems	202

10	**Sludge Treatment and Disposal**	**205**
	Sources of Sludge	205
	Characteristics of Sludges	207
	Sludge Treatment	210
	Ultimate Disposal	228
	Conclusion	230
	Problems	231

11	**Nonpoint Source Water Pollution**	**233**
	Sediment Erosion and the Pollutant Transport Process	235
	Prevention and Mitigation of Nonpoint Source Pollution	241
	Conclusion	248
	Problems	248

12	**Solid Waste**	**251**
	Quantities and Characteristics of Municipal Solid Waste	252
	Collection	254
	Disposal Options	259
	Litter	261
	Conclusion	261
	Problems	261

13	**Solid Waste Disposal**	**263**
	Disposal of Unprocessed Refuse in Sanitary Landfills	263
	Volume Reduction Before Disposal	269
	Conclusion	270
	Problems	270

14 Reuse, Recycling, and Resource Recovery — 273
- Recycling — 273
- Recovery — 274
- Conclusion — 292
- Problems — 292

15 Hazardous Waste — 295
- Magnitude of the Problem — 295
- Waste Processing and Handling — 298
- Transportation of Hazardous Wastes — 299
- Recovery Alternatives — 301
- Hazardous Waste Management Facilities — 303
- Conclusion — 310
- Problems — 311

16 Radioactive Waste — 313
- Radiation — 313
- Health Effects — 321
- Sources of Radioactive Waste — 325
- Movement of Radionuclides Through the Environment — 333
- Radioactive Waste Management — 334
- Transportation of Radioactive Waste — 337
- Conclusion — 337
- Problems — 338

17 Solid and Hazardous Waste Law — 341
- Nonhazardous Solid Waste — 342
- Hazardous Waste — 345
- Conclusion — 350
- Problems — 350

18 Meteorology and Air Pollution — 351
- Basic Meteorology — 351
- Horizontal Dispersion of Pollutants — 352
- Vertical Dispersion of Pollutants — 355
- Atmospheric Dispersion — 361
- Cleansing the Atmosphere — 368
- Conclusion — 371
- Problems — 371

19 Measurement of Air Quality — 375
- Measurement of Particulate Matter — 375
- Measurement of Gases — 377
- Reference Methods — 380

	Grab Samples	381
	Stack Samples	381
	Smoke and Opacity	382
	Conclusion	382
	Problems	382
20	**Air Pollution Control**	**385**
	Source Correction	385
	Collection of Pollutants	385
	Cooling	386
	Treatment	387
	Control of Gaseous Pollutants	399
	Control of Moving Sources	404
	Control of Global Climate Change	407
	Conclusion	407
	Problems	408
21	**Air Pollution Law**	**411**
	Air Quality and Common Law	411
	Statutory Law	413
	Moving Sources	418
	Tropospheric Ozone	419
	Acid Rain	419
	Problems of Implementation	420
	Conclusion	421
	Problems	421
22	**Noise Pollution**	**423**
	The Concept of Sound	423
	Sound Pressure Level, Frequency, and Propagation	426
	Sound Level	430
	Measuring Transient Noise	434
	The Acoustic Environment	436
	Health Effects of Noise	437
	The Dollar Cost of Noise	440
	Noise Control	441
	Conclusion	443
	Problems	444

Appendices

	A	**Conversion Factors**	**447**
	B	**Elements of the Periodic Table**	**451**

C	**Physical Constants**	**455**
D	**List of Symbols**	**457**
E	**Bibliography**	**465**
Index		*471*

Preface

Everything seems to matter in environmental engineering. The social sciences and humanities, as well as the natural sciences, can be as important to the practice of environmental engineering as classical engineering skills. Many environmental engineers find this combination of skills and disciplines, with its inherent breadth, both challenging and rewarding. In universities, however, inclusion of these disciplines often requires the environmental engineering student to cross discipline and department boundaries. Deciding what to include in an introductory environmental engineering book is critical but difficult, and this difficulty has been enhanced by the growth of environmental engineering since the first edition of this book.

The text is organized into areas important to all environmental engineers: water resources, air quality, solid and hazardous wastes (including radioactive wastes), and noise. Chapters on environmental impact assessment and on risk analysis are also included. Any text on environmental engineering is somewhat dated by the time of publication, because the field is moving and changing rapidly. We have included those fundamental topics and principles on which the practice of environmental engineering is grounded, illustrating them with contemporary examples. We have incorporated emerging issues, such as global climate change and the controversy over the linear nonthreshold theory, whenever possible.

This book is intended for engineering students who are grounded in basic physics, chemistry, and biology, and who have already been introduced to fluid mechanics. The material presented can readily be covered in a one-semester course.

The authors are indebted to Professor P. A. Vesilind of Bucknell University and Professor J. J. Peirce of Duke University, the authors of the original *Environmental Engineering*. Without their work, and the books that have gone before, this edition would never have come to fruition.

Ruth F. Weiner
Robin Matthews

Chapter 1

Environmental Engineering

Environmental engineering is a relatively new profession with a long and honorable history. The descriptive title of "environmental engineer" was not used until the 1960s, when academic programs in engineering and public health schools broadened their scope and required a more accurate title to describe their curricula and their graduates. The roots of this profession, however, go back as far as recorded history. These roots reach into several major disciplines including civil engineering, public health, ecology, chemistry, and meteorology. From each foundation, the environmental engineering profession draws knowledge, skill, and professionalism. From ethics, the environmental engineer draws concern for the greater good.

CIVIL ENGINEERING

Throughout western civilization settled agriculture and the development of agricultural skills created a cooperative social fabric and spawned the growth of communities, as well as changed the face of the earth with its overriding impact on the natural environment. As farming efficiency increased, a division of labor became possible, and communities began to build public and private structures that engineered solutions to specific public problems. Defense of these structures and of the land became paramount, and other structures subsequently were built purely for defensive purposes. In some societies the conquest of neighbors required the construction of machines of war. Builders of war machines became known as engineers, and the term "engineer" continued to imply military involvement well into the eighteenth century.

In 1782 John Smeaton, builder of roads, structures, and canals in England, recognized that his profession tended to focus on the construction of public facilities rather than purely military ones, and that he could correctly be designated a civil engineer. This title was widely adopted by engineers engaged in public works (Kirby *et al.* 1956).

The first formal university engineering curriculum in the United States was established at the U.S. Military Academy at West Point in 1802. The first engineering course outside the Academy was offered in 1821 at the American Literary, Scientific, and Military Academy, which later became Norwich University. The Renssalaer Polytechnic Institute conferred the first truly civil engineering degree in 1835. In 1852, the American Society of Civil Engineers was founded (Wisely 1974).

Water supply and wastewater drainage were among the public facilities designed by civil engineers to control environmental pollution and protect public health. The availability of water had always been a critical component of civilizations. Ancient Rome, for example, had water supplied by nine different aqueducts up to 80 km (50 miles) long, with cross sections from 2 to 15 m (7 to 50 ft). The purpose of the aqueducts was to carry spring water, which even the Romans knew was better to drink than Tiber River water.

As cities grew, the demand for water increased dramatically. During the eighteenth and nineteenth centuries the poorer residents of European cities lived under abominable conditions, with water supplies that were grossly polluted, expensive, or nonexistent. In London the water supply was controlled by nine different private companies and water was sold to the public. People who could not afford to pay for water often begged or stole it. During epidemics of disease the privation was so great that many drank water from furrows and depressions in plowed fields. Droughts caused water supplies to be curtailed and great crowds formed to wait their "turn" at the public pumps (Ridgway 1970).

In the New World the first public water supply system consisted of wooden pipes, bored and charred, with metal rings shrunk on the ends to prevent splitting. The first such pipes were installed in 1652, and the first citywide system was constructed in Winston-Salem, NC, in 1776. The first American water works was built in the Moravian settlement of Bethlehem, PA. A wooden water wheel, driven by the flow of Monocacy Creek, powered wooden pumps that lifted spring water to a hilltop wooden reservoir from which it was distributed by gravity (American Public Works Association 1976). One of the first major water supply undertakings was the Croton Aqueduct, started in 1835 and completed six years later. This engineering marvel brought clear water to Manhattan Island, which had an inadequate supply of groundwater (Lankton 1977).

Although municipal water systems might have provided adequate quantities of water, the water quality was often suspect. One observer noted that the poor used the water for soup, the middle class dyed their clothes in it, and the very rich used it for top-dressing their lawns.

The earliest known acknowledgment of the effect of impure water is found in Susruta Samhitta, a collection of fables and observations on health, dating back to 2000 BCE, which recommended that water be boiled before drinking. Water filtration became commonplace toward the middle of the nineteenth century. The first successful water supply filter was in Parsley, Scotland, in 1804, and many less successful attempts at filtration followed (Baker 1949). A notable failure was the New Orleans system for filtering water from the Mississippi River. The water proved to be so muddy that the filters clogged too fast for the system to be workable. This problem was not alleviated until aluminum sulfate (alum) began to be used as a pretreatment to filtration. The use of alum to clarify water was proposed in 1757, but was not convincingly demonstrated until 1885. Disinfection of water with chlorine began in Belgium in 1902 and in America, in Jersey City, NJ, in 1908. Between 1900 and 1920 deaths from infectious disease dropped dramatically, owing in part to the effect of cleaner water supplies.

Human waste disposal in early cities presented both a nuisance and a serious health problem. Often the method of disposal consisted of nothing more than flinging

Figure 1-1. Human excreta disposal, from an old woodcut (source: W. Reyburn, *Flushed with Pride*. McDonald, London, 1969).

the contents of chamberpots out the window (Fig. 1-1). Around 1550, King Henri II repeatedly tried to get the Parliament of Paris to build sewers, but neither the king nor the parliament proposed to pay for them. The famous Paris sewer system was built under Napoleon III, in the nineteenth century (De Camp 1963).

Stormwater was considered the main "drainage" problem, and it was in fact illegal in many cities to discharge wastes into the ditches and storm sewers. Eventually, as water supplies developed,[1] the storm sewers were used for both sanitary waste and stormwater. Such "combined sewers" existed in some of our major cities until the 1980s.

---

[1] In 1844, to hold down the quantity of wastewater discharge, the city of Boston passed an ordinance prohibiting the taking of baths without doctor's orders.

4   ENVIRONMENTAL ENGINEERING

The first system for urban drainage in America was constructed in Boston around 1700. There was surprising resistance to the construction of sewers for waste disposal. Most American cities had cesspools or vaults, even at the end of the nineteenth century. The most economical means of waste disposal was to pump these out at regular intervals and cart the waste to a disposal site outside the town. Engineers argued that although sanitary sewer construction was capital intensive, sewers provided the best means of wastewater disposal in the long run. Their argument prevailed, and there was a remarkable period of sewer construction between 1890 and 1900.

The first separate sewerage systems in America were built in the 1880s in Memphis, TN, and Pullman, IL. The Memphis system was a complete failure. It used small pipes that were to be flushed periodically. No manholes were constructed and cleanout became a major problem. The system was later removed and larger pipes, with manholes, were installed (American Public Works Association 1976).

Initially, all sewers emptied into the nearest watercourse, without any treatment. As a result, many lakes and rivers became grossly polluted and, as an 1885 Boston Board of Health report put it, "larger territories are at once, and frequently, enveloped in an atmosphere of stench so strong as to arouse the sleeping, terrify the weak and nauseate and exasperate everybody."

Wastewater treatment first consisted only of screening for removal of the large floatables to protect sewage pumps. Screens had to be cleaned manually, and wastes were buried or incinerated. The first mechanical screens were installed in Sacramento, CA, in 1915, and the first mechanical comminutor for grinding up screenings was installed in Durham, NC. The first complete treatment systems were operational by the turn of the century, with land spraying of the effluent being a popular method of wastewater disposal.

Civil engineers were responsible for developing engineering solutions to these water and wastewater problems of these facilities. There was, however, little appreciation of the broader aspects of environmental pollution control and management until the mid-1900s. As recently as 1950 raw sewage was dumped into surface waters in the United States, and even streams in public parks and in U.S. cities were fouled with untreated wastewater. The first comprehensive federal water pollution control legislation was enacted by the U.S. Congress in 1957, and secondary sewage treatment was not required at all before passage of the 1972 Clean Water Act. Concern about clean water has come from the public health professions and from the study of the science of ecology.

PUBLIC HEALTH

Life in cites during the middle ages, and through the industrial revolution, was difficult, sad, and usually short. In 1842, the Report from the Poor Law Commissioners on an Inquiry into the Sanitary Conditions of the Labouring Population of Great Britain described the sanitary conditions in this manner:

> Many dwellings of the poor are arranged around narrow courts having no other opening to the main street than a narrow covered passage. In these courts there are several occupants, each

of whom accumulated a heap. In some cases, each of these heaps is piled up separately in the court, with a general receptacle in the middle for drainage. In others, a plot is dug in the middle of the court for the general use of all the occupants. In some the whole courts up to the very doors of the houses were covered with filth.

The great rivers in urbanized areas were in effect open sewers. The River Cam, like the Thames, was for many years grossly polluted. There is a tale of Queen Victoria visiting Trinity College at Cambridge, and saying to the Master, as she looked over the bridge abutment, "What are all those pieces of paper floating down the river?" To which, with great presence of mind, he replied, "Those, ma'am, are notices that bathing is forbidden" (Raverat 1969).

During the middle of the nineteenth century, public health measures were inadequate and often counterproductive. The germ theory of disease was not as yet fully appreciated, and epidemics swept periodically over the major cities of the world. Some intuitive public health measures did, however, have a positive effect. Removal of corpses during epidemics, and appeals for cleanliness, undoubtedly helped the public health.

The 1850s have come to be known as the "Great Sanitary Awakening." Led by tireless public health advocates like Sir Edwin Chadwick in England and Ludwig Semmelweiss in Austria, proper and effective measures began to evolve. John Snow's classic epidemiological study of the 1849 cholera epidemic in London stands as a seminally important investigation of a public health problem. By using a map of the area and identifying the residences of those who contracted the disease, Snow was able to pinpoint the source of the epidemic as the water from a public pump on Broad Street. Removal of the handle from the Broad Street pump eliminated the source of the cholera pathogen, and the epidemic subsided.[2] Waterborne diseases have become one of the major concerns of the public health. The control of such diseases by providing safe and pleasing water to the public has been one of the dramatic successes of the public health profession.

Today the concerns of public health encompass not only water but all aspects of civilized life, including food, air, toxic materials, noise, and other environmental insults. The work of the environmental engineer has been made more difficult by the current tendency to ascribe many ailments, including psychological stress, to environmental origins, whether or not there is any evidence linking cause and effect. The environmental engineer faces the rather daunting task of elucidating such evidence relating causes and effects that often are connected through years and decades as human health and the environment respond to environmental pollutants.

ECOLOGY

The science of ecology defines "ecosystems" as interdependent populations of organisms interacting with their physical and chemical environment. The populations of the

---

[2]Interestingly, it was not until 1884 that Robert Koch proved that *Vibrio comma* was the microorganism responsible for the cholera.

6   ENVIRONMENTAL ENGINEERING

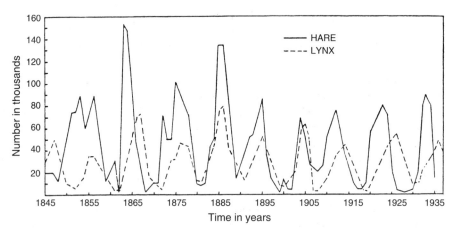

Figure 1-2. The hare and lynx homeostasis (source: D.A. MacLurich, "Fluctuations in the Numbers of Varying Hare," University of Toronto Studies, Biological Sciences No. 43, Reproduced in S. Odum, *Fundamentals of Ecology*, 3rd ed., W.B. Saunders, Philadelphia, 1971).

species in an ecosystem do not vary independently but rather fluctuate in an approximate steady state in response to self-regulating or negative feedback (*homeostasis*). Homeostatic equilibrium is dynamic, however, because the populations are also governed by positive feedback mechanisms that result from changes in the physical, chemical, and biological environment (*homeorhesis*).

Homeostatic mechanisms can be illustrated by a simple interaction between two populations, such as the hare and the lynx populations pictured in Fig. 1-2. When the hare population is high the lynx have an abundant food supply and procreate. The lynx population increases until the lynx outstrip the available hare population. Deprived of adequate food, the lynx population then decreases, while the hare population increases because there are fewer predators. This increase, in turn, provides more food for the lynx population, and the cycle repeats. The numbers of each population are continually changing, making the system dynamic. When studied over a period of time, the presence of this type of self-regulating feedback makes the system appear to be in a steady state, which we call homeostasis.

In reality, populations rarely achieve steady state for any extended period of time. Instead, populations respond to physical, chemical, and biological changes in the environment along a positive feedback trajectory that will eventually settle into a new, but again temporary, homeostasis. Some of these changes are natural (e.g., a volcanic eruption that covers the lynx and hare habitat with ash or molten rock); many are caused by humans (e.g., destruction or alteration of habitat, introduction of competing species, trapping or hunting).

Ecosystem interactions obviously can also include more than two species; consider, for example, the sea otter, the sea urchin, and kelp in a homeostatic interaction. The kelp forests along the Pacific coast consist of 60-m (200-ft) streamers fastened to the ocean floor. Kelp can be economically valuable, since it is the source of algin used in foods, paints, and cosmetics. In the late 1900s kelp began to disappear mysteriously, leaving a barren ocean floor. The mystery was solved when it was recognized that sea urchins feed on the kelp, weaken the stems, and cause them to detach and float away. The sea urchin population had increased because the population of the predators, the sea otters, had been reduced drastically. The solution was protection of the sea otter and increase in its population, resulting in a reduction of the sea urchin population and maintenance of the kelp forests.

Some ecosystems are fragile, easily damaged, and slow to recover; some are resistant to change and are able to withstand even serious perturbations; and others are remarkably resilient and able to recover from perturbation if given the chance. Engineers must consider that threats to ecosystems may differ markedly from threats to public health; for example, acid rain poses a considerable hazard to some lake ecosystems and agricultural products, but virtually no direct hazard to human health. A converse example is that carcinogens dispersed in the atmospheric environment can enter the human food chain and be inhaled, putting human health at risk, but they could pose no threat to the ecosystems in which they are dispersed.

Engineers must appreciate the fundamental principles of ecology and design in consonance with these principles in order to reduce the adverse impacts on fragile ecosystems. For example, since the deep oceans are among the most fragile of all ecosystems this fragility must be part of any consideration of ocean disposal of waste. The engineer's job is made even harder when he or she must balance ecosystem damage against potential human health damage. The inclusion of ecological principles in engineering decisions is a major component of the environmental engineering profession.

ETHICS

Historically the engineering profession in general and environmental engineering in particular did not consider the ethical implications of solutions to problems. Ethics as a framework for making decisions appeared to be irrelevant to engineering since the engineer generally did precisely what the employer or client required.

Today, however, the engineer is no longer free from concern for ethical questions. Scientists and engineers look at the world objectively with technical tools, but often face questions that demand responses for which technical tools may be insufficient. In some cases all the alternatives to a particular engineering solution include "unethical" elements. Engineers engaged in pollution control, or in any activity that impinges on the natural environment, interface with environmental ethics.[3] An environmental ethic concerns itself with the attitude of people toward other living things and toward the natural environment, as well as with their attitudes toward each other. The search

---

[3] See, for example, *Environmental Ethics*, a professional journal published quarterly by the University of Georgia, Athens, GA.

for an environmental ethic raises the question of the origin of our attitude toward the environment.

It is worth noting that the practice of settled agriculture has changed the face of the earth more than any other human activity; yet the Phaestos Disk—the earliest Minoan use of pictographs—elevates to heroism the adventurer who tries to turn North Africans from hunting and gathering to settled agriculture. The tradition of private ownership of land and resources, which developed hand-in-hand with settled agriculture, and the more recent tradition of the planned economies that land and resources are primarily instruments of national policy have both encouraged the exploitation of these resources. Early European settlers arriving in the New World from countries where all land was owned by royalty or wealthy aristocrats considered it their right to own and exploit land.[4] An analogous situation occurred with the Soviet development of Siberia and the eastern lands of the former Soviet Union (now the Russian Federation): land once under private ownership now belonged to the state. Indeed, in both America and Russia, natural resources appeared to be so plentiful that a "myth of superabundance" grew in which the likelihood of running out of any natural resource, including oil, was considered remote (Udall 1968). These traditions are contrary to the view that land and natural resources are public trusts for which people serve the role of stewards.

Nomadic people and hunter-gatherers practiced no greater stewardship than the cultures based on settled agriculture. In the post-industrial revolution world, the less industrialized nations did less environmental damage than industrialized nations only because they could not extract resources as quickly or efficiently. The Navajo sheepherders of the American southwest allowed overgrazing and consequent erosion and soil loss to the same extent as the Basque sheepherders of southern Europe. Communal ownership of land did not guarantee ecological preservation.

Both animistic religion and early improvements in agricultural practices (e.g., terracing, allowing fallow land) acted to preserve resources, particularly agricultural resources. Arguments for public trust and stewardship were raised during the nineteenth century, in the midst of the ongoing environmental devastation that followed the industrial revolution. Henry David Thoreau, Ralph Waldo Emerson, and later John Muir, Gifford Pinchot, and President Theodore Roosevelt all contributed to the growth of environmental awareness and concern. One of the first explicit statements of the need for an environmental ethic was penned by Aldo Leopold (1949). Since then, many have contributed thoughtful and well-reasoned arguments toward the development of a comprehensive and useful ethic for judging questions of conscience and environmental value.

Since the first Earth Day in 1970 environmental and ecological awareness has been incorporated into public attitudes and is now an integral part of engineering

---

[4]An exception is found in states within the boundaries of the Northwest Purchase, notably Wisconsin. Included in the Purchase agreement between France and America is the condition that state constitutions must ensure that the water and air must be held in trust by the state for the people for "as long as the wind blows and the water flows." Wisconsin provides a virtually incalculable number of public accesses to lakes and rivers.

processes and designs. Environmental awareness and concern became an essentially permanent part of the U.S. public discourse with the passage of the National Environmental Policy Act of 1970. Today, every news magazine, daily newspaper, and radio and TV station in the United States has staff who cover the environment and publish regular environmental features. Candidates for national, state, and local elective office run on environmental platforms. Since passage of the National Environmental Policy Act, no federal public works project is undertaken without a thorough assessment of its environmental impact and an exploration of alternatives (as is discussed in the following chapter). Many state and local governments have adopted such requirements as well, so that virtually all public works projects include such assessments. Engineers are called on both for project engineering and for assessing the environmental impact of that engineering. The questions that engineers are called upon to answer have increased in difficulty and complexity with the development of a national environmental ethic.

The growing national environmental ethic, coupled (unfortunately) with a general lack of scientific understanding, is at the root of public response to reports of "eco-disasters" like major oil spills or releases of toxic or radioactive material. As a result, this public response includes a certain amount of unproductive hand-wringing, occasional hysteria, and laying of blame for the particular disaster. The environmental engineer is often called on in such situations to design solutions and to prevent future similar disasters, and is able to respond constructively.

In recent years, and particularly after the accident at the Three Mile Island nuclear plant in 1979, the release of methyl isocyanate at the chemical plant in Bhopal, India, in 1984, and the disastrous nuclear criticality and fire at the Chernobyl nuclear power plant in 1986, general appreciation of the threats to people and ecosystems posed by toxic or polluting substances has increased markedly. In 1982 the U.S. Environmental Protection Agency (EPA) began to develop a system of "risk-based" standards for carcinogenic substances. The rationale for risk-based standards is the theory, on which regulation is based, that there is no threshold for carcinogenesis. The U.S. Nuclear Regulatory Commission is also considering risk-based standards. As a result of the consequent increase of public awareness of risk, some members of the public appear to be unwilling to accept any risk in their immediate environment to which they are exposed involuntarily. It has become increasingly difficult to find locations for facilities that can be suspected of producing any toxic, hazardous, or polluting effluent: municipal landfills, radioactive waste sites, sewage treatment plants, or incinerators. Aesthetically unsuitable developments, and even prisons, mental hospitals, or military installations, whose lack of desirability is social rather than environmental, are also difficult to site. Popper (1985) refers to such unwanted facilities as *l*ocally *u*ndesirable *l*and *u*ses, or LULUs.

Local opposition to LULUs is generally focused on the site of the facility and in particular on the proximity of the site to the residences of the opponents, and can often be characterized by the phrase "not in my back yard." Local opponents are often referred to by the acronym for this phrase, NIMBY. The NIMBY phenomenon has also been used for political advantage, resulting in unsound environmental decisions. The environmental engineer is cautioned to identify the fine line between real concern about environmental degradation and an almost automatic "not in my back yard" reaction. He or she recognizes, as many people do not, that virtually all human activity entails some environmental alteration and some risk, and that a risk-free environment is impossible to achieve. The balance between risk and benefit to various segments of the population often involves questions of environmental ethics.

Is it ethical to oppose a particular location of an undesirable facility because of its proximity to ecologically or politically sensitive areas, rather than working to mitigate the undesirable features of the facility? Moreover, is it ethical to locate such a facility where there is less local opposition, perhaps because employment is needed, instead of in the environment where it will do the least damage? The enactment of pollution control legislation in the United States has had a sort of NIMBY by-product: the siting of U.S.-owned plants with hazardous or toxic effluents, like oil desulfurization and copper smelting, in countries that have little or no pollution control legislation. The ethics of such "pollution export" deserve closer examination than they have had.

ENVIRONMENTAL ENGINEERING AS A PROFESSION

The general mission of colleges and universities is to allow students to mature intellectually and socially and to prepare for careers that are rewarding. The chosen vocation is ideally an avocation as well. It should be a job that is enjoyable and one approached with enthusiasm even after experiencing many of the ever-present bumps in the road. Designing a water treatment facility to provide clean drinking water to a community can serve society and become a personally satisfying undertaking to the environmental engineer. Environmental engineers now are employed in virtually all heavy industries and utility companies in the United States, in any aspect of public works construction and management, by the EPA and other federal agencies, and by the consulting firms used by these agencies. In addition, every state and most local governments have agencies dealing with air quality, water quality and water resource management, soil quality, forest and natural resource management, and agricultural management that employ environmental engineers. Pollution control engineering has also become an exceedingly profitable venture.

Environmental engineering has a proud history and a bright future. It is a career that may be challenging, enjoyable, personally satisfying, and monetarily rewarding. Environmental engineers are committed to high standards of interpersonal and environmental ethics. They try to be part of the solution while recognizing that all people including themselves are part of the problem.

ORGANIZATION OF THIS TEXT

The second chapter, Assessing Environmental Impact, gives an overview of the tools needed to assess environmental impact of engineering projects. The concept of risk is introduced and coupled with the concept of environmental ethics. The third chapter deals with some details of risk analysis. The text that follows is organized into four sections:

- Water resources, water quality, and water pollution assay and control
- Air quality and air pollution assay and control
- Solid waste: municipal solid waste, chemically hazardous waste, and radioactive waste
- Noise

These sections include parts of chapters that deal with the relevant pollution control laws and regulations. All sections include problems to be addressed individually by the reader or collectively in a classroom setting.

Chapter 2

Assessing Environmental Impact

Environmental engineering requires that the impact and interaction of engineered structures on and with the natural environment be considered in any project. In 1970 this principle was enacted into federal law in the United States. The National Environmental Policy Act (NEPA) requires that environmental impact be assessed whenever a federal action will have an environmental impact, as well as requiring that alternatives be considered. Many states have enacted similar legislation to apply to state or state-licensed actions. Since 1990, a number of "programmatic" environmental impact statements have been drafted.

Environmental impact of federal projects currently is performed in several stages: environmental assessment, a finding of no significant impact (FONSI) if that is appropriate, an environmental impact statement (if no FONSI is issued), and a record of the decision (ROD) made following the environmental assessment. In this chapter we consider the methods for making an environmental assessment as well as introducing the economic and ethical implications of environmental engineering.

Engineers ideally approach a problem in a sequence suggested to be rational by the theories of public decisionmaking: (1) problem definition, (2) generation of alternative solutions, (3) evaluation of alternatives, (4) implementation of a selected solution, and (5) review and appropriate revision of the implemented solution. This step-by step approach is essentially the NEPA process defined by the federal and state governments. This chapter presents an overview of environmental impact analysis. The specific analytical tools as well as the specific impacts and mitigation measures are discussed in detail in the other chapters of the book. However, impact assessment provides an integrated view of the problems of environmental engineering.

ENVIRONMENTAL IMPACT

On January 1, 1970, President Richard Nixon signed NEPA into law, setting a national policy to encourage "productive and enjoyable harmony" between people and their environment. This law established the Council on Environmental Quality (CEQ), which monitors the environmental effects of all federal activities, assists the President in evaluating environmental problems, and determines solutions to these problems.

14   ENVIRONMENTAL ENGINEERING

However, few people realized in 1970 that NEPA contained a "sleeper," Section 102(2)(C), that requires federal agencies to evaluate with public input the consequences of any proposed action on the environment:

> Congress authorizes and directs that, to the fullest extent possible: (1) the policies, regulations, and public laws of the United States shall be interpreted and administered in accordance with the policies set forth in this chapter, and (2) all agencies of the Federal Government shall include in every recommendation or report on proposals for legislation and other major Federal actions significantly affecting the quality of the human environment, a detailed statement by the responsible official on—
>
> (i) the environmental impact of the proposed action,
> (ii) any adverse environmental effects that cannot be avoided should the proposal be implemented,
> (iii) alternatives to the proposed action,
> (iv) the relationship between local short-term uses of man's environment and the maintenance and enhancement of long-term productivity, and
> (v) any irreversible and irretrievable commitments of resources that would be involved in the proposed action should it be implemented.

In other words each project funded by the federal government or requiring a federal permit must be accompanied by an environmental assessment. This assessment results in issuance of one of three documents:

(1) *Finding of No Significant Impact (FONSI)*. Such a stand-alone finding results when potential environmental impacts are compared to a checklist of significant impacts, with the result that no significant impact can be identified.
(2) *Environmental Assessment (EA)*. A detailed assessment of potential environmental impact resulting in one of two conclusions: either the EA must be expanded to a full-scale environmental impact statement or a FONSI results from the EA.
(3) *Environmental Impact Statement (EIS)*. An EIS must assess in detail the potential environmental impacts of a proposed action and alternative actions. Additionally, the agencies must generally follow a detailed public review of each EIS before proceeding with the project or permit. It should be noted that both positive and negative impacts are included; i.e., "impact" does not imply "adverse impact."

These impact statements are assessments and contain no judgments about the positive or negative value of the project in question. An EIS publication sequence is prescribed by law. First, a draft EIS (DEIS) is issued by the appropriate federal agency. After mandated public hearings and incorporation of comments, the federal agency issues a final EIS (FEIS). A Record of Decision (ROD), which includes the final decision about the project, the alternative chosen, and any value judgments, is also issued.

The purpose of environmental assessments was not to justify or fault projects, but to introduce environmental factors into the decision-making machinery and have them discussed in public before decisions about a project are made. However, this objective is difficult to apply in practice. Alternatives may be articulated by various interest groups in and out of government, or the engineer may be left to create his or her own alternatives. In either case, there are normally one or two plans that, from the outset, seem eminently more feasible and reasonable, and these are sometimes legitimized by juggling, for example, selected time scales or standards of enforcement patterns just slightly and calling them alternatives, as they are in a limited sense. As a result, "nondecisions" are made (Bachrach and Baratz 1962), i.e., wholly different ways of perceiving the problems and conceiving the solutions have been overlooked, and the primary objective of the EIS has been circumvented. Over the past few years, court decisions and guidelines by various agencies have, in fact, helped to mold this procedure for the development of environmental impact statements.

As the environmental assessment procedure has evolved, assessment of socioeconomic impact of the project has played an increasing role. In addition to direct economic impact (number of jobs, total household income, property values, etc.), socioeconomic impact includes impacts on archaeological and historical sites, impacts on sites that have cultural significance and on cultural practices, and environmental justice impacts (assessments of excessive impacts on minority populations). As impact assessment moves into successively "softer" science, overlap with questions of ethics and values increases, and the engineer must take care to differentiate between quantitatively measurable impacts and qualitative assessments that might be influenced by value judgments. Risk assessment has also become increasingly important in environmental assessment.

This text focuses on the "hard science" and risk assessment aspects of environmental assessment. The socioeconomic aspects of environmental assessment are usually analyzed by experts in the social sciences and economics, and so are not discussed in detail. Assessment of future impacts frequently requires probabilistic risk analyses instead of deterministic analyses. Risk analysis is discussed later in this chapter and in Chap. 3.

An environmental assessment must be thorough, interdisciplinary, and as quantitative as possible. The writing of an environmental assessment involves four distinct phases: scoping, inventory, assessment, and evaluation. The first phase defines the scope or extent of the assessment. For example, if the project involves transporting construction materials to a site, the scope may or may not include the environmental impacts of that transportation. At least one public hearing is generally held on scoping.

The second phase is a cataloging of environmentally susceptible areas and activities, including socioeconomically impacted areas. The third phase is the process of estimating the impact of the alternatives, including cumulative impacts, and the impacts of a "no action" alternative. The last phase is the interpretation of these findings, which is often done concurrently with estimating impacts.

All federal agencies are currently required to assess the environmental impacts of extended projects and programs as well as individual projects under their jurisdiction. The impact assessment for an extended or multifaceted project is frequently called a generic environmental impact statement or GEIS, while an assessment for an entire program is referred to as a programmatic environmental impact statement, or PEIS.

For example, in 1980 the U.S. Department of Energy issued a GEIS on the impact of disposition of commercially generated nuclear fuel. In 1984, the Bonneville Power Administration issued a PEIS on its proposed energy conservation program.

Environmental Inventories

The first step in evaluating the environmental impact of a project's alternatives is to inventory factors that may be affected by the proposed action. Existing conditions are measured and described, but no effort is made to assess the importance of a variable. Any number and many kinds of variables may be included, such as:

1. the "ologies": hydrology, geology, climatology, anthropology, and archaeology;
2. environmental quality: land, surface and subsurface water, air, noise, and transportation impacts;
3. plant and animal life;
4. economic impact on the surrounding community: number of jobs, average family income, etc.;
5. analysis of the risks to both people and the natural environment from accidents that may occur during the life of the project; and
6. other relevant socioeconomic parameters, like future land use, expansion or diminution of the population of urban areas and exurbs, the impacts of nonresident populations, and environmental justice considerations.

Environmental Assessment

The process of calculating projected effects that a proposed action or construction project will have on environmental quality is called environmental assessment. A methodical, reproducible, and reasonable method is needed to evaluate both the effect of the proposed project and the effects of alternatives that may achieve the same ends but that may have different environmental impacts. A number of semiquantitative approaches, among them the checklist, the interaction matrix, and the checklist with weighted rankings, have been used.

Checklists are lists of potential environmental impacts, both primary and secondary. Primary effects occur as a direct result of the proposed project, such as the effect of a dam on aquatic life. Secondary effects occur as an indirect result of the action. For example, an interchange for a highway may not directly affect wildlife, but indirectly it will draw such establishments as service stations and quick food stores, thus changing land use patterns.

The checklist for a highway project could be divided into three phases: planning, construction, and operation. During planning, consideration is given to the environmental effects of the highway route and the acquisition and condemnation of property. The construction phase checklist will include displacement of people, noise, soil erosion, air and water pollution, and energy use. Finally, the operation phase will list direct impacts owing to noise, water pollution resulting from runoff, energy use, etc., and indirect impacts owing to regional development, housing, lifestyle, and economic development.

The checklist technique thus lists all of the pertinent factors; then the magnitude and importance of the impacts are estimated. Estimated importance of impact may be quantified by establishing an arbitrary scale, such as:

$0 =$ no impact
$1 =$ minimal impact
$2 =$ small impact
$3 =$ moderate impact
$4 =$ significant impact
$5 =$ severe impact

The numbers may then be combined, and a quantitative measurement of the severity of the environmental impact for any given alternative be estimated.

In the checklist technique most variables must be subjectively valued. Further, it is difficult to predict further conditions such as land-use pattern changes or changes in lifestyle. Even with these drawbacks, however, this method is often used by engineers because of its simplicity. Impact assessments of controversial projects often do not use the checklist technique because the numerical ranking implies a subjective judgment by the environmental assessment team. A checklist remains a convenient method for developing a FONSI, although a FONSI requires subjective selection of the number judged to be the lowest value of significance.

EXAMPLE 2.1. A landfill is to be placed in the floodplain of a river. Estimate the impact by using the checklist technique.

First the items to be impacted are listed; then a quantitative judgement concerning both importance and magnitude of the impact is made. In Table 2-1, the items are only a sample of the impacts one would normally consider. The importance and magnitude are then multiplied and the sum obtained.

Table 2-1.

Potential impact	Importance × magnitude
Groundwater contamination	$5 \times 5 = 25$
Surface contamination	$4 \times 3 = 12$
Odor	$1 \times 1 = 1$
Noise	$1 \times 2 = 2$
Jobs provided	$-2 \times 3 = -6$
Total	34

This total of 34 may then be compared with totals calculated for alternative courses of action. Note that jobs are a positive impact, as distinct from the negative environmental

18   ENVIRONMENTAL ENGINEERING

impacts of the other variables, and is arbitrarily assigned a negative value, so that its impact is subtracted from the other impacts.

In order to determine a FONSI in this example, a value that indicates the lower limit of significance can be assigned to either importance or magnitude, or to their product. Thus, if the absolute value of the product of importance and magnitude of any variable is considered significant only if that product is 10% of the total, odor and noise would not be deemed significant and need not be considered further in assessing environmental impact. Alternatively, if the absolute value of the product of importance and magnitude of any variable is considered significant only if that product is greater than 5, the same conclusion is reached.

The interaction matrix technique is a two-dimensional listing of existing characteristics and conditions of the environment and detailed proposed actions that may affect the environment. This technique is illustrated in Example 2.2. For example, the characteristics of water might be subdivided into:

- surface
- ocean
- underground
- quantity
- temperature
- groundwater
- recharge
- snow, ice, and permafrost.

Similar characteristics must also be defined for air, land, socioeconomic conditions, and so on. Opposite these listings in the matrix are lists of possible actions. In our example, one such action is labeled resource extraction, which could include the following actions:

- blasting and drilling
- surface extraction
- subsurface extraction
- well drilling
- dredging
- timbering
- commercial fishing and hunting.

The interactions, as in the checklist technique, are measured in terms of magnitude and importance. The magnitudes represent the extent of the interaction between the environmental characteristics and the proposed actions and typically may be measured. The importance of the interaction, on the other hand, is often a judgment call on the part of the engineer.

If an interaction is present, for example, between underground water and well drilling, a diagonal line is placed in the block. Values may then be assigned to the interaction, with 1 being a small and 5 being a large magnitude or importance, and these

Assessing Environmental Impact   19

are placed in the blocks with the magnitude above and importance below. Appropriate blocks are filled in, using a great deal of judgement and personal bias, and then are summed over a line, thus giving a numerical grade for either the proposed action or environmental characteristics.

EXAMPLE 2.2. Lignite (brown) coal is to be surface-mined in the Appalachian Mountains. Construct an interaction matrix for the water resources (environmental characteristics) vs resource extraction (proposed actions).

Table 2-2.

Environmental characteristics	Proposed action							Total
	Blasting + drilling	Surface excavation	Subsurface excavation	Well drilling	Dredging	Timbering	Commercial fishing	
Surface water	3/2	5/5						8/7
Ocean water								
Underground water		3/3						3/3
Quantity								
Temperature		1/2						1/2
Recharge								
Snow, ice								
Total	3/2	9/10						12/12

In Table 2-2, we see that the proposed action would have a significant effect on surface water quality, and that the surface excavation phase will have a large impact. The value of the technique is seen when the matrix is applied to alternative solutions. The individual elements in the matrix, as well as row and column totals, can be compared.

Example 2.2 is trivial, and cannot fully illustrate the advantage of the interaction technique. With large projects having many phases and diverse impacts, it is relatively easy to pick out especially damaging aspects of the project, as well as the environmental characteristics that will be most severely affected.

The search for a comprehensive, systematic, interdisciplinary, and quantitative method for evaluating environmental impact has led to the checklist-with-weighted-rankings technique. The intent here is to use a checklist as before to ensure that all aspects of the environment are covered, as well as to give these items a numerical rating in common units.

An EA or EIS is usually organized into the following sections:

Introduction

The introduction provides an overview of the proposed project, alternative actions, and the assessment methods that will be used. It includes a statement of purpose: why the assessment is being done. It often includes a summary of the most critical and important results of the assessment. The introduction can often serve as an executive summary of the EA or EIS.

Description of the Proposed Action and Alternatives

This section describes the proposed project and all of the alternatives that need to be considered, including the "no action" alternative. The last is a description of projections of future scenarios if the proposed project is not done. All possible alternatives need not be included; inclusion depends on the project being undertaken. For example, the EIS for the proposed high-level radioactive waste repository at Yucca Mountain, NV, was not required to consider alternate waste repository sites.

Description of the Environment Affected by the Proposed Action

This description is best organized by listing environmental parameters that could be impacted by the proposed alternative, grouping them into logical sets. One listing might be:

- Ecology

    —Species and populations
    —Habitats and communities
    —Ecosystems
    —Wetlands

- Aesthetics

    —Land
    —Air
    —Water
    —Biota
    —Human-made objects
    —Objects of historical or cultural significance

- Environmental Pollution and Human Health

    —Water
    —Air
    —Land
    —Noise

- Economics

    —Jobs created or lost
    —Property values
    —Jobs

Each title might have several specific subtopics to be studied; for example, under Aesthetics, "Air" may include odor, sound, and visual impacts as items in the checklist.

Numerical ratings may be assigned to these items. One procedure is to first estimate the ideal or natural levels of environmental quality (without anthropogenic pollution) and take a ratio of the expected condition to the ideal. For example, if the ideal dissolved oxygen in the stream is 9 mg/L, and the effect of the proposed action is to lower the dissolved oxygen to 3 mg/L, the ratio would be 0.33. This is sometimes called the environmental quality index (EQI). Another option to this would be to make the relationship nonlinear, as shown in Fig. 2-1. Lowering the dissolved oxygen by a few milligrams per liter will not affect the EQI nearly as much as lowering it, for example, below 4 mg/L, since a dissolved oxygen below 4 mg/L definitely has a severe adverse effect on the fish population.

EQIs may be calculated for all checklist items that have a natural quantitative scale. In order to assess those items that do not have a quantitative scale, like aesthetics or historical objects, a scale based on qualitative considerations may be generated by an expert in the particular area. For example, impact on a historic building might be measured by the cost of recovering from certain amounts of damage. Something like visual aesthetics can simply be assigned a scale.

The EQI values are then tabulated for each parameter. Next, weights may be attached to the items, usually by distributing 1000 parameter importance units (PIU) among the items. Assigning weights is a subjective exercise and is usually done by the decision makers: those individuals who are going to make decisions about the project. The product of EQI and PIU, called the environmental impact unit (EIU), is thus the

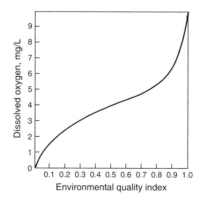

Figure 2-1. Projected environmental quality index curve for dissolved oxygen.

magnitude of the impact multiplied by the importance:

$$EIU = PIU \times EQI.$$

This method has several advantages. We may calculate the sum of EIUs and evaluate both the cumulative impact of the proposed project and the "worth" of many alternatives, including the do-nothing alternative. We may also detect points of severe impact, for which the EIU after the project may be much lower than before, indicating severe degradation in environmental quality. Its major advantage, however, is that it makes it possible to input data and evaluate the impact on a much less qualitative and a much more objective basis.

EXAMPLE 2.3. Evaluate the effect of a proposed lignite strip mine on a local stream. Use 10 PIU and linear functions for EQI.
  The first step is to list the areas of potential environmental impact. These may be:

- appearance of water
- suspended solids
- odor and floating materials
- aquatic life
- dissolved oxygen.

Other factors could be listed, but these will suffice for this example. Next, we need to assign EQIs to the factors. Assuming a linear relationship, we can calculate them as in Table 2-3.

Table 2-3.

Item	Condition before project	Condition after project	EQI
Appearance of water	10	3	0.33
Suspended solids	10 mg/L	1000 mg/L	0.01
Odor	10	5	0.50
Aquatic life	10	2	0.20
Dissolved oxygen	9 mg/L	8 mg/L	0.88

Note that we had to put in subjective quantities for three of the items — "Appearance of water," "Odor," and "Aquatic life" — based on an arbitrary scale of decreasing quality from 10 to 1. The actual magnitude is not important since a ratio is calculated. Also note that the sediment ratio had to be inverted to make its EQI indicate environmental degradation, i.e., EQI < 1. The EQI indices are weighted by the 10 available PIU, and the EIU are calculated.
  In Table 2-4, The EIU total of 2.72 for this alternative is then compared with the total EIU for other alternatives.

Table 2-4.

Item	Project PIU	After project EQI × PIU = EIU
Appearance of water	1	0.33 × 1 = 0.33
Suspended solids	2	0.01 × 2 = 0.02
Odor	1	0.5 × 1 = 0.5
Aquatic life	5	0.2 × 5 = 1.0
Dissolved oxygen	1	0.88 × 1 = 0.88
Total	10	2.73

Multiattribute utility analysis, a more sophisticated ranking and weighting method, has been developed by Keeney (Keeney and Raiffa 1993; Bell *et al.* 1989) and is now used by many federal agencies. The details of the method are unfortunately beyond the scope of this text.

Evaluation

The final part of the environmental impact assessment, which is reflected in the record of decision, is the evaluation of the results of the preceding studies. Typically the evaluation phase is out of the hands of the engineers and scientists responsible for the inventory and assessment phases. The responsible governmental agency ultimately uses the environmental assessment to justify the record of decision.

USE OF RISK ANALYSIS IN ENVIRONMENTAL ASSESSMENT

The rationale for including risk analysis in environmental impact assessment is threefold:

(1) Risk analysis provides a method for comparing low-probability, high-consequence impacts with high-probability, low-consequence impacts.
(2) Risk analysis allows assessment of future uncertain impacts, and incorporates uncertainty into the assessment.
(3) The United States and international agencies concerned with regulating environmental impact are adopting risk-based standards in place of consequence-based standards.

The following example incorporates risk analysis into environmental impact assessment: In 1985, the U.S. EPA promulgated a regulation for radioactive waste disposal which allowed a 10% probability of a small release of radioactive material, and a 0.1% probability of release of ten times that amount (USEPA 1985). This standard is shown in the stair-step of the diagram of Fig. 2-2.

24 ENVIRONMENTAL ENGINEERING

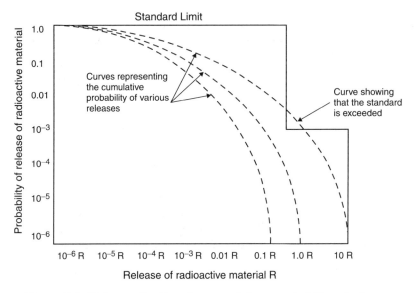

Figure 2-2. Release of radioactive materials vs. probability of release.

The curved lines are complementary cumulative distribution functions and represent the risks of releases for three different alternatives being assessed. This is a typical representation of the probability of release of material from a hazardous or radioactive waste landfill. The alternatives might be three different sites, three different surface topographies, or three different engineered barriers to release. In the case of the EPA standard, the curves represented three different geological formations. Risk analysis is particularly useful in assessing future or projected impacts and impacts of unlikely (low-probability) events, like transportation accidents that affect cargo, or earthquakes and other natural disasters.

Recently much has been written about the difficulty of communicating risk to a large public, and the perception of risk as it differs from assessed risk (see, for example, Slovic 1985; Weiner 1994). The engineer must remember that risk assessment, when used in environmental impact assessment, is independent of risk as perceived by, or presented to, the public. He or she should assay risk as quantitatively as possible. The details of performing a risk analysis are presented in Chap. 3.

SOCIOECONOMIC IMPACT ASSESSMENT

Historically, the President's Council on Environmental Quality has been responsible for overseeing the preparation of environmental assessments, and CEQ regulations have listed what should be included in all environmental assessments developed by federal agencies. For the proposed projects discussed earlier in this chapter the primary issues are public health dangers and environmental degradation. Under original NEPA and CEQ regulations, both issues must be addressed whenever alternatives are developed and compared.

Federal courts have ruled (cf. for example, State of Nevada vs. Herrington, 1987) that consideration of public health and environmental protection alone are not sufficient grounds on which to evaluate a range of alternative programs. Socioeconomic considerations such as population increases, need for public services like schools, and increased or decreased job availability are also included under NEPA considerations. Very recently (O'Leary 1995), federal agencies have also mandated inclusion of environmental justice considerations in environmental impact assessments. Frequently, public acceptability is also a necessary input to an evaluation process. Although an alternative may protect public health and minimize environmental degradation, it may not be generally acceptable. Factors that influence public acceptability of a given alternative are generally discussed in terms of economics and broad social concerns. Economics includes the costs of an alternative, including the state, regional, local, and private components; the resulting impacts on user charges and prices; and the ability to finance capital expenditures. Social concerns include public preferences in siting (e.g., no local landfills in wealthy neighborhoods) and public rejection of a particular disposal method (e.g., incineration of municipal solid waste rejected on "general principle"). Moreover, as budgets become tighter, the cost/benefit ratio of mitigating a particular impact is increasing in importance. Consequently, each alternative that is developed to address the issues of public health and environmental protection must also be analyzed in the context of rigid economic analyses and broad social concerns.

Financing of Capital Expenditures

A municipality's or industry's inability to finance large capital expenditures will necessarily affect choice among alternatives and possibly affect their ability to comply with environmental regulations. Traditional economic impact assessments examine the amortized capital, operation and maintenance (O&M) costs of a project, and the community's ability to pay, but the analyses typically overlook the problems involved in raising the initial capital funds required for implementation. Financing problems face municipalities and industries of all sizes, but may be particularly troublesome for small communities and firms that face institutional barriers to financing. The following discussion examines only financing capability of compliance with water quality regulations, but parallel issues arise for other types of public and private projects.

For relatively small capital needs, communities may make use of bank borrowing or capital improvement funds financed through operating revenues. However, local shares of wastewater treatment facility capital costs are generally raised by long-term borrowing in the municipal bond market. In the absence of other sources of funds, both the availability of funds through the bond market and the willingness of the community to assume the costs of borrowing may affect the availability of high-cost programs. The availability and cost of funds are, in turn, affected by how the financing is arranged.

Bonds issued by a municipality for the purpose of raising capital for a wastewater treatment plant typically are general obligation (GO) bonds or revenue bonds. Both have fixed maturities and fixed rates of interest, but they differ in the security pledge by the issuing authority to meet the debt service requirements, i.e., payments for principal plus interest. GO bonds are backed by the basic taxing authority of the issuer; revenue bonds are backed solely by the revenue for the service provided by the specific project.

The GO bond is generally preferred since the overhead costs of financing GO bonds are lower and their greater security allows them to be offered at a lower rate of interest. Some states are constitutionally prevented from issuing GO bonds or are limited in the quantity they can issue, and cities may have to resort to revenue bonds. Nevertheless, most bond issues for wastewater treatment projects are GOs.

Most municipal bonds carry a credit rating from at least one of the private rating agencies, Standard & Poor's or Moody's Investor Service. Both firms attempt to measure the credit worthiness of borrowers, focusing on the potential for decrease on bond quality by subsequent debt and on the risk of default. Although it is not the sole determinant, the issuer's rating helps to determine the interest costs of borrowing, since individual bond purchasers have little else to guide them, and commercial banks wishing to purchase bonds are constrained by federal regulations to favor investments in the highest rating categories.

Of the rating categories used, only the top ones are considered to be of investment quality. Even among these grades, the difference in interest rates may impose a significantly higher borrowing cost on communities with low ratings. During the 1970s, for example, the interest rate differential between the highest grade (Moody's Aaa) and the lowest investment grade (Moody's Baa) bonds averaged 1.37%. Such a differential implies a substantial variation in financing costs for facilities requiring extensive borrowing.

To highlight the importance of financing costs, consider the example of a city planning a $2 million expenditure on an incinerator to serve a publicly owned (wastewater) treatment works (POTW) with a capacity of 15 million gallons per day (mgd). Under one set of assumptions, such a facility would support a population of roughly 75,000 that would bear anywhere from 12.5 to 100% of the total cost, depending on what portion of the capital expense state and federal agencies agreed to pay. Under the more conservative assumption, this would amount to $250,000. Assuming that the capital is raised through one Aaa bond issue amortized over 25 years at an interest rate of 5.18%, the interest payment over the entire borrowing period would total $212,500. If the Baa rate were 6.34%, the interest payments would amount to $262,000. As a rule of thumb, total interest payments are roughly equal to the principal and are sensitive to the interest rate.

The rating may also determine the acceptability of bonds on the market. A city with a low or nonexistent rating might find that credit is simply not available. The fiscal problems of small communities are particularly sensitive to the rating process, which places lower ratings on small towns on the basis of an ill-defined "higher risk of default," despite the lack of supporting evidence. Hence, small towns must often endure higher interest payments. The problem is compounded by the higher average cost of small bond issues owing to certain fixed underwriting fees associated with buying and selling the bonds.

Despite the availability of GO financing and good credit, municipalities may face difficulties in raising capital because of the state of the market, while some borrowers may be precluded from borrowing altogether. Two trends are of particular significance.

Periods of high rates of inflation are expected to lead to higher interest rates for all municipal borrowers. The impact is significant for borrowers with marginal credit ratings, who will be squeezed out as the difference in interest rates between high- and low-rated bond issues widens. Second, the combination of expanding bond supply owing to increased municipal borrowing and shrinking demand in periods of slow economic growth is likely to drive interest rates higher, increasing costs for all municipal borrowers.

Clearly, the availability of funding to finance capital expenditures is far from certain. Credit restrictions are not limited to large cities that often suffer from severe budget problems. The financial health of a local government may be analyzed. A two-step procedure is useful as engineers consider the impacts of proposed projects:

1. Define capital requirements of each alternative.
2. Apply financial classification criteria.

Although rating agency analysts use a variety of factors in assessing a city's fiscal strength, current outstanding debt is the principal determinant of the bond rating. Debt is generally expressed as a proportion of assets or on a per capita basis to make comparisons among cities. Three of these municipal debt ratios are of particular significance in determining market acceptance of bond offerings. Sample ratios and estimates of market-acceptable thresholds are:

- Debt/capita ≤ 300
- Debt/capita as percentage of per capita income $\leq 7\%$
- Debt/full property value $\leq 4.5\%$.

Empirical studies have shown that bond offerings by cities that exceed all three debt ratios are not likely to be successful.

Increases in User Charges

A second component of an economic impact assessment is the analyses of projected increases in user charges. For example, if a city constructs a large wastewater treatment facility, what is the likely change in household sewer bills in that city? Based on the number of residential users supporting a given facility, the proportion of wastewater generated by them, and the data on alternatives, a projected cost per household may be computed. This cost may then be compared with the existing charge per household to determine the percent increase attributable to the alternative.

The results of these types of user charge analyses often indicate that engineers may find it difficult to make a distinction among alternatives based solely on increases in user charges. For example, an average difference of only 36 cents per year per household is projected between alternatives 1 and 2 in Fig. 2-3. Respectively, these two alternatives represent a less stringent and a more stringent regulation of effluent water quality. Figure 2-3 summarizes projected impacts on the user charges in 350 cities. By almost any standard, 36 cents per year is insignificant. The findings suggest that the selection

28 ENVIRONMENTAL ENGINEERING

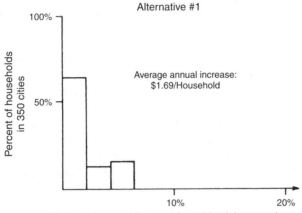

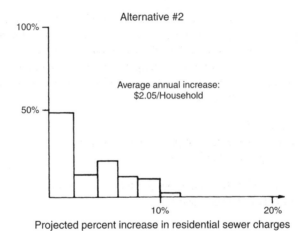

Figure 2-3. Projected increases in sewer use charges per household in a sample community if a new wastewater treatment facility is constructed.

process, in this instance between alternative 1 (little required reduction in pollution) and alternative 2 (large requirements in pollution reduction), must be based on such other criteria as environmental impact and public health concerns, not on changes in user charges.

Sociological Impacts

Large changes in the population of a community, such as influxes of temporary construction workers or establishment of a military base with immigration of the associated personnel and their families, may have a number of impacts, both positive and adverse. New service jobs may well be created, particularly in small communities, but there may also be increases in the crime rate, the need for police officers on the street, the need for fire protection, etc. Study of such "boom town" phenomena has led to the inclusion of these assessments in any environmental assessment.

The business of preparing EISs, EAs, and FONSIs is truly a growth business worldwide. Besides becoming a cornerstone of environmental regulation in the United States and Europe, the environmental assessment process is now found throughout the world. In Brazil, for example, the siting and design of submarine wastewater outfalls are among the activities that require an EIA according to Brazilian legislation (Jordao and Leitao 1990). While acting to protect the environment, such roles and procedure have acted to slow development while creating the ever-expanding job market for those qualified to conduct and report within the environmental assessment systems of many countries.

Ethical Considerations

A properly done environmental impact assessment is independent of any ethical system and is value-free. However, ethical questions can arise in formulating a record of decision based on an environmental assessment. Some of these questions are:

Is it ethical to limit resource extraction, with its concomitant environmental damage, by raising the resource price and thereby limiting its use to those who can afford it?

Is it ethical to eliminate jobs in an area in order to protect the environment for a future generation?

Conversely, is it ethical to use up a resource so that future generations do not have it at all?

Given limited financial resources, is it ethical to spend millions mitigating a high-consequence impact that is extremely unlikely to occur (a low-probability, high-consequence event)?

Is it ethical to destroy a watershed by providing logging jobs for 50 years?

Conversely, is it ethical to close down a lumber mill, eliminating jobs for an entire small community, in order to save an old-growth forest?

The engineer should remember that none of these considerations are part of environmental assessment, nor are they addressed by NEPA. Addressing them may be part of a record of decision, and is the responsibility of the decision makers.

CONCLUSION

Engineers are required to develop, analyze, and compare a range of solutions to any given environmental pollution problem. This range of alternatives must be viewed in terms of their respective environmental impacts and economic assessments. A nagging question exists throughout any such viewing: Can individuals really measure, in the strict "scientific" sense, degradation of the environment? For example, can we place a value on an unspoiled wilderness area? Unfortunately, qualitative judgments are required to assess many impacts of any project.

PROBLEMS

2.1 Develop and apply an interaction matrix for the following proposed actions designed to clean municipal wastewater in your home town: (a) construct a large activated wastewater treatment plant, (b) require septic tanks for households and small-scale package treatment plants for industries, (c) construct decentralized, small-scale treatment facilities across town, (d) adopt land application technology, and (e) continue direct discharge of untreated wastewater into the river. Draw conclusions from the matrix. Which alternatives appears to be superior? Which environmental characteristics appear to be the most important? What should the town do?

2.2 What conclusions may be drawn about the abilities of the three towns to finance a proposed water treatment facility:

City	Debt per capita		Debt per capita as % per capita income		Debt per full property valuation	
	Current	Projected	Current	Projected	Current	Projected
A	$946	$950	22.3	22.3	15.0	15.1
B	$335	$337	9.1	9.1	7.8	7.9
C	$6	$411	0.1	10.7	0.1	8.6

Discuss the current financial picture of each town as well as the projected incremental impact of this new proposed construction.

2.3 Discuss the advantages and disadvantages of a benefit/cost ratio in deciding whether a town should build a wastewater treatment facility. Focus on the valuation problems associated with analyzing the impacts of such a project.

2.4 Determine the average residential sewer charge for the three cities listed below. What is the incremental impact on these charges if the cost of sludge disposal increases as noted?

	Annual		Projected	
	Wastewater flow (mgd)	% Flow residential	Wastewater treatment	Sludge disposal budget increase
A	20.0	60	$5,100,000	$1,500,000
B	5.0	90	$3,600,000	$1,000,000
C	0.1	100	$50,000	$20,000

Discuss your assumptions.

2.5 A major oil company wishes to establish a port for large tankers in Washington State in Northern Puget Sound, in a community of 40,000 people, 5,000 of whom are members of two Native American tribes. Construction of the port would supply 500 temporary jobs, and the port's operation would supply 300 permanent jobs. The community now is home to an oil refinery, a pulp mill, and a state-funded university with 8,000 students. Part of the oil unloaded at the port would supply the local refinery, but a sizable part would be shipped to Minnesota by pipeline. Construction of the port facility and pipeline extension requires federal permits.

(a) Would you require an EA or an EIS? Give reasons for your answer.
(b) Construct an environmental checklist, including socioeconomic factors.
(c) Write down at least six ethical questions that would arise.

2.6 Extensive environmental impact analysis has been performed for the proposed high-level radioactive waste repository at Yucca Mountain, NV. Part of this impact analysis is the impact of transporting construction materials to the site, transporting site-generated nonradioactive waste from the site to an appropriate landfill, and commuter-transporting workers to the site. Truck and rail transportation of materials and automobile and bus transportation of commuters pose a risk of fatal traffic accidents in direct proportion to the distance traveled. The average risk is about 3.5 deaths per 100 million km traveled by rail and truck, and about one death per 100 million km traveled by automobile. Trucks and trains carrying materials would travel a total of about 280 million km, and commuters a total of about 800 million km. Do you think that the projected traffic fatalities for any of this transportation should be part of the EIS? Please give reasons for your answer. Note that there is no right or wrong answer to this question.

2.7 A microchip manufacturer wishes to open a plant in the community where your university is located. Air and water discharge permits would be needed, which require environmental assessments. Given your community, would you require an EIS, or would an EA suffice? Outline the topics that would be included in an EIS, or EA. In your community, would you file a FONSI? Note that this problem is best worked as a team or class effort, and the results will differ from community to community.

Chapter 3

Risk Analysis

One of the tasks of the environmental engineer is to understand and reduce the risks from hazards from environmental pollution to the environment and to public health in both long and short term. In particular the environmental engineer frequently is asked to estimate or project future risks, then to use science, engineering, and technology to design facilities and/or processes to prevent or mitigate those risks. To accomplish this objective the risks associated with various hazards must initially be evaluated and quantified.

Risk analysis is introduced here as a tool of the environmental engineer that crosses the boundaries of science, engineering, and risk analysis. This chapter is not a comprehensive treatise on risk analysis; rather, it includes those elements of risk analysis that an environmental engineer most importantly must understand and use.

RISK

Most pollution control and environmental laws were enacted in the early 1970s in order to protect public health and welfare.[1] Throughout this text a substance is considered a pollutant if it has been perceived to have an adverse effect on human health or the environment. In recent years increasing numbers of substances appear to pose such threats; for example the Clean Air Act listed seven hazardous substances between 1970 and 1989 and now lists approximately 300! The environmental engineer thus has one more task to face: determine the comparative risks from various environmental pollutants and, further, which risk(s) is it most important to decrease or eliminate.

Adverse effects on human health are often difficult to identify and to determine. Even when an adverse effect on health has been identified, it is still difficult to recognize those components of the individual's environment associated with the adverse effect. Risk analysts refer to these components as *risk factors*. In general a risk factor should meet the following conditions:

- Exposure to the risk factor precedes appearance of the adverse effect.
- The risk factor and the adverse effect are consistently associated. That is, the adverse effect is not usually observed in the absence of the risk factor.

---

[1] See Chapters 11, 17, and 21 for the details of air, water, and land laws and regulations in the United States.

34 ENVIRONMENTAL ENGINEERING

- The more of the risk factor there is, or the greater its intensity, the greater the adverse effect, although the functional relationship need not be linear or monotonic.
- The occurrence or magnitude of the adverse effect is statistically significantly greater in the presence of the risk factor than in its absence.

Identification of a risk factor for a particular adverse effect may be done with confidence only if the relationship is consonant with, and does not contradict, existing knowledge of the cellular and organismic mechanisms producing the adverse effect.

Identification of the risk factor is more difficult than identification of an adverse effect. For example, we are now fairly certain that cigarette smoke is unhealthy, both to the smoker (primary smoke risk) and to those around the smoker (secondary smoke risk). Specifically, lung cancer, chronic obstructive pulmonary disease, and heart disease occur much more frequently among habitual smokers than among nonsmokers or even in the whole population including smokers. In the interest of simplifying the problem, we are defining "habitual" smoking as two packs or more per day. The increased frequency of occurrence of these diseases is statistically significant.[2] Cigarette smoke is thus a risk factor for these diseases; smokers and people exposed to secondhand smoke are at increased risk for them.

Note, however, that we do not say that cigarette smoking *causes* lung cancer, chronic obstructive pulmonary disease, or heart disease, because we have not identified the actual causes, or etiology, of any of them. How, then, has cigarette smoking been identified as a risk factor if it cannot be identified as the cause? This observation was not made, and indeed could not be made, until the middle of the twentieth century, when the lifespan in at least the developed countries of the world was long enough to observe the diseases that have been correlated with exposure to cigarette smoke. In the first half of the twentieth century, infectious diseases were a primary cause of death. With the advent of antibiotics and the ability to treat infectious diseases, the lifespan in the developed nations of the world lengthened, and cancer and heart disease became the leading causes of death. From the early 1960s, when the average lifespan in the United States was about 70, lifelong habitual cigarette smokers were observed to die from lung cancer at ages between 55 and 65. This observation, which associated early death with cigarette smoke, identified cigarette smoke as a risk factor.

ASSESSMENT OF RISK

Risk assessment is a system of analysis that includes four tasks:

- identification of a substance (a toxicant) that may have adverse health effects,
- scenarios for exposure to the toxicant,
- characterization of health effects, and
- an estimate of the probability (risk) of occurrence of these health effects.

---

[2] It should be noted that cigarette smoke is not regulated in the way that other pollutants are regulated.

The decision that the concentration of a certain toxicant in air, water, or food is acceptable or unacceptable is usually based on a risk assessment.

Toxicants are usually identified when an associated adverse health effect is noticed. In most cases, the first intimation that a substance is toxic is its association with an unusual number of deaths. Mortality risk, or risk of death, is easier to determine for populations, especially in the developed countries, than morbidity risk (risk of illness). All deaths and their apparent causes are reported on death certificates, while recording of disease incidence began in the relatively recent past, and is done only for a very few diseases. Death certificate data may be misleading: an individual who suffers from high blood pressure but is killed in an automobile accident becomes an accident statistic rather than a cardiovascular disease statistic. In addition, occupational mortality risks are well documented only for men; until recently, too few women worked outside the home all their lives to form a good statistical base.

These particular uncertainties may be overcome in assessing risk from a particular cause or exposure to a toxic substance by isolating the influence of that particular cause. Such isolation requires studying two populations whose environment is virtually identical except that the risk factor in question is present in the environment of one population but not of the other. Such a study is called a *cohort study* and may be used to determine morbidity as well as mortality risk. One cohort study showed that residents of copper-smelting communities, who were exposed to airborne arsenic, had a higher incidence of a certain type of lung cancer than residents of similar industrial communities where there was no airborne arsenic.

Retrospective cohort studies are almost impossible to perform because of uncertainties on data, habits, other exposures, etc. Cohorts must be well matched in cohort size, age distribution, lifestyle, and other environmental exposures, and must be large enough for an effect to be distinguishable from the deaths or illnesses that occur anyway.

PROBABILITY

Extensive discussion of probability is beyond the scope of this text, and the reader is encouraged to consult a statistics text. Some basic considerations are in order, however. Probability is often confused with frequency, because frequencies are used to estimate probabilities in risk assessment. One such use is the estimate of the risk of being in an automobile accident. We have an excellent idea of the frequency in the United States of automobile accidents of varying severity, and we use these frequencies to predict automobile death risks: the consequence is death, and the probability is the frequency of fatal car accidents. Frequency presents a more reliable estimate of probability than most other estimates. Estimates of probabilities change as observed frequencies change.

Note that probability is a dimensionless number and is always less than unity. A probability of 1 means that the event has a 100% chance of occurring, or is a certainty. A frequency, on the other hand, has dimensions and can have a value larger than unity, depending on how the frequency is defined. Consider the following example:

The observed average truck accident frequency (rate) in the United States is 3.5×10^{-7} accidents per truck-kilometer (Saricks and Tompkins 1999). This is an observed *frequency*. However, it allows us to estimate that the average annual *probability* of a truck accident in the United States is 3.5×10^{-7}. If there are 20,000 truck shipments of hazardous materials per year for 10 years, and the average distance traveled is 1000 km per shipment, there would be

$$(10 \text{ years}) \times (3.5 \times 10^{-7} \text{ accidents/year-shipment-km})$$
$$\times (2 \times 10^4 \text{ shipments}) \times (1000 \text{ km/shipment})$$
$$= 70 \text{ accidents involving trucks carrying hazardous materials.}$$

This result can be expressed by at least four different expressions:

- The frequency of accidents involving trucks carrying hazardous materials is 70 accidents in 10 years, or
- The 10-year *average* frequency of accidents involving trucks carrying hazardous materials is 7 accidents per year, or
- The 10-year *average* frequency of accidents involving trucks carrying hazardous materials is 0.007 accidents per kilometer per year, or
- The *average* frequency of accidents involving trucks carrying hazardous materials is 0.07 accidents per kilometer for this 10-year period.

The *probability* of an accident involving a truck carrying hazardous materials is 3.5×10^{-7}, as *estimated* from the observed accident frequency.

The unconditional probabilities of a set of alternative outcomes of the same event always add to unity. That is, assume that an automobile trip, when a driver and one passenger occupy the car, has ten possible outcomes:

1. The car will arrive safely at the destination without incident.
2. The car will be involved in an accident but no occupant will be injured.
3. The car will be involved in an accident and an occupant other than the driver will be injured. The driver will be uninjured.
4. The car will be involved in an accident and the driver will be injured, but the occupant will be uninjured.
5. The car will be involved in an accident and an occupant other than the driver will be killed, while the driver is uninjured.
6. The car will be involved in an accident and the driver will be killed while the occupant will be uninjured.
7. The car will be involved in an accident and an occupant other than the driver will be killed, while the driver is injured.
8. The car will be involved in an accident and the driver will be killed while the occupant is injured.
9. The car will be involved in an accident and driver and occupant will both be injured.
10. The car will be involved in an accident and driver and occupant will both be killed.

The sum of the probabilities of each of these ten outcomes is unity. Therefore, only nine of the ten probabilities are independent.

A more precise representation of these probabilities is that some are independent, or unconditional, and some are conditional. Probabilities 2 through 10 are conditional on the car being in an accident. A more precise representation of the sum is then

$$P(\text{no accident}) + P(\text{accident}) = 1$$

$$P(\text{accident}) = P(2) + P(3) + P(4) + P(5) + P(6) + P(7)$$
$$+ P(8) + P(9) + P(10) = 1 - P(\text{no accident}).$$

A conditional probability is the *product* of the probabilities. If the probability of an accident is 10% (or 0.1), and the probability of an injury to either of two occupants of the vehicle conditional on that accident is 10% (or 0.1), then the overall probability of an injury to either occupant is

$$P(\text{nonfatal injury}) = P(\text{accident}) \times P(\text{injury}|\text{accident}) = 0.1 \times 0.1 = 0.01$$

or 1%. The sum of the conditional probabilities must also be unity. Thus if the conditional probability of an injury is 1% (0.01), then the conditional probability of an accident involving no injury is $1 - 0.01 = 0.99$. The overall probability of an accident with neither an injury nor a death is then $0.1 \times 0.99 = 0.099$, or 9.9%.

This problem may be illustrated using an *event tree*, as shown in Fig. 3-1, where the net probabilities are listed on the right-hand side of the figure. Note that these are

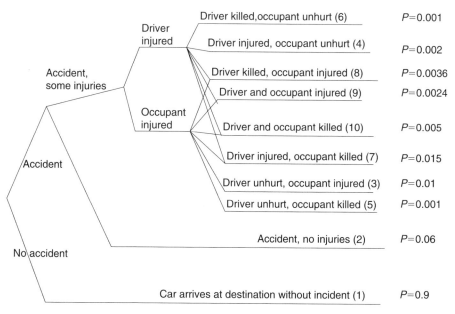

Figure 3-1. Event tree for automobile accident example, Numbers in parentheses refer to the scenarios on page 36.

38   ENVIRONMENTAL ENGINEERING

not actual frequencies or probabilities, but do reflect, in general, the probability of each branch of the event tree. An event tree is helpful in identifying all possible outcomes of any situation.

In general, the risks attendant on exposure to hazardous substances are conditional probabilities with multiple conditions; that is, the probability of cancer fatality is the product of three probabilities: the probability of exposure, the probability that the exposure will result in cancer, and the probability that the resulting cancer will be the cause of death. All three of these probabilities are considerably less than unity: "exposure," for example, means that the carcinogen finds a sensitive target.

DOSE–RESPONSE EVALUATION

Dose–response evaluation is required both in determining exposure scenarios for the pollutant in question and in characterizing a health effect. The response of an organism to a pollutant always depends in some way on the amount or dose of pollutant to the organism. The magnitude of the dose, in turn, depends on the exposure pathway. The same substance may have a different effect depending on whether it is inhaled, ingested, or absorbed through the skin, or whether the exposure is external. The exposure pathway determines the biochemistry of the pollutant in the organism. In general, the human body detoxifies an ingested pollutant more efficiently than an inhaled pollutant.

The relationship between the dose of a pollutant and the organism's response can be expressed in a dose–response curve, as shown in Fig. 3-2. The figure shows four basic types of dose–response curves possible for a dose of a specific pollutant and the respective responses. For example, such a curve may be drawn for various concentrations of carbon monoxide, the dose, plotted against the associated blood concentrations of carboxylated hemoglobin, the response. Some characteristic features of the dose–response relationship are as follows:

1. *Threshold.* The existence of a threshold in health effects of pollutants has been debated for many years. A threshold dose is the lowest dose at which there is

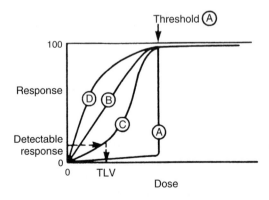

Figure 3-2. Possible dose–response curves.

an observable effect. Curve A in Fig. 3-2 illustrates a threshold response: there is no observed effect until a particular concentration is reached. This concentration is designated as the threshold concentration. Curve B shows a linear response with no threshold; i.e., the intensity of the effect is directly proportional to the pollutant dose, and an effect is observed for any detectable concentration of the pollutant in question. Curve C, sometimes called *sublinear*, is a sigmoidal dose–response curve, and is characteristic of many pollutant dose–response relationships. Although Curve C has no clearly defined threshold, the lowest dose at which a response can be detected is called the *threshold limit value* (TLV). Occupational exposure guidelines are frequently set at the TLV. Curve D displays a *supralinear* dose–response relationship, which is found when low doses of a pollutant appear to provoke a disproportionately large response.

2. *Total Body Burden.* An organism, or a person, can be exposed simultaneously to several different sources of a given pollutant. For example, we may inhale about 50 μg/day of lead from the ambient air and ingest about 300 μg/day in food and water. The concentration of lead in the body is then the sum of what is inhaled and ingested and what remains in the body from prior exposure, less what has been eliminated from the body. This sum is the total body burden of the pollutant.

3. *Physiological Half-Life.* The physiological half-life of a pollutant in an organism is the time needed for the organism to eliminate half of the internal concentration of the pollutant, through metabolism or other normal physiological functions.

4. *Bioaccumulation and Bioconcentration.* Bioaccumulation occurs when a substance is concentrated in one organ or type of tissue of an organism. Iodine, for example, bioaccumulates in the thyroid gland. The *organ dose* of a pollutant can thus be considerably greater than what the total body burden would predict. Bioconcentration occurs with movement up the food chain. A study of Lake Michigan ecosystem (Hickey *et al.* 1966) found the following bioconcentration of DDT:

0.014 ppm (wet weight) in bottom sediments
0.41 ppm in bottom-feeding crustacea
3 to 6 ppm in fish
2400 ppm in fish-eating birds

Pollution control criteria for which an engineer designs must take both bioconcentration and bioaccumulation into account.

5. *Exposure Time and Time vs Dosage.* Most pollutants need time to react; the exposure time is thus as important as the level of exposure. Because of the time–response interaction, ambient air quality standards are set at maximum allowable concentrations for a given time.

6. *Synergism.* Synergism occurs when two or more substances enhance each other's effects, and when the resulting effect of the combination on the organism is greater than the additive effects of the substances separately. For example, black lung disease in miners occurs more often in miners who smoke than in those who do not. The synergistic effect of breathing coal dust *and* smoking puts miners at high risk. The opposite of synergism is *antagonism*, a phenomenon that occurs when two substances counteract each other's effects.

40   ENVIRONMENTAL ENGINEERING

7. *LC_{50} and LD_{50}*. Dose–response relationships for human health are usually determined from health data or epidemiological studies. Human volunteers obviously cannot be subjected to pollutant doses that produce major or lasting health effects, let alone fatal doses. Toxicity can be determined, however, by subjecting nonhuman organisms to increasing doses of a pollutant until the organism dies. The LD_{50} is the dose that is lethal for 50% of the experimental animals used; LC_{50} refers to lethal concentration rather than lethal dose. LD_{50} values are most useful in comparing toxicities, as for pesticides and agricultural chemicals; no direct extrapolation is possible, either to humans or to any species other than that used for the LD_{50} determination. LD_{50} can sometimes be determined retrospectively when a large population has been exposed accidentally, although estimates of dose in such cases is very uncertain.

8. *Hormesis.* Some substances appear to have a beneficial effect in small doses and a detrimental effect when there is greater exposure. For example, soft (low-energy) X-rays in very small doses are believed to stimulate the healing of broken bones. To date, observations of hormesis have been rare and uncertain.

POPULATION RESPONSES

Individual responses to a particular pollutant may differ widely; dose–response relationships differ from one individual to another. In particular, thresholds differ. Threshold values in a population, however, generally follow a Gaussian distribution.

Individual responses and thresholds also depend on age, sex, and general state of physical and emotional health. Healthy young adults are on the whole less sensitive to pollutants than are elderly people, those who are chronically or acutely ill, and children. Allowable releases of pollutants are, in theory, restricted to amounts that ensure protection of the health of the entire population, including its most sensitive members. In many cases, however, such protection would mean zero release.

The levels of release actually allowed taking technical and economic control feasibility into account. Even so, regulatory agencies try to set such levels below threshold level for 95% or more of the U.S. population. For nonthreshold pollutants, however, no such determination can be made. In these instances, there is no release level for which protection can be ensured for everyone, so a comparative risk analysis is necessary. Carcinogens are all considered to be in this category of nonthreshold pollutants.

EXPOSURE AND LATENCY

Characterization of some health risks can take a very long time. Many cancers grow very slowly and are noticed (expressed) many years, or even decades, after exposure to the potentially responsible carcinogen. Current medical thought suggests that some carcinogens act by damaging a tumor-suppressing factor, and that expression occurs when the tumor-suppressing factor is overwhelmed. The length of time between exposure to a risk factor and expression of the adverse effect is called the *latency period*. Cancers in adults have apparent latency periods of between 10 and 40 years. Relating a cancer to a particular exposure is fraught with inherent inaccuracy; it is exceedingly difficult to isolate the effect of a single carcinogen when examining thirty or forty years of a person's life. Many carcinogenic effects are not identifiable in the lifetime of a single individual. There are a few instances in which a particular cancer is found only on exposure to a particular agent (e.g., a certain type of hemangioma is found only on exposure to vinyl chloride monomer) but for most cases, the connection between exposure and effect is far from clear. Many carcinogens are identified through animal studies, but one cannot always extrapolate from animal results to human results. The U.S. Environmental Protection Agency (EPA) classifies known animal carcinogens for which there is inadequate evidence for human carcinogenicity as probable human carcinogens.

There is a growing tendency to regulate any substance for which there is any evidence, even inconclusive evidence, of adverse health effects. This is considered a conservative assumption, but may not be valid in all cases. Such a conservative posture toward regulation and control is the result of the cumulative uncertainty surrounding the epidemiology of pollutants.

EXPRESSION OF RISK

In order to use risks in determining pollution standards, as EPA does, it is necessary to develop quantitative expressions for risk. The quantitative expressions reflect both the proportionality of the risk factor to the adverse effect and the statistical significance of the effect.

Risk is defined as the product of probability and consequence, and is expressed as the probability or frequency of occurrence of an undesirable event. For example, if 10% of the students in a course were randomly given an "F," the "risk" of getting "F" is 0.1 "F" per total number of grades assigned. The probability is 0.1 and the consequence is "F." The units in which risk is expressed incorporate both the probability and some measure of consequence. In discussing human health or environmental risk, the consequences are adverse health effects or adverse effects on some species of plant or animal. The frequency of occurrence of adverse health effects in a population is written as

$$F = \frac{X}{N}, \qquad (3.1)$$

where

F = frequency,
X = number of adverse health effects, and
N = number of individuals in the population.

42   ENVIRONMENTAL ENGINEERING

This frequency is often called a probability P, and written without units. Because of this common (though confusing) usage, we will refer to "probability" throughout the remainder of this chapter.

If the adverse effect is death from cancer, and the cancer occurs after a long latency period, the adverse health effects are called latent cancer fatalities, or LCF.

Relative risk is the ratio of the probabilities that an adverse effect will occur in two different populations. For example, the relative risk of fatal lung cancer in smokers may be expressed as

$$\frac{P_s}{P_n} = \frac{(X_s/N_s)}{(X_n/N_n)} \tag{3.2}$$

where

P_s = probability of fatal lung cancer in smokers,
P_n = probability of fatal lung cancer in nonsmokers,
X_s = fatal lung cancers in smokers,
X_n = fatal lung cancers in nonsmokers,
N_s = total number of smokers, and
N_n = total number of nonsmokers.

Relative risk of death is also called the *standard mortality ratio* (SMR), which is also written as

$$\text{SMR} = \frac{D_s}{D_n} = \frac{P_s}{P_n}, \tag{3.3}$$

where

D_s = observed lung cancer deaths in a population of habitual smokers, and
D_n = expected lung cancer deaths in a nonsmoking population of the same size.

In this particular instance, the SMR is approximately 11/1, and is significantly greater than 1.[3]

Three important characteristics of epidemiological reasoning are illustrated by this example:

- Everyone who smokes heavily will not die of lung cancer.
- Some nonsmokers die of lung cancer.

---

[3] Determination of statistical significance is beyond the scope of this text. In determining statistical significance, a test of significance (like the Fisher's test or the Student's *t*-test) appropriate to the population under consideration is applied.

- Therefore, one cannot unequivocally relate any given individual lung cancer death to cigarette smoking.[4]

Risk may be expressed in several ways:

- *Deaths per 100,000 persons.* In 1998, in the United States, 260,000 smokers died as a result of lung cancer and chronic obstructive pulmonary disease (COPD), according to the National Center for Health Statistucs. In that year, the United States had a population of 270.3 million. The risk of death (from these two factors) associated with habitual smoking may thus be expressed as deaths per 100,000 population, or

$$\frac{(260,000)(100,000)}{270.3 \times 10^6} = 96.$$

In other words, a habitual smoker in the United States has an annual risk of 96 in 100,000, or about one in a thousand, of dying of lung cancer or COPD. The probability is 9.6×10^{-4}; the consequence is death from lung cancer or COPD. Table 3-1 presents some typical statistics for the United States (National Center for Health Statistics, 2000).

- *Deaths per 1,000 deaths.* Using 1998 data again, there were 2,337,256 deaths in the United States that year. Of these, 260,000, or 111 deaths per 1,000 deaths, were related to habitual smoking.
- *Loss of years of life or for occupational risks, loss of work days or work years.* Loss of years of life depends on life expectancy, which differs considerably from one country to another. Average life expectancy in the United States is now 75 years; in Canada, it is 76.3 years and in Ghana, 54 years (World Resources Institute, 1987). Table 3-2 (from National Center for Health Statistics) gives the loss of life expectancy from various causes of death in the United States.

These figures show that meaningful risk analyses can be conducted only with very large populations. Health risk that is considerably lower than the risks cited in Tables 3-1 and 3-2 may not be observed in small populations. Chapters 19 and 20 cite several examples of statistically valid risks from air pollutants.

EXAMPLE 3.1. A butadiene plastics manufacturing plant is located in Beaverville, and the atmosphere is contaminated by butadiene, a suspected carcinogen. The cancer death rate in the community of 8000 residents is 36 people per year, and the total death rate is 106 people per year. Does Beaverville appear to be a healthy place to live, or is the cancer risk unusually high?

---

[4]Despite this principle of risk analysis, in 1990 the family of Rose Cipollino successfully sued cigarette manufacturers and advertisers, claiming that Ms. Cipollino had been enticed to smoke by advertising, and that the cigarette manufacturers had concealed known adverse health effects. Ms. Cipollino died of lung cancer at the age of 59.

Table 3-1. Adult Deaths in the United States/100,000 Population (National Center for Health Statistics: www.census.gov)

Cause of death	Deaths/10^5 population
Cardiovascular disease	328
Cancer (all)	200
Chronic obstructive pulmonary disease	40
Motor vehicle accidents	16.2
Alcohol-related disease	19.0
Other causes	594
All causes	1197

Table 3-2. Years of Life Lost per 100,000 Population, before Age 75, from Various Causes of Death in the United States (National Center for Health Statistics: www.census.gov)

Cause of death	Loss of life expectancy (years/100,000 persons)
Cardiovascular disease	1343
Cancer — all types	1716
Respiratory system cancer	458
Chronic obstructive pulmonary disease	188
Motor vehicle accidents	596
Alcohol-related disease	159
Pneumonia and influenza	123
HIV	177

From Table 3-1, we see that the annual cancer death rate in the United States is 200 deaths/10^5 persons, and the death rate from all causes is 1197 deaths/10^5 persons. The *expected* annual death rate in Beaverville from cancer is thus

$$\left(\frac{200 \text{ deaths}}{10^5 \text{ persons}}\right)(8000 \text{ persons}) = 16.0 \text{ deaths},$$

and the expected death rate from all causes is

$$\left(\frac{1197 \text{ deaths}}{10^5 \text{ persons}}\right)(8000 \text{ persons}) = 95.8 \text{ deaths}.$$

The annual SMR for cancer is thus

$$\text{SMR(cancer)} = \frac{36}{16.0} = 2.25.$$

For all causes of death, the annual SMR is

$$\text{SMR(total)} = \frac{106}{95.8} = 1.1.$$

Without performing a test of statistical significance, we may assume that the annual SMR for cancer is significantly greater than 1, and thus Beaverville residents display excess cancer deaths. Moreover, a Beaverville resident is only a little more likely to die in any given year from any cause as the average resident of the United States. We may also calculate whether cancer deaths per 1000 deaths are higher in Beaverville than in the United States as a whole. In the United States, cancer deaths per 1000 deaths are

$$\left(\frac{200}{1197}\right)(1000) = 167,$$

while in Beaverville, cancer deaths per thousand deaths are

$$\left(\frac{36}{106}\right)(1000) = 340.$$

We may thus conclude further that a death in Beaverville in any given year is about twice as likely to be a cancer death than is the case in the United States as a whole.

Risk assessment is usually used to compare risks, because the absolute value of a risk is not very meaningful. EPA has adopted the concept of unit risk in discussion of potential risk. *Unit risk* for airborne and waterborne pollutants is defined as the risk to an individual from exposure to a concentration of 1 $\mu g/m^3$ of an airborne pollutant, or 10^{-9} g/L of a waterborne pollutant. *Unit lifetime risk* is the risk to an individual from exposure to these concentrations for 70 years (a lifetime, as EPA defines it). *Unit occupational lifetime risk* implies exposure for 8 h per day and 22 days per month every year, or 2000 h per year, for 47 years (a working lifetime).

EPA's concern with respect to somatic risk from a number of hazardous substances is the carcinogenic potential of these substances, so the "consequence" part of the risk is given as latent cancer fatalities (LCF). We can then write equations for the different expressions for unit risk, and use these to calculate the estimated risk. In the example below, these calculations assume that risk increases linearly with time and concentration. EPA considers this a conservative assumption for low exposure to a carcinogen over a period of years. Nonlinear dose–response relationships imply more complex relationships between risk, concentration, and exposure time; examples of such more complex relationships will not be considered here. For waterborne pollutants:

$$\text{Unit annual risk} = \frac{\text{LCF/year}}{10^{-9} \text{ g/L}} \qquad (3.4a)$$

$$\text{Unit lifetime risk} = \frac{\text{LCF}}{(10^{-9} \text{ g/L})(70 \text{ yrs})} \qquad (3.4b)$$

$$\text{Unit lifetime occupational risk} = \frac{\text{LCF}}{(10^{-9} \text{ g/L})(47 \text{ yrs})(2000/8760)}. \qquad (3.4c)$$

The factor 2000/8760 in Eq. (3.4c) is the fraction of hours per year spent in the workplace. For airborne pollutants:

$$\text{Unit annual risk} = \frac{\text{LCF/year}}{10^{-6}\,\text{g/m}^3} \tag{3.5a}$$

$$\text{Unit lifetime risk} = \frac{\text{LCF}}{(10^{-6}\,\text{g/m}^3)(70\,\text{yrs})} \tag{3.5b}$$

$$\text{Unit lifetime occupational risk} = \frac{\text{LCF}}{(10^{-6}\,\text{g/m}^3)(47\,\text{yrs})(2000/8760)}. \tag{3.5c}$$

EXAMPLE 3.2. The EPA has calculated that unit lifetime risk from exposure to ethylene dibromide (EDB) in drinking water is 0.85 LCF per 10^5 persons. What risk is experienced by drinking water with an average EDB concentration of 5 pg/L for five years?

The risk may be estimated using either unit annual risk or unit lifetime risk. Since the unit lifetime risk is given, we may write

$$\text{Risk} = \frac{(5 \times 10^{-12}\,\text{g/L})(0.85\,\text{LCF})(5\,\text{yrs})}{(10^5)(10^{-9}\,\text{g/L})(70\,\text{yrs})} = 3.0 \times 10^{-9}\,\text{LCF}.$$

The estimated risk is that about three fatal cancers would be expected in a population of a billion people who drink water containing 5 pg/L EDB for five years. Although there is a popular tendency to translate this to an "individual risk" of "a chance of three in a billion of having a fatal cancer," this statement of risk is less meaningful than the statement of population risk.

RISK PERCEPTION

Since 1983, a great deal has been written about risk perception, and there has been considerable research and speculation as to why risks are perceived very differently from their estimates. For example, the risks attendant on transporting radioactive waste are perceived by some people as far greater than the risks attendant on transporting gasoline in tank trucks, when in fact the latter are considerably larger. The reader is referred to the works of Paul Slovic and Hank Jenkins-Smith, listed in the bibliography, for some detailed discussions of risk perception. In general, the factors that appear to influence risk perception are the familiarity of the risk, knowledge about the risk, whether the risk is undertaken (or perceived as undertaken) voluntarily, and the estimated and perceived benefits of the activity that offsets the attendant risk. While engineers deal with mitigation of estimated risks, they must often take risk perception into account.

ECOSYSTEM RISK ASSESSMENT

Regulation of toxic or hazardous substances often requires an assessment of hazard or risk to some living species other than *homo sapiens*, or assessment of risk to an entire ecosystem. Methods for ecosystem risk assessment are now being developed by Suter and others (Suter 1990). Ecosystem risk assessment is done in the same general way as human health risk assessment, except that identification of the species at risk and the exposure pathway is a far more complex process than in human health risk assessment. Assessment *endpoints* are values of the ecosystem to be protected and are identified early in the analysis; these endpoints may include numbers of different species, life-cycle stages for a given species, reproductive patterns, or growth patterns. Identification of specific endpoints implies choices among potential target species. Ecosystem risk assessment is as yet in its infancy, and details of its practice are beyond the scope of this textbook.

CONCLUSION

The best available control for nonthreshold pollutants will still entail a residual risk. Our industrial society needs accurate quantitative risk assessment to evaluate the protection afforded by various levels of pollution control. We must also remain aware that determination of safe levels of pollutants based on risk analysis is a temporary measure until the mechanism of the damage done by the pollutant is elucidated and understood. At present, we can only identify apparent associations between most pollutants and a given health effect. We should note that analysis of epidemiological data and determination of significance of effects require application of a test of statistical significance. There are a number of such tests in general use, but their application is not central to the scope of this text and will not be considered here.

Almost all of our knowledge of adverse health effects comes from occupational exposure, which is orders of magnitude higher than exposure of the general public. Doses to the public are usually so low that excess mortality, and even excess morbidity, are not identifiable. However, development of pollution control techniques continues to reduce risk. The philosophy, regulatory approaches, and engineering design of environmental pollution control compose the remainder of this book.

PROBLEMS

3.1 Using the data given in the chapter, calculate the expected deaths (from all diseases) for heavy smoking in the United States.

3.2 Calculate the relative risks of smoking and alcoholism in the United States. Do you think a regulatory effort should be made to limit consumption of alcohol?

3.3 The International Commission on Radiation Protection has determined the lifetime unit risk for cancer for low-energy ionizing radiation to be 5.4×10^{-4} per rem of absorbed radiation. The allowed level of ionizing radiation (the EPA standard) above background is 25 mrem per source per year. Average background is about 360 mrem per year. How many fatal cancers attributable to ionizing radiation would result in the United States each year if the entire population were exposed at the level of the EPA standard? How many cancers may be attributed to background? If only 10% of the cancers were fatal each year, what percentage of the annual cancer deaths in the United States would be attributed to exposure to background radioactivity? (Note what "unit risk" means in this problem.)

3.4 The allowed occupational dose for ionizing radiation is 1 rem per year. By what factor does a worker exposed to this dose over a working lifetime increase her risk of cancer?

3.5 Workers in a chemical plant producing molded polyvinyl chloride plastics suffered from hemangioma, a form of liver cancer that is usually fatal. During the 20 years of the plant's operation, 20 employees out of 350 — the total number of employees at the plant during those years — developed hemangioma. Does working in the plant present an excess cancer risk? Why? What assumptions need to be made?

3.6 Additional, previously unavailable data on hemangioma incidence indicate that, among people who have never worked in the plastics industry, there are only 10 deaths per 100,000 persons per year from hemangioma. How does this change your answer to Problem 3.5?

3.7 The unit lifetime risk from airborne arsenic is 9.2×10^{-3} latent cancer fatalities (LCF). EPA regards an acceptable annual risk from any single source to be 10^{-6}. A copper smelter emits arsenic into the air, and the average concentration within a 2-mile radius of the smelter is 5.5 $\mu g/m^3$. Is the risk from smelter arsenic emissions acceptable to EPA?

3.8 In the community of Problem 3.7, approximately 25,000 people live within a 2-mile radius of the smelter. Assuming that the residents live there throughout their lifetimes, how many excess LCF would be expected per year in this population?

3.9 Using the data of Problems 3.7 and 3.8, estimate an acceptable workplace standard for ambient airborne arsenic.

3.10 Plutonium has a physiological half-life of one year. The isotope Pu-239 has a radiological half-life of 24,600 years, and the isotope Pu-238, a radiological half-life of 87 years. If a volunteer eats 5.0 mg of Pu-239, how much is left in his body after three years? If he eats 5.0 mg of Pu-238, how much is left in his body after three years?

3.11 During WWII, a U.S. pilot flying a single bombing mission over Europe faced a 4% probability of being shot down and not returning. After successfully completing (and returning from) 25 missions, a pilot was not required to fly any more missions. What was the probability that a pilot would be shot down on his 25th mission? (The answer is not 100%!)

3.12 Can the event tree of Fig. 3-1 be drawn in different ways? Draw this event tree in a different way, and estimate the probabilities in each branch.

3.13 It is widely assumed that the Love Canal and Times Beach incidents, in which the resident population was unwittingly exposed to toxic chemicals in the environment over a period of years, resulted in adverse health effects to that population. Design an epidemiological study that would determine whether there actually were adverse health effects resulting from these incidents. Assume that toxic material leaked into the soil, air, and drinking water over a period of 5 years.

3.14 A couple with two children ages 18 months and 4 years is buying an older house. The house has aluminum wiring, 10-ft-high asbestos "popcorn" ceilings in which the asbestos has been painted over, and high radon readings in the basement. The couple can only afford to mitigate one of these risks (let us assume that mitigation costs the same for all three). Which risk should be mitigated, in your opinion? Give reasons for your answer.

3.15 How would you rank the following risks (start with highest risk):

- smoking
- exposure to second-hand smoke
- riding in an automobile
- drinking water from a mountain stream
- technical mountain climbing
- bicycling on bike lanes on city streets
- bicycling on city streets
- downhill skiing
- bungee jumping
- flying in a commercial jet
- flying in a small twin-engine plane
- living 5 miles from a nuclear power plant
- living 5 miles from a coal-burning power plant.

This problem is best done by a team or class. It is also interesting to revisit this problem at the end of the course.

Chapter 4

Water Pollution

Although people intuitively relate filth to disease, the transmission of disease by pathogenic organisms in polluted water was not recognized until the middle of the nineteenth century. The Broad Street pump handle incident demonstrated dramatically that water could carry diseases.

In 1854, a public health physician named John Snow, assigned to try to control the spread of cholera, noticed a curious concentration of cholera cases in one part of London. Almost all of the people affected drew their drinking water from a community pump in the middle of Broad Street. However, people who worked in an adjacent brewery were not affected. Snow recognized that the brewery workers' apparent immunity to cholera occurred because the brewery drew its water from a private well and not from the Broad Street pump (although the immunity might have been thought due to the health benefits of beer). Snow's evidence convinced the city council to ban the polluted water supply, which was done by removing the pump handle so that the pump was effectively unusable. The source of infection was cut off, the cholera epidemic subsided, and the public began to understand the importance of having clean drinking water supplies.

Until recently, water pollution was viewed primarily as a threat to human health because of the transmission of bacterial and viral waterborne diseases. In less developed countries, and in almost any country in time of war, waterborne diseases remain a major public health threat. In the United States and other developed countries, however, water treatment and distribution methods have almost eradicated microbial contamination in drinking water. We now recognize that water pollution constitutes a much broader threat and continues to pose serious health risks to the public as well as aquatic life. In this chapter we discuss the sources of water pollution and the effect of this pollution on streams, lakes and oceans.

SOURCES OF WATER POLLUTION

Water pollutants are categorized as *point source* or *nonpoint source*, the former being identified as all dry weather pollutants that enter watercourses through pipes or channels. Storm drainage, even though the water may enter watercourses by way of pipes or channels, is considered nonpoint source pollution. Other nonpoint source pollution comes from agricultural runoff, construction sites, and other land disturbances, as discussed in Chap. 11. Point source pollution comes mainly from industrial facilities and municipal wastewater treatment plants. The range of pollutants is vast, depending only on what gets "thrown down the drain."

Oxygen demanding substances such as might be discharged from milk processing plants, breweries, or paper mills, as well as municipal wastewater treatment plants, compose one of the most important types of pollutants because these materials decompose in the watercourse and can deplete the water of dissolved oxygen.

Sediments and suspended solids may also be classified as a pollutant. Sediments consists of mostly inorganic material washed into a stream as a result of land cultivation, construction, demolition, and mining operations. Sediments interfere with fish spawning because they can cover gravel beds and block light penetration, making food harder to find. Sediments can also damage gill structures directly, smothering aquatic insects and fishes. Organic sediments can deplete the water of oxygen, creating anaerobic (without oxygen) conditions, and may create unsightly conditions and cause unpleasant odors.

Nutrients, mainly nitrogen and phosphorus, can promote accelerated eutrophication, or the rapid biological "aging" of lakes, streams, and estuaries. Phosphorus and nitrogen are common pollutants in residential and agricultural runoff, and are usually associated with plant debris, animal wastes, or fertilizer. Phosphorus and nitrogen are also common pollutants in municipal wastewater discharges, even if the wastewater has received conventional treatment. Phosphorus adheres to inorganic sediments and is transported with sediments in storm runoff. Nitrogen tends to move with organic matter or is leached from soils and moves with groundwater.

Heat may be classified as a water pollutant when it is caused by heated industrial effluents or from *anthropogenic* (human) alterations of stream bank vegetation that increase the stream temperatures due to solar radiation. Heated discharges may drastically alter the ecology of a stream or lake. Although localized heating can have beneficial effects like freeing harbors from ice, the ecological effects are generally deleterious. Heated effluents lower the solubility of oxygen in the water because gas solubility in water is inversely proportional to temperature, thereby reducing the amount of dissolved oxygen available to *aerobic* (oxygen-dependent) species. Heat also increases the metabolic rate of aquatic organisms (unless the water temperature gets too high and kills the organism), which further reduces the amount of dissolved oxygen because respiration increases.

Municipal wastewater often contains high concentrations of organic carbon, phosphorus, and nitrogen, and may contain pesticides, toxic chemicals, salts, inorganic solids (e.g., silt), and pathogenic bacteria and viruses. A century ago, most discharges from municipalities received no treatment whatsoever. Since that time, the population and the pollution contributed by municipal discharge have both increased, but treatment has increased also.

We define a population equivalent of municipal discharge as equivalent of the amount of untreated discharge contributed by a given number of people. For example, if a community of 20,000 people has 50% effective sewage treatment, the population equivalent is $0.5 \times 20,000$ or 10,000. Similarly, if each individual contributes 0.2 lb of solids per day into wastewater, and an industry discharges 1,000 lb/day, the industry has a population equivalent of 1,000/0.2, or 5,000. The current estimate of the population equivalent of municipal discharges into U.S. surface water is about 100 million, for a population of nearly 300 million. The contribution of municipal discharges to water pollution has not decreased significantly in the past several decades, nor has it significantly increased; at least we are not falling behind.

The sewerage systems in older U.S. cities have aggravated the wastewater discharge situation. When these cities were first built, engineers realized that sewers were necessary to carry off both storm water and sanitary wastes, and they usually designed a single system to carry both discharges to the nearest appropriate body of water. Such systems are known as combined sewers.

Almost all of the cities with combined sewers have treatment plants that can only handle dry weather flow (i.e., no storm water runoff). When it rains, the flow in the combined sewer system increases to many times the dry weather flow and most of it must be bypassed directly into a river, lake, or bay. The overflow will contain raw sewage as well as storm water, and can be a significant pollutant to the receiving water. Attempts to capture and store the excess flow for subsequent treatment are expensive, and the cost of separating combined sewer systems may be prohibitive.

As years passed, city populations increased, and the need for sewage treatment became apparent. Separate sewer systems were built: one system to carry sanitary sewage to the treatment facility and the other to carry storm water runoff. This change improved the overall treatment of sewage by decreasing the frequency of bypasses and allowing additional levels of sewage treatment, such as phosphorus removal, to be added at the wastewater treatment plant. It left unresolved the treatment of storm water runoff, which is now one of the major sources of water pollution in the United States.

Agricultural wastes that flow directly into surface waters have a collective population equivalent of about two billion. Agricultural wastes are typically high in nutrients (phosphorus and nitrogen), biodegradable organic carbon, pesticide residues, and *fecal coliform* bacteria (bacteria that normally live in the intestinal tract of warm-blooded animals and indicate contamination by animal wastes). Feedlots where large numbers of animals are penned into relatively small spaces provide an efficient way to raise animals for food. They are usually located near slaughterhouses, and thus near cities. Feedlot drainage (and drainage from intensive poultry cultivation) creates an extremely high potential for water pollution. Aquaculture has a similar problem because wastes are concentrated in a relatively small space. Even relatively low densities of animals can significantly degrade water quality if the animals are allowed to trample the stream bank, or runoff from manure-holding ponds is allowed to overflow into nearby waterways. Both surface and groundwater pollution are common in agricultural regions because of the extensiveness of fertilizer and pesticide application.

Pollution from petroleum compounds ("oil pollution") first came to public attention with the Torrey Canyon disaster in 1967. The huge tanker loaded with crude oil plowed into a reef in the English Channel. Despite British and French attempts to burn the oil, almost all of it leaked out and fouled French and English beaches. Eventually, straw was used to soak up the oil and detergents were applied to disperse the oil (detergents were later found to be harmful to the coastal ecology).

By far the most notorious recent incident has been the *Exxon Valdez* spill in Prince William Sound in Alaska. Oil in Alaska is produced in the Prudhoe Bay region in northern Alaska and piped down to the tanker terminal in Valdez on the southern coast. On March 24, 1989, the *Exxon Valdez*, a huge oil tanker loaded with crude oil veered off course and hit a submerged reef, spilling about 11 million gallons of oil, into Prince William Sound. The effect was devastating to the fragile ecology. About 40,000 birds died, including about 150 bald eagles. The final toll on wildlife will never be known, but the effect of the spill on the local fishing economy can be calculated and it exceeds $100 million. The cleanup by Exxon cost at least $2 billion, and the legal responsibility is still being debated.

While oil spills as large as the *Exxon Valdez* spill get a lot of publicity, it is estimated that there are about 10,000 serious oil spills in the United States every year, and many more minor spills from routine operations that do not make headlines. The effect of some of these spills may never be known. In addition to oil spills, petroleum hydrocarbons from atmospheric sources (e.g., automobile exhaust fumes) are deposited daily on road surfaces. When it rains, these oily deposits wash into nearby streams and lakes.

The acute effect of oil on birds, fish, and other aquatic organisms is well cataloged; the subtle effects of oil on aquatic life is not so well understood and is potentially more harmful. For example, anadromous fish that find their home stream by the smell or taste of the water can become so confused by the presence of strange hydrocarbons that they will refuse to enter their spawning stream.

Acids and bases from industrial and mining activities can alter the water quality in a stream or lake to the extent that it kills the aquatic organisms living there, or prevents them from reproducing. Acid mine drainage has polluted surface waters since the beginning of ore mining. Sulfur-laden water leached from mines, including old and abandoned mines as well as active ones, contains compounds that oxidize to sulfuric acid on contact with air. Deposition of atmospheric acids originating in industrial regions has caused lake acidification throughout vast areas of Canada, Europe, and Scandinavia.

Synthetic organics and *pesticides* can adversely affect aquatic ecosystems as well as making the water unusable for human contact or consumption. These compounds may come from point source industrial effluents or from nonpoint source agricultural and urban runoff.

The effects of water pollution can be best understood in the context of an aquatic ecosystem, by studying one or more specific interactions of pollutants with that ecosystem.

ELEMENTS OF AQUATIC ECOLOGY

Plants and animals in their physical and chemical environment make up an *ecosystem*. The study of ecosystems is termed *ecology*. Although we often draw lines around a specific ecosystem to be able to study it more fully (e.g., a farm pond) and thereby assume that the system is completely self-contained, this is obviously not true. One of the tenets of ecology is that "everything is connected with everything else."

Three categories of organisms make up an ecosystem. The *producers* use energy from the sun and nutrients like nitrogen and phosphorus from the soil to produce high-energy chemical compounds by the process of photosynthesis. The energy from the sun is stored in the molecular structure of these compounds. Producers are often referred to as being in the first *trophic* (growth) level and are called *autotrophs*. The second category of organisms in an ecosystem includes the *consumers*, who use the energy stored during photosynthesis by ingesting the high-energy compounds. Consumers in the second trophic level use the energy of the producers directly. There may be several more trophic levels of consumers, each using the level below it as an energy source. A simplified ecosystem showing various trophic levels is illustrated in Fig. 4-1, which also shows the progressive use of energy through the trophic levels. The third category of organisms, the *decomposers* or decay organisms, use the energy in animal wastes, along with dead animals and plants, converting the organic compounds to stable inorganic compounds (e.g., nitrate) that can be used as nutrients by the producers.

Ecosystems exhibit a flow of both energy and nutrients. The original energy source for nearly all ecosystems is the sun (the only notable exception is oceanic hydrothermal vent communities, which derive energy from geothermal activity). Energy flows in only one direction: from the sun and through each trophic level. Nutrient flow, on the other hand, is cyclic: nutrients are used by plants to make high-energy molecules that are eventually decomposed to the original inorganic nutrients, ready to be used again.

Most ecosystems are sufficiently complex that small changes in plant or animal populations will not result in long-term damage to the ecosystem. Ecosystems are constantly changing, even without human intervention, so ecosystem *stability* is best defined by its ability to return to its original rate of change following a disturbance. For example, it is unrealistic to expect to find the exact same numbers and species of aquatic invertebrates in a "restored" stream ecosystem as were present before any disturbance. Stream invertebrate populations vary markedly from year to year, even in undisturbed streams. Instead, we should look for the return of similar *types* of invertebrates, in about the same relative proportion as would be found in undisturbed streams.

The amount of perturbation that an ecosystem can absorb is called *resistance*. Communities dominated by large, long-lived plants (e.g., old growth forests) tend to be fairly resistant to perturbation (unless the perturbation is a chain saw!). Ecosystem resistance is partially based on which species are most sensitive to the particular disturbance. Even relatively small changes in the populations of "top of the food chain" predators (including humans) or critical plant types (e.g., plants that provide irreplaceable habitat) can have a substantial impact on the structure of the ecosystem. The ongoing attempt to limit the logging of old-growth forests in the Pacific Northwest is an attempt to preserve critical habitat for species that depend on old growth, such as the spotted owl and the marbled murrelet.

The rate at which the ecosystem recovers from perturbation is called *resilience*. Resilient ecosystems are usually populated with species that have rapid colonization and growth rates. Most aquatic ecosystems are very resilient (but not particularly resistant). For example, during storm events, the stream bottom is scoured, removing most of the attached algae that serve as food for small invertebrates. The algae grow

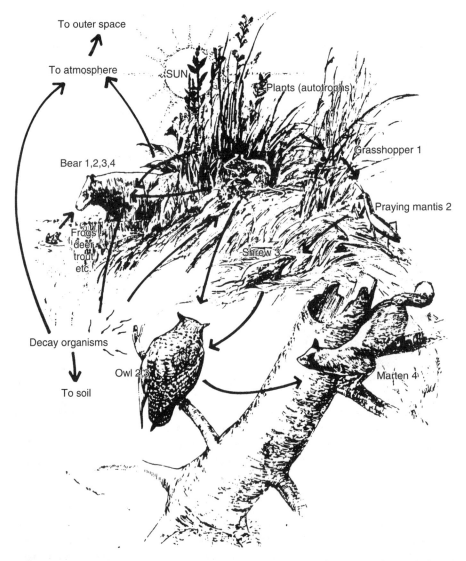

Figure 4-1. A typical terrestrial ecosystem. The numbers refer to trophic level above the autotrophic, and the arrows show progressive loss of energy. (Source: Turk, A. et al., *Environmental Science*. Philadelphia: W.B. Saunders, 1974.)

quickly after the storm flow abates, so the invertebrates do not starve. In contrast, the deep oceanic ecosystem is extraordinarily fragile, not resilient, and not resistant to environmental disturbances. This must be considered before the oceans are used as waste repositories.

Although inland waters (streams, lakes, wetlands, etc.) tend to be fairly resilient ecosystems, they are certainly not totally immune to destruction by outside perturbations. In addition to the direct effect of toxic materials like metals, pesticides, and synthetic organic compounds, one of the most serious effects of pollutants on inland waters is depletion of dissolved oxygen. All higher forms of aquatic life exist only in the presence of oxygen, and most desirable microbiologic life also requires oxygen. Natural streams and lakes are usually aerobic. If a watercourse becomes anaerobic, the entire ecology changes and the water becomes unpleasant and unsafe. The dissolved oxygen concentration in waterways and the effect of pollutants are closely related to the concept of decomposition and biodegradation, part of the total energy transfer system that sustains life.

BIODEGRADATION

Plant growth, or photosynthesis, may be represented by the equation:

$$6CO_2 + 12H_2O + \text{light} \rightarrow C_6H_{12}O_6 + 6H_2O + 6O_2 \quad (4.1)$$

In this simplified example, glucose ($C_6H_{12}O_6$), water (H_2O), and oxygen (O_2) are produced from carbon dioxide (CO_2) and water, with sunlight as the source of energy and chlorophyll as a catalyst. Photosynthesis is basically a redox reaction where CO_2 is reduced to glucose, or other high-energy carbon compounds of the general form *HCOH*, using water as the hydrogen donor. When glucose is metabolized (used as food), by the plant cell or by an animal consumer, energy is released in much the same way as in burning any fuel, with the end products being heat, carbon dioxide, and water.

In addition to sunlight, CO_2, and H_2O, plants require inorganic nutrients, particularly nitrogen and phosphorous, to grow. Equation (4.2) shows that production of algal protoplasm (the living portion of an algae cell) requires a ratio of 106 units of carbon and 16 units of nitrogen for every unit of phosphorus:

$$106CO_2 + 16NO_3^- + HPO_4^{2-} + 122H_2O + 18H^+ \rightarrow \underset{\text{(algal protoplasm)}}{[(CH_2O)_{106}(NH_3)_{16}(H_3PO_4)]} \quad (4.2)$$

As discussed previously, plants (producers) use inorganic nutrients and sunlight as an energy source to build high-energy compounds. Consumers eat and metabolize these compounds, releasing some of the energy for the consumer to use. The end product of metabolism (excrement) becomes food for decomposers and is degraded further, but at a much slower rate because many of the readily digestible compounds have already been consumed. After several such steps, only very low-energy compounds remain, and decomposers can no longer use the residue as food. Plants then use these compounds to build more high-energy compounds by photosynthesis, and the process starts over. The process is shown symbolically in Fig. 4-2.

Many organic materials responsible for water pollution enter watercourses at a high energy level. The biodegradation, or gradual use of energy, of the compounds by a chain of organisms causes many water pollution problems.

58 ENVIRONMENTAL ENGINEERING

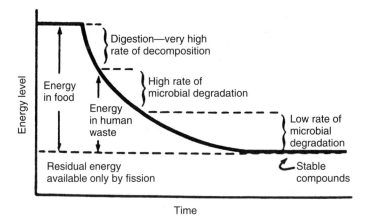

Figure 4-2. Energy loss in biodegradation. (After McGauhey, P.H., *Engineering Management of Water Quality*. New York: McGraw–Hill, 1968.)

AEROBIC AND ANAEROBIC DECOMPOSITION

Decomposition or biodegradation may take place in one of two distinctly different ways: aerobic (using free oxygen) and anaerobic (in the absence of free oxygen). The basic equation for *aerobic decomposition* of complex organic compounds is:

$$HCOH + O_2 \rightarrow CO_2 + H_2O + \text{energy} \tag{4.3}$$

The biological *respiration* or decomposition of glucose (the reverse of Eq. (4.1)) under aerobic conditions would result in the release of CO_2, H_2O, and energy that can be used for metabolism:

$$C_6H_{12}O_6 + 6H_2O + 6O_2 \rightarrow 6CO_2 + 12H_2O + \text{energy} \tag{4.4}$$

Carbon dioxide and water are always two of the end products of aerobic decomposition. Both are stable, low in energy, and used by plants in photosynthesis (plant photosynthesis is a major CO_2 sink for the earth).

Nitrogen, phosphorus, and sulfur compounds are often included in the general discussion of decomposition because the breakdown and release of these compounds during decomposition of organic matter can contribute to water quality problems. In aerobic environments, sulfur compounds are oxidized to the sulfate ion (SO_4^{2-}) and phosphorus is oxidized to phosphate (PO_4^{3-}). Any phosphate not rapidly taken up by microorganisms is bound by physical or chemical attraction to suspended sediments and metal ions, making it unavailable to most aquatic organisms. Nitrogen is oxidized through a series of steps in the progression:

$$\text{Organic N} \rightarrow NH_3 (\text{ammonia}) \rightarrow NO_2^- (\text{nitrite}) \rightarrow NO_3^- (\text{nitrate})$$

Because of this distinctive progression, various forms of nitrogen are used as indicators of water pollution. A schematic representation of the aerobic cycle for carbon,

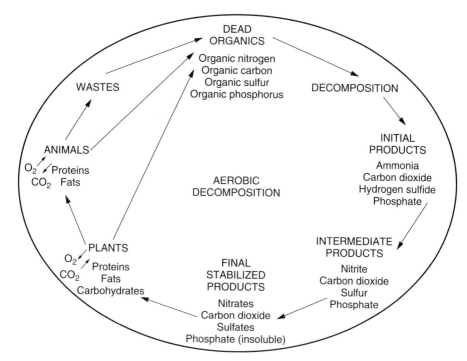

Figure 4-3. Aerobic carbon, nitrogen, phosphorus, and sulfur cycles. (After McGauhey, P.H., *Engineering Management of Water Quality*. New York: McGraw–Hill, 1968.)

nitrogen, sulfur, and phosphorus is shown in Fig. 4-3. This figure shows only the basic phenomena and greatly simplifies the actual steps and mechanisms.

Anaerobic decomposition is usually performed by a completely different set of microorganisms, to which oxygen may even be toxic. (Microorganisms that can only survive in anaerobic environments are called obligate anaerobes; facultative anaerobes can survive in aerobic or anaerobic environments.) The basic equation for anaerobic biodegradation is:

$$2HCOH \rightarrow CH_4 + CO_2 + \text{energy} \tag{4.5}$$

Figure 4-4 is a schematic representation of anaerobic decomposition. Note that the left half of the cycle, photosynthesis by plants, is identical to the aerobic cycle. Many of the end products of anaerobic decomposition are biologically unstable. Methane (CH_4), for example, is a high-energy gas commonly called "marsh gas" (or "natural gas" when burned as fuel). Although methane is physically stable (does not break down spontaneously), it can be oxidized and used as an energy source (food) by a variety of aerobic bacteria. Ammonia (NH_3) can also be oxidized by aerobic bacteria or used by plants as a nutrient. Sulfur is anaerobically biodegraded to evil smelling

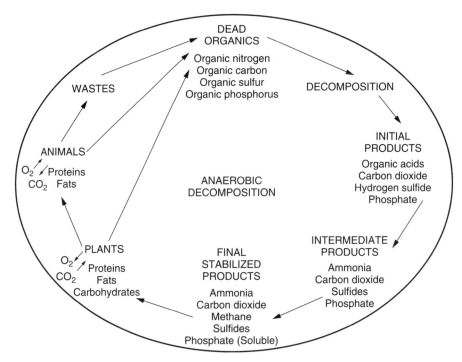

Figure 4-4. Anaerobic carbon, nitrogen, phosphorus, and sulfur cycles. (After McGauhey, P.H., *Engineering Management of Water Quality*. New York: McGraw–Hill, 1968.)

sulfhydryl compounds like hydrogen sulfide (H_2S), and can be used as an energy source by aerobic bacteria. Phosphates released during anaerobic decomposition are very soluble in water and do not bind to metal ions or sediments. Soluble phosphate is easily taken up by plants and used as a nutrient.

Biologists often speak of certain compounds as *hydrogen acceptors*. When energy is released from high-energy compounds a C=H or N=H bond is broken, and the freed hydrogen must be attached somewhere. In aerobic decomposition, oxygen serves the purpose of a hydrogen scavenger or hydrogen acceptor, and forms water. In anaerobic decomposition, oxygen is not available. The next preferred hydrogen acceptor is nitrate (NO_3^-) or nitrite (NO_2^-), forming ammonia (NH_3). If no appropriate nitrogen compound is available, sulfate (SO_4^{2-}) accepts hydrogen to form sulfur ($S°$) and H_2S, the compound responsible for the notorious rotten egg smell.

EFFECT OF POLLUTION ON STREAMS

The effect of pollution on streams depends on the type of pollutant. Some compounds are acutely toxic to aquatic life (e.g., heavy metals), and will cause dead zones downstream from the pollutant source. Some types of pollutants are health concerns to humans, but have little impact on stream communities. For example, coliform bacteria are an indicator of animal waste contamination, and are therefore an important human health concern, but most aquatic organisms are not harmed by the presence of coliforms.

One of the most common types of stream pollutants is the introduction of biodegradable organic material. When a high-energy organic material such as raw sewage is discharged into a stream, a number of changes occur downstream from the point of discharge. As the organic components of the sewage are oxidized, oxygen is used at a rate greater than that upstream from the sewage discharge, and the dissolved oxygen in the stream decreases markedly. The rate of reaeration, or solution of oxygen from the air, also increases, but is often not enough to prevent total depletion of oxygen in the stream. If the dissolved oxygen is totally depleted, the stream becomes anaerobic. Often, however, the dissolved oxygen does not drop to 0 and the stream recovers without a period of anaerobiosis. Both of these situations are depicted graphically in Fig. 4-5. The dip in dissolved oxygen is referred to as a dissolved oxygen sag curve.

The effect of a biodegradable organic waste on a stream's oxygen level may be estimated mathematically. Let

$z(t)$ = the amount of oxygen still required at time t, in milligrams per liter (mg/L), and
k_1' = the deoxygenation constant, in days^{-1}.

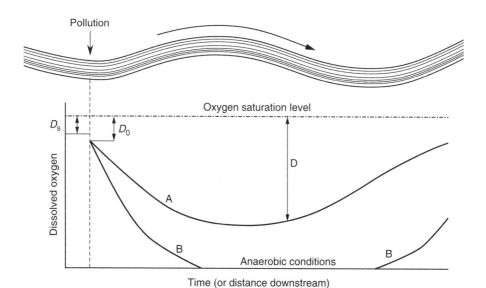

Figure 4-5. Dissolved oxygen downstream from a source of organic pollution. Curve A depicts an oxygen sag without anaerobic conditions; curve B shows an oxygen sag curve when pollution is concentrated enough to create anaerobic conditions, D_0 is the oxygen deficit in the stream after the stream has mixed with the pollutant, and D_s is the oxygen deficit of the upstream water.

62   ENVIRONMENTAL ENGINEERING

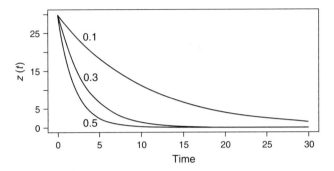

Figure 4-6. Amount of oxygen required at any time $t(z(t))$ for various deoxygenation constants (k_1') when the ultimate carbonaceous oxygen demand (L_0) is 30 mg/L.

The deoxygenation constant k_1' will depend on the type of waste, the temperature, the stream velocity, etc. The rate of change of z over time is proportional to k_1':

$$\left(\frac{d}{dt} z(t)\right) = k_1' z(t). \tag{4.6}$$

This differential equation has a simple solution:

$$z(t) = L_0 e^{-k_1' t}, \tag{4.7}$$

where L_0 is the *ultimate carbonaceous oxygen demand*, in milligrams per liter (mg/L), or the amount of oxygen needed to degrade the carbonaceous organic material in the wastewater at the point where the effluent first enters into and mixes with the stream (see next chapter). This equation is plotted in Fig. 4-6 for various values of k_1', and with $L_0 = 30$ mg/L.

Since the ultimate oxygen requirement is L_0 and the amount of oxygen still needed at any given time is z, the amount of oxygen used after time t, the *biochemical oxygen demand* (BOD), is simply the difference between L_0 and $z(t)$:

$$\text{BOD}(t) = L_0 - z(t) = L_0(1 - e^{-k_1' t}). \tag{4.8}$$

This relationship is plotted in Fig. 4-7, and it can be seen that the BOD asymptotically approaches L_0 as time passes.

Contrasting with this increase in BOD over time is the reoxygenation of the stream by natural forces. This will depend on the difference between the current amount of dissolved oxygen, and the maximum amount of oxygen the water can hold at saturation. In other words, if d is the actual amount of dissolved oxygen in the water, and d_s is the amount of dissolved oxygen at saturation, then

$$\frac{d}{dt} d(t) = k_2'(d_s - d(t)) = k_2' D(t), \tag{4.9}$$

where $D(t)$ is the oxygen deficit at time t, in milligrams per liter (mg/L), and k_2' is the reoxygenation constant, in days^{-1}.

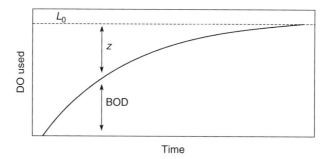

Figure 4-7. Dissolved oxygen used (BOD) at any time t plus the dissolved oxygen still needed at time $t(z(t))$ is equal to the ultimate oxygen demand (L_0).

Table 4-1. Reaeration Constants

Type of watercourse	k_2' at 20°C[a] (days^{-1})
Small ponds or backwaters	0.10–0.23
Sluggish streams	0.23–0.35
Large streams, low velocity	0.35–0.46
Large streams, normal velocity	0.46–0.69
Swift streams	0.69–1.15
Rapids	>1.15

[a] For temperatures other than 20°C, $k_2'(T) = k_2'(20°C)(1.024)^{T-20}$.

The value of k_2' is obtained by studying the stream using a tracer. If this cannot be done, a generalized expression (O'Connor 1958) may be used:

$$k_2' = \frac{3.9v^{1/2}\sqrt{(1.037)^{(T-20)}}}{H^{3/2}}, \quad (4.10)$$

where T is the temperature of the water in degrees Celsius, H is the average depth of flow in meters, and v is the mean stream velocity in meters per second (m/s).

Alternatively, k_2' values may be estimated from a table like Table 4-1.

For a stream loaded with organic material, the simultaneous deoxygenation and reoxygenation of the water forms the dissolved oxygen sag curve, first developed by Streeter and Phelps in 1925 (Streeter and Phelps 1925). The shape of the oxygen sag curve, as shown in Fig. 4-5, is the sum of the rate of oxygen use and the rate of oxygen supply. Immediately downstream from a source of organic pollution the rate of use will often exceed the reaeration rate and the dissolved oxygen concentration will fall sharply. As the discharged organic matter is oxidized, and fewer high-energy organic compounds are left, the rate of use will decrease, the supply will begin to catch up with the use, and the dissolved oxygen will once again reach saturation.

This may be expressed mathematically as:

$$\frac{d}{dt}D(t) = k_1' z(t) - k_2' D(t), \tag{4.11}$$

which can be solved to give:

$$D(t) = \frac{k_1' L_0}{k_2' - k_1'}(e^{-k_1' t} - e^{-k_2' t}) + D_0 e^{-k_2' t}, \tag{4.12}$$

where D_0 is the initial oxygen deficit in the stream at the point of wastewater discharge, after the stream flow has mixed with the wastewater, in milligrams per liter (mg/L).

The deficit equation can also be expressed in common logarithms:

$$D = \frac{k_1 L_0}{k_2 - k_1}(10^{-k_1 t} - 10^{-k_2 t}) + D_0 10^{-k_2 t}, \tag{4.13}$$

since:

$$e^{-k't} = 10^{-kt} \text{ when } k = 0.43 k'.$$

The initial oxygen deficit (D_0) is calculated as a flow-weighted proportion of the initial stream oxygen deficit and the wastewater oxygen deficit:

$$D_0 = \frac{D_s Q_s + D_p Q_p}{Q_s + Q_p}, \tag{4.14}$$

where D_s is the oxygen deficit in the stream directly upstream from the point of discharge, in milligrams per liter (mg/L); Q_s is the stream flow upstream from the wastewater discharge, in cubic meters per second (m³/s); D_p is the oxygen deficit in the wastewater being added to the stream, in milligrams per liter (mg/L); and Q_p is the flow rate of wastewater, in cubic meters per second (m³/s).

Similarly, the ultimate carbonaceous BOD (L_0) is:

$$L_0 = \frac{L_s Q_s + L_p Q_p}{Q_s + Q_p}, \tag{4.15}$$

where L_s is the ultimate BOD in the stream immediately upstream from the point of wastewater discharge, in milligrams per liter (mg/L); Q_s is the stream flow upstream from the wastewater discharge, in cubic meters per second (m³/s); L_p is the ultimate BOD of the wastewater, in milligrams per liter (mg/L); and Q_p is the flow rate of the wastewater, in cubic meters per second (m³/s).

The most serious water quality concern is the downstream location where the oxygen deficit will be the greatest, or where the dissolved oxygen concentration is the lowest. By setting $dD/dt = 0$, we can solve for the time when this minimum dissolved oxygen occurs, the *critical time*, as

$$t_c = \frac{1}{k'_2 - k'_1} \ln \left[\frac{k'_2}{k'_1} \left(1 - \frac{D_0 \left(k'_2 - k'_1\right)}{k'_1 L_0} \right) \right], \qquad (4.16)$$

where t_c is the time downstream when the dissolved oxygen concentration is the lowest.

An actual dissolved oxygen sag curve is shown in Fig. 4-8. Note that the stream becomes anaerobic at about mile 3.5, recovers, then drops back to 0 after receiving effluents from a city and a paper mill. Also shown in the figure is the expected dissolved oxygen sag if 95% of the demand for oxygen is removed from all discharges.

EXAMPLE 4.1. Assume that a large stream has a reoxygenation constant k'_2 of 0.4/day, a flow velocity of 5 miles/h, and at the point of pollutant discharge, the stream is saturated with oxygen at 10 mg/L. The wastewater flow rate is very small compared with the stream flow, so the mixture is assumed to be saturated with dissolved oxygen and to have an oxygen demand of 20 mg/L. The deoxygenation constant k'_1 is 0.2/day. What is the dissolved oxygen level 30 miles downstream?

Stream velocity = 5 miles/h, hence it takes 30/5 or 6 h to travel 30 miles. Therefore, $t = 6$ h/24 h/day $= 0.25$ day, and $D_0 = 0$ because the stream is saturated.

$$D = \frac{(0.2)(20)}{0.4 - 0.2} \left(e^{-(0.2)(0.25)} - e^{-(0.4)(0.25)} \right) = 1.0 \text{ mg/L}.$$

The dissolved oxygen 30 miles downstream will be the saturation level minus the deficit, or $10 - 1.0 = 9.0$ mg/L.

Stream flow is variable, of course, and the critical dissolved oxygen levels can be expected to occur when the flow is the lowest. Accordingly, most state regulatory agencies base their calculations on a statistical low flow, such as a 7-day, 10-year low flow: the 7 consecutive days of lowest flow that may be expected to occur once in 10 years. This is calculated by first estimating the lowest 7-day flow for each year, then assigning ranks: $m = 1$ for the least flow (most severe) to $m = n$ for the greatest flow (least severe), where n is the number of years considered. The probability of occurrence of a flow greater than or equal to a particular low flow is

$$P = \frac{m}{n+1}. \qquad (4.17)$$

When P is graphed against the flow using probability paper, the result is often a straight line (log-probability sometimes gives a better fit). The 10-year low flow can be estimated from the graph at $m/(n+1) = 0.1$ (or 1 year in 10). The data from Example 4.2 are plotted in Fig. 4-9; the minimum 7-day, 10-year low flow is estimated to be 0.5 m^3/s.

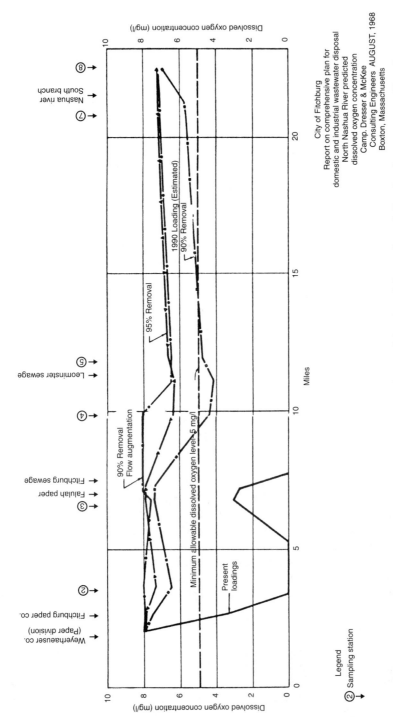

Figure 4-8. Actual dissolved oxygen sag (going to 0; anaerobic conditions) and the projected dissolved oxygen sags if various types of wastewater treatment are provided. (Courtesy of Camp Dresser & McKee, Boston, MA.)

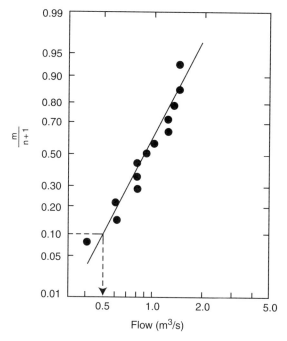

Figure 4-9. Plot of 7-day, 10-year low flows, for Example 4.2.

EXAMPLE 4.2. Calculate the 7-day, 10-year low flow given the data below.

Year	Lowest flow 7 consecutive days (m³/s)	Ranking (m)	m/(n + 1)	Lowest flow in order of severity (m³/s)
1965	1.2	1	1/14 = 0.071	0.4
1966	1.3	2	2/14 = 0.143	0.6
1967	0.8	3	3/14 = 0.214	0.6
1968	1.4	4	4/14 = 0.285	0.8
1969	0.6	5	5/14 = 0.357	0.8
1970	0.4	6	6/14 = 0.428	0.8
1971	0.8	7	7/14 = 0.500	0.9
1972	1.4	8	8/14 = 0.571	1.0
1973	1.2	9	9/14 = 0.642	1.2
1974	1.0	10	10/14 = 0.714	1.2
1975	0.6	11	11/14 = 0.786	1.3
1976	0.8	12	12/14 = 0.857	1.4
1977	0.9	13	13/14 = 0.928	1.4

68 ENVIRONMENTAL ENGINEERING

When the rate of oxygen use overwhelms the rate of oxygen reaeration, the stream may become anaerobic. An anaerobic stream is easily identifiable by the presence of floating sludge, bubbling gas, and a foul smell. The gas is formed because oxygen is no longer available to act as the hydrogen acceptor, and NH_3, H_2S, and other gases are formed. Some of the gases dissolve in water, but others can attach themselves as bubbles to sludge (solid black or dark benthic deposits) and buoy the sludge to the surface. In addition, the odor of H_2S will advertise the anaerobic condition for some distance, the water is usually black or dark, and filamentous bacteria (sewage "fungus") grow in long slimy filaments that cling to rocks and wave graceful streamers downstream.

Other adverse effects on aquatic life accompany the unpleasant physical appearance of an anaerobic stream. The types and numbers of species change drastically downstream from the pollution discharge point. Increased turbidity, settled solid matter, and low dissolved oxygen all contribute to a decrease in fish life. Fewer and fewer species of fish are able to survive, but those species that do survive find food plentiful, and often multiply in large numbers. Carp and catfish can survive in water that is quite foul and can even gulp air from the surface. Trout, on the other hand, need very pure, cold, oxygen-saturated water and are notoriously intolerant of pollution.

The numbers of other aquatic species are also reduced under anaerobic conditions, as shown in Fig. 4-10. The remaining species, like sludge worms, bloodworms, and rat-tailed maggots, abound, often in staggering numbers — as many as 50,000 sludge worms per square foot. The diversity of species may be quantified by using an index, such as the Shannon–Weaver diversity index (Shannon and Weaver 1949),

$$H' = \sum_{i=1}^{S} \left(\frac{n_i}{n}\right) * \ln\left(\frac{n_i}{n}\right), \tag{4.18}$$

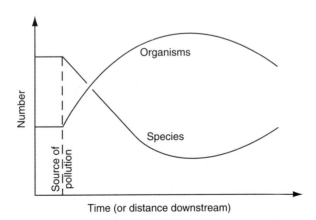

Figure 4-10. The number of species and the total number of organisms downstream from a point of organic pollution.

where H' is the diversity index, n_i is the number of individuals in the ith species, and n is the total number of individuals in all S species.

Diversity indices can be quite difficult to interpret because they are composed of two different measurements: species richness (how many different kinds of organisms are present?) and species equitability (how evenly are the individuals distributed among the species?). One way to overcome this problem is to convert the diversity index into an *equitability* index, such as Pielou's J (Pielou 1975):

$$J = \frac{H'}{\ln S}. \tag{4.19}$$

Pielou's J is a measure of how close H' is to its maximum value for any given sample, approaching 1.0 at maximum equitability. Although still widely used for general comparisons, both H' and J have been replaced with more complex indices that take into account the relative abundance of pollution-tolerant or -intolerant species. Table 4-2 shows a simplified example of biotic diversity and equitability upstream and downstream from a pollution outfall.

As mentioned earlier, nitrogen compounds may be used as indicators of pollution. The changes in the various forms of nitrogen with distance downstream are shown in Fig. 4-11. The first transformation, in both aerobic and anaerobic decomposition, is the formation of ammonia; thus the concentration of ammonia increases as organic nitrogen decreases. As long as the stream remains aerobic, the concentration of nitrate will increase to become the dominant form of nitrogen.

These reactions of a stream to pollution occur when a rapidly decomposable organic material is the waste. The stream will react much differently to inorganic waste, as from a metal-plating plant. If the waste is toxic to aquatic life, both the kind and total number of organisms will decrease downstream from the outfall. The dissolved oxygen will not fall, and might even rise. There are many types of pollution, and a stream will react differently to each. When two or more wastes are involved, the situation is even more complicated.

Table 4-2. Diversity and Equitability of Aquatic Organisms

Species	Pollution tolerance	No. of individuals in sample	
		Upstream from outfall	Downstream from outfall
Mayflies	Intolerant	20	5
Rat-tailed maggots	Tolerant	0	500
Trout	Intolerant	5	0
Carp	Tolerant	1	20
	H' diversity =	0.96	0.22
	Equitability (J) =	0.87	0.20

70 ENVIRONMENTAL ENGINEERING

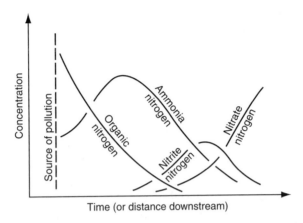

Figure 4-11. Typical variations in nitrogen compounds downstream from a point of organic pollution.

EFFECT OF POLLUTION ON LAKES

The effect of pollution on lakes differs in several respects from the effect on streams. Water movement in lakes is slower than in streams, so reaeration is more of a problem in lakes than streams. Because of the slow movement of water in a lake, sediments, and pollutants bound to sediments, tend to settle out of the water column rather than being transported downstream.

Light and temperature have important influences on a lake, and must be included in any *limnological* analysis (limnology is the study of lakes). Light is the source of energy in the photosynthetic reaction, so the penetration of light into the lake water determines the amount of photosynthesis that can occur at various depths in the lake. Light penetration is logarithmic and a function of wavelength. Short wavelengths (blue, ultraviolet) penetrate farther than long wavelengths (red, infrared). Light penetration at all wavelengths is less in lakes with high concentrations of dissolved organic matter. In pristine lakes, 60–80% of the incident blue/UV light, and 10–50% of the red/IR may penetrate beyond the first 3 ft; in *humic* (boggy) lakes, the presence of large amounts of organic matter causes 90–99% of all wavelengths to be absorbed within the first 3 ft. Because of this, algal growth is concentrated near the surface of a lake, in the *photic zone*, which is limited to the maximum depth where there is still enough light to support photosynthesis.

Temperature and heat often have a profound effect on a lake. Water is at a maximum density at 4°C; warmer or colder water (including ice) is less dense, and will float. Water is also a poor conductor of heat and retains heat quite well.

Lake water temperature usually varies seasonally (see Fig. 4-12). During the winter, if the lake does not freeze, the temperature is relatively constant with depth. As the weather warms in the spring the top layers of water begin to warm. Since warmer water is less dense, and water is a poor conductor of heat, the lake eventually stratifies into a warm, less dense, surface layer called the *epilimnion* and a cooler,

Water Pollution   71

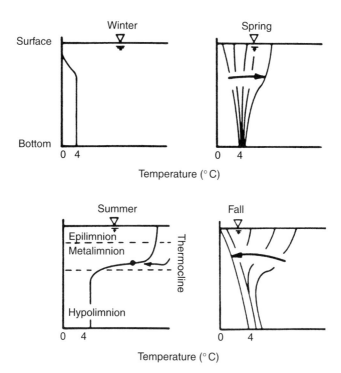

Figure 4-12. Typical temperature–depth relationships in lakes.

denser, bottom layer, the *hypolimnion*. A thermal gradient, the *metalimnion*, is present between these two layers. The inflection point in the temperature gradient is called the *thermocline* (early limnologists used "thermocline" to describe the entire thermal gradient). Circulation of water occurs only within a stratum, and thus there is only limited transfer of biological or chemical material (including dissolved oxygen) between the epilimnion and the hypolimnion. As colder weather approaches, the top layer cools, becomes denser, and sinks. This creates circulation within the lake, known as fall turnover. If the lake freezes over in the winter, the lake surface temperature will be less than 4°C, and the ice will float on top of the slightly denser (but still cold!) underlying water. When spring comes, the lake surface will warm slightly and there will be a spring turnover as the ice thaws.

The biochemical reactions in a natural lake are represented schematically in Fig. 4-13. A river feeding the lake would contribute carbon, phosphorus, and nitrogen, either as high-energy organics or as low-energy compounds. The phytoplankton (free-floating algae) take carbon, phosphorus, and nitrogen, and, using sunlight as an energy source, make high-energy compounds. Algae are eaten by zooplankton (tiny aquatic animals), which are in turn eaten by larger aquatic life such as fish. All of these life forms defecate or excrete waste products, contributing a pool of dissolved organic carbon. This pool is further fed by the death of aquatic life, and by the near-constant leakage of soluble organic compounds from algae into the water. Bacteria use dissolved

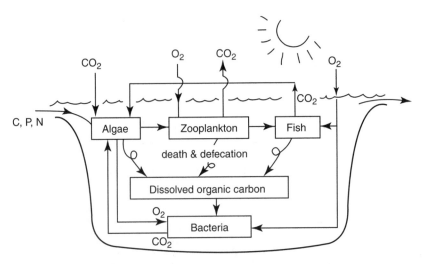

Figure 4-13. Schematic representation of lake ecology. (With thanks to Don Francisco.)

organic carbon and produce carbon dioxide, which is in turn used by algae. Carbon dioxide is also provided by respiration of fish and zooplankton, as well as dissolving into the water directly from the air.

The growth of algae in most lakes is limited by the availability of phosphorus; if phosphorus is in sufficient supply, nitrogen is usually the next *limiting nutrient*. (A limiting nutrient is an essential element or compound that controls the rate of algal growth because the nutrient is not readily available.) Some algal species have special growth requirements that result in co-limitation by other nutrients (e.g., silica is required for diatom growth).

When phosphorus and nitrogen are introduced into the lake, either naturally from storm runoff, or from a pollution source, the nutrients promote rapid growth of algae in the epilimnion. When the algae die, they drop to the lake bottom (the hypolimnion) and become a source of carbon for decomposing bacteria. Aerobic bacteria will use all available dissolved oxygen in the process of decomposing this material, and the dissolved oxygen may be depleted enough to cause the hypolimnion to become anaerobic. As more and more algae die, and more and more dissolved oxygen is used in their decomposition, the metalimnion may also become anaerobic. When this occurs, aerobic biological activity is restricted to the epilimnion.

The increasing frequency of this condition over the years is called eutrophication. Eutrophication is the continually occurring natural process of lake aging and occurs in three stages:

- the oligotrophic stage, which is characterized by low levels of biological productivity and high levels of oxygen in the hypolimnion;

- the mesotrophic stage, which is characterized by moderate levels of biological productivity and the beginnings of declining oxygen levels following lake stratification; and
- the eutrophic stage, at which point the lake is very productive, with extensive algal blooms, and increasingly anaerobic conditions in the hypolimnion.

Natural eutrophication may take thousands of years. If enough nutrients are introduced into a lake system, as may happen as a result of human activity, the eutrophication process may be shortened to as little as a decade.

Because phosphorus is usually the nutrient that limits algal growth in lakes, the addition of phosphorus, in particular, can speed eutrophication. If only phosphorus is introduced into a lake, it will cause some increase in algal growth, but nitrogen quickly becomes a limiting factor for most species of algae. One group of photosynthetic organisms, however, is uniquely adapted to take advantage of high phosphorus concentrations: the *cyanobacteria*, or blue green "algae." Cyanobacteria are autotrophic bacteria that can store excess phosphorus inside their cells in a process called *luxury consumption*. The bacteria use the excess phosphorus to support future cell growth (up to about 20 cell divisions). The cyanobacteria also have the ability to use dissolved N_2 gas as a nitrogen source, which is rapidly replenished by atmospheric N_2. Most other aquatic autotrophs cannot use N_2 as a nitrogen source. As a result, cyanobacteria thrive in environments where nitrogen has become limiting to other algae, and can sustain their growth using cellular phosphorus for long periods of time. Not surprisingly, cyanobacteria are often water quality indicators of phosphorus pollution.

Where do these nutrients originate? One source is excrement, since all human and animal wastes contain organic carbon, nitrogen, and phosphorus. Synthetic detergents and fertilizers are a much greater source. About half of the phosphorus in U.S. lakes is estimated to come from agricultural runoff, about one-fourth from detergents, and the remaining one-fourth from all other sources.

Phosphate concentrations between 0.01 and 0.1 mg/L appear to be enough to accelerate eutrophication. Sewage treatment plant effluents may contain 5–10 mg/L of phosphorus as phosphate, and a river draining farm country may carry 1–4 mg/L. Residential and urban runoff may carry up to 1 mg/L, mostly from pet wastes, detergents, and fertilizer. In moving water, the effects of elevated phosphorus are usually not apparent because the algae are continually flushed out and do not accumulate. Eutrophication occurs mainly in lakes, ponds, estuaries, and sometimes in very sluggish rivers.

Actual profiles in a lake for a number of parameters are shown in Fig. 4-14. The foregoing discussion clarifies why a lake is warmer on top than at lower depths, how dissolved oxygen can drop to 0, and why nitrogen and phosphorus are highly concentrated in the lake depths while algae bloom on the surface.

EFFECT OF POLLUTION ON GROUNDWATER

A popular misconception is that all water that moves through the soil will be purified "naturally" and will emerge from the ground in a pristine condition. Unfortunately,

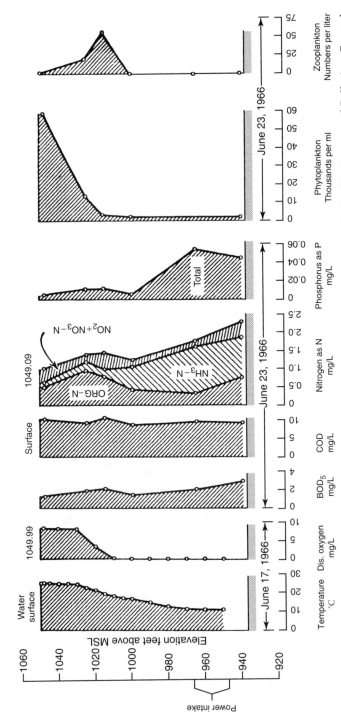

Figure 4-14. Water quality profiles for a water supply reservoir. (From Berthouex, P. and D. Rudd, *Strategy of Pollution Control*. New York: Wiley, 1977.)

there are limits to what soil can remove, and groundwater pollution is becoming an increasing concern throughout the world.

Many soils do have the ability to remove certain types of pollutants, including phosphorus, heavy metals, bacteria, and suspended solids. Pollutants that dissolve in water, like nitrate and ammonia, may pass through soils into the groundwater. In agricultural regions, the nitrogen and other soluble chemicals in fertilizers or animal wastes can seep into the groundwater and show up in alarmingly high concentrations in local drinking water wells. A recent study of the Abbotsford/Sumas aquifer (a water-bearing zone of rock, sand, gravel, etc.), which supplies water to more than 100,000 people in the western portion of Canada and Washington, indicated that 40% of the wells tested had nitrate levels above 10 mg/L (EPA maximum recommended drinking water level), and 60% had nitrate levels above 3 mg/L (a general warning level for nitrate in drinking water).

The agricultural community is becoming more aware of the connection between agricultural practices and groundwater pollution. Many states have begun working with dairy owners and farmers to develop farm management plans that restrict fertilizer applications to periods of active plant growth, which helps prevent groundwater pollution by sequestering nitrate into growing vegetation. These farm plans also include surface water pollution prevention techniques such as restricting animal access to stream banks, setting maximum animal density goals, requiring manure-holding ponds, and revegetating *riparian* (stream side) areas.

Other potential sources of groundwater pollution include leaking underground storage tanks, solid waste landfills, improperly stored hazardous waste, careless disposal of solvents and hazardous chemicals on ground surfaces, and road salts and deicing compounds. Many of the current U.S. Superfund sites (see Chap. 17, "Solid and Hazardous Waste Law") are concerned with the cleanup of materials that have contaminated, or have the potential to contaminate, groundwater.

EFFECT OF POLLUTION ON OCEANS

Not many years ago, the oceans were considered infinite sinks; the immensity of the seas and oceans seemed impervious to assault. Now we know that the seas and oceans are fragile environments and we are able to measure detrimental effects.

Ocean water is a complicated chemical solution, and appears to have changed very little over millions of years. Because of this constancy, however, marine organisms have become specialized and intolerant to environmental change. Oceans are thus fragile ecosystems, quite susceptible to pollution.

A relief map of the ocean bottom reveals two major areas: the continental shelf and the deep oceans. The continental shelf, especially near major estuaries, is the most productive in terms of food supply. Because of its proximity to human activity, it receives the greatest pollution load. Many estuaries have become so badly polluted that they are closed to commercial fishing. The Baltic and Mediterranean Seas are in danger of becoming permanently damaged.

Ocean disposal of untreated wastewater is severely restricted in the United States, but many major cities all over the world still discharge untreated sewage into the ocean. Although the sewage is carried a considerable distance from shore by pipeline and discharged through diffusers to achieve maximum dilution, the practice remains controversial, and the long-term consequences are much in doubt. Even in the United States, most sewage effluents receive only secondary treatment (see Chap. 7, "Water Treatment"), which is not effective at removing certain types of pollutants, including phosphorus.

HEAVY METALS AND TOXIC SUBSTANCES

In 1970, Barry Commoner (Commoner 1970) and other scientists alerted the nation to the growing problem of mercury contamination of lakes, streams, and marine waters. The manufacture of chlorine and lye from brine, called the chlor-alkali process, was identified as a major source of mercury contamination. Elemental mercury is methylated by aquatic organisms (usually anaerobic bacteria), and methylated mercury finds its way into fish and shellfish and thus into the human food chain. Methylmercury is a powerful neurological poison. Methylmercury poisoning was first identified in Japan in the 1950s as "Minamata disease." Mercury-containing effluent from the Minamata Chemical Company was found to be the source of mercury in food fish. Mercury contamination in oceanic fishes is currently widespread, and of sufficient concern that the U.S. Food and Drug Administration issued a consumer alert on March 9, 2001, advising that pregnant women, women of childbearing age, nursing mothers, and young children should avoid eating shark, swordfish, king mackerel, and tilefish. Many states in the United States have issued similar warnings about potentially hazardous levels of mercury or other bioaccumulated toxins in freshwater sport fish.

Arsenic, copper, lead, and cadmium are often deposited in lakes and streams from the air near emitting facilities. These substances may also enter waterways from runoff from slag piles, mine drainage, and industrial effluent. Effluents from electroplating contain a number of heavy metal constituents. Heavy metals, copper in particular, may be toxic to aquatic species as well as harmful to human health.

In the past quarter century, a considerable number of incidents of surface water contamination by hazardous and carcinogenic organic compounds were reported in the United States. The sources of these include effluent from petrochemical industries and agricultural runoff, which contains both pesticide and fertilizer residues. Trace quantities of chlorinated hydrocarbon compounds in drinking water may also be attributed to the chlorination of organic residues by chlorine added as a disinfectant. The production of these disinfection by-products is difficult to eliminate in the drinking water treatment process, but maintaining clean, unpolluted, source water is the first step.

CONCLUSION

Water pollution stems from many sources and causes, only a few of which are discussed here. Rivers and streams demonstrate some capacity to recover from the effects of certain pollutants, but lakes, bays, ponds, sluggish rivers, and oceans have little resistance to the effects of water pollution. We have a long history of introducing pollutants into aquatic environments, and have had only partial success at repairing the damage that has already been done and curbing the activities that result in environmental degradation. Nonpoint source pollution continues to be a serious threat to receiving waters, as does the continued release of sewage and industrial effluents throughout the world. As we have seen with mercury contamination in fishes, environmental pollution can have widespread and lasting consequences.

PROBLEMS

4.1 Some researchers have suggested that the empirical analysis of some algae gives it the chemical composition $C_{106}H_{181}O_{45}N_{16}P$. Suppose that analysis of lake water yields the following: C = 62 mg/L, N = 1.0 mg/L, P = 0.01 mg/L. Which element would be the limiting nutrient for the growth of algae in this lake?

4.2 A stream feeding a lake has an average flow of 1 ft^3/s and a phosphate concentration of 10 mg/L. The water leaving the lake has a phosphate concentration of 5 mg/L.

a. What weight of phosphate is deposited in the lake each year?
b. Where does this phosphorus go, since the outflow concentration is less than the inflow concentration?
c. Would the average phosphate concentration be higher near the surface of the lake or near the bottom?
d. Would you expect eutrophication of the lake to be accelerated? Why?

4.3 The following water quality data were collected at two stream sites. What can you say about the conditions at each site?

	Dissolved oxygen (mg/L)	Ammonia (mg/L)	Phosphorus (mg/L)
Site 1	1.8	1.5	15
Site 2	12.5	<0.01	<0.01

4.4 Show how the compound thiodiazine — $C_{21}H_{26}N_2S_2$ — decomposes anaerobically and how these end products in turn decompose aerobically to stabilized sulfur and nitrogen compounds.

4.5 If an industrial plant discharges an effluent with a solids concentration of 5,000 lb/day, and if each person contributes 0.2 lb/day, what is the population equivalent of the waste?

78 ENVIRONMENTAL ENGINEERING

4.6 The temperature soundings and dissolved oxygen measurements for a lake are as follows:

Depth (ft)	Temperature (°F)	Oxygen (mg/L)
0 (surface)	70	9.5
10	70	9.5
20	70	9.5
30	50	4.0
40	40	2.0
50	40	0.0
60	40	0.0

Plot depth vs temperature and depth vs oxygen. Label the hypolimnion, epilimnion, and thermocline. Is this lake oligotrophic or eutrophic?

4.7 Draw the dissolved oxygen sag curves you would expect in a stream from the following wastes. Assume the stream flow equals the flow of wastewater. (Do not calculate.)

Waste source	BOD (mg/L)	Susp. solids (mg/L)	Phosphorus (mg/L)
Dairy	2000	100	40
Brick manufacturing	5	100	10
Fertilizer manufacturing	25	5	200
Electroplating plant	0	100	10

4.8 Starting with nitrogenous dead organic matter, follow N around the aerobic and anaerobic cycles by writing down all the various forms of nitrogen.

4.9 Describe what is occurring in the following process. Is this process aerobic or anaerobic? Where might you expect this process to occur?

$$2NH_4 + 3O_2 \rightarrow 2NO_2^- + H_2O + 4H^+$$

$$2NO_2^- + O_2 \rightarrow 2NO_3^-$$

4.10 Suppose a stream with a velocity of 1 ft/s, a flow of 10 million gallons per day (mgd), and an ultimate BOD of 5 mg/L was hit with treated sewage at 5 mgd with an ultimate BOD (L_0) of 60 mg/L. The temperature of the stream water is 20°C, at the point of the sewage discharge the stream is 90% saturated with oxygen, and the wastewater is at 30°C and has no oxygen (see Table 4-1). Measurements show the deoxygenation constant $k_1' = 0.5$ and the reoxygenation $k_2' = 0.6$, both as days^{-1}. Calculate: (a) the oxygen deficit 1 mile downstream, (b) the minimum dissolved oxygen (the lowest part of the sag curve), and (c) the minimum dissolved oxygen (or maximum deficit) if the ultimate BOD of the treatment plant's effluent was 10 mg/L. You may write a computer program or use a spreadsheet program to solve this problem.

4.11 An industry discharges sufficient quantities of organic wastes to depress the oxygen sag curve to 2 mg/L 5 miles downstream, at which point the dissolved oxygen begins to increase. The state regulatory agency finds the industry out of compliance because the minimum dissolved oxygen violates state standards, and wants to impose a fine. A wastewater treatment plant is too expensive, so the industry looks around for another solution. A salesman suggests that a special kind of freeze-dried bacteria, particularly adapted to the plant's organic waste, be dumped into the river at the point of discharge: the freeze-dried bacteria will come to life, break down the organic material, and solve the problem. Will this approach work? Why or why not? Draw the dissolved oxygen sag curve before and after the bacteria are added and explain what happens.

4.12 An industry discharges into a river with a velocity of 0.1 m/s, a temperature of 26°C, and an average depth of 4 m. The ultimate oxygen demand immediately downstream from the effluent discharge point where the waste has mixed with the river water (L_0) is 14.5 mg/L and the dissolved oxygen is 5.4 mg/L. Laboratory studies show that $k'_1 = 0.11$ day^{-1}. Find the point of critical dissolved oxygen and the minimum dissolved oxygen. You may solve this problem by writing a computer program or by using a spreadsheet program.

Chapter 5

Measurement of Water Quality

Quantitative measurements of pollutants are obviously necessary before water pollution can be controlled. Measurement of these pollutants is, however, fraught with difficulties. Sometimes specific materials responsible for the pollution are not known. Moreover, these pollutants are generally present at low concentrations, and very accurate methods of detection are required.

Only a few of the analytical tests available to measure water pollution are discussed in this chapter. A complete volume of analytical techniques used in water and wastewater engineering is compiled as *Standard Methods for the Examination of Water and Wastewater* (Clesceri *et al.* 1998). This volume is updated every few years so that it can incorporate current information on standardized testing techniques. It is considered definitive in its field and has the weight of legal authority.

Many water pollutants are measured in terms of milligrams of the substance per liter of water (mg/L). In older publications pollutant concentrations were often expressed as parts per million (ppm), a weight/weight parameter. If the only liquid involved is water, ppm is identical with mg/L, since one liter (L) of water weighs 1000 grams (g). For many aquatic pollutants, ppm is approximately equal to mg/L; however, because of the possibility that some wastes have specific gravity different from water, mg/L is preferred to ppm.

SAMPLING

Some tests require the measurement to be conducted at the site because the process of obtaining a sample may change the measurement. For example, to measure the dissolved oxygen in a stream or lake, either the measurement should be conducted at the site or the sample must be extracted with great care to ensure that there has been no loss or addition of oxygen as the sample is exposed to the air. Similarly, it is better to measure pH at the site if you are sampling water that is poorly buffered from pH changes (see discussion on alkalinity).

Most tests may be performed on a water sample taken from the stream. The process by which the sample is obtained, however, may greatly influence the result. The three basic types of samples are grab samples, composite samples, and flow-weighted composite samples.

The grab sample, as the name implies, measures water quality at only one sampling point. Grab samples accurately represent the water quality at the moment of sampling, but say nothing about the quality before or after the sampling. A composite sample is obtained by taking a series of grab samples and mixing them together. The flow-weighted composite is obtained by taking each sample so that the volume of the sample is proportional to the flow at that time. The last method is especially useful when daily loadings to wastewater treatment plants are calculated. Whatever the technique or method, however, the analysis can only be as accurate as the sample, and often the sampling methods are far more sloppy than the analytical determination.

DISSOLVED OXYGEN

One of the most important measures of water quality is dissolved oxygen. Oxygen, although poorly soluble in water, is fundamental to aquatic life. Without free dissolved oxygen, streams and lakes become uninhabitable to aerobic organisms, including fish and most invertebrates. Dissolved oxygen is inversely proportional to temperature, and the maximum amount of oxygen that can be dissolved in water at 0°C is 14.6 mg/L. The saturation value decreases rapidly with increasing water temperature, as shown in Table 5-1. The balance between saturation and depletion is therefore tenuous.

The amount of oxygen dissolved in water is usually measured either with an oxygen probe or by iodometric titration. The latter method, known as the *Winkler* test,

Table 5-1. Solubility of Oxygen in Water

Water temperature (°C)	Saturation concentration of oxygen in water (mg/L)
0	14.6
2	13.8
4	13.1
6	12.5
8	11.9
10	11.3
12	10.8
14	10.4
16	10.0
18	9.5
20	9.2
22	8.8
24	8.5
26	8.2
28	8.0
30	7.6

was developed about 100 years ago and is the standard against which all other measurements are compared. The chemical reactions of the Winkler test are as follows:

Manganous sulfate ($MnSO_4$) and a mixture of potassium hydroxide and potassium iodide (KOH and KI) are added to a water sample. If there is no oxygen present, the $MnSO_4$ will react with the KOH to form a white precipitate, manganous hydroxide ($Mn(OH)_2$). If oxygen is present, the $Mn(OH)_2$ will react further to form a brown precipitate, manganic oxide ($MnO(OH)_2$):

$$MnSO_4 + 2KOH \rightarrow Mn(OH)_2 + K_2SO_4 \tag{5.1}$$

$$2Mn(OH)_2 + O_2 \rightarrow 2MnO(OH)_2. \tag{5.2}$$

Sulfuric acid is added, which dissolves the manganic oxide and, in conjunction with the KI added earlier, forms iodine (I_2), which imparts a yellowish orange color to the sample:

$$2MnO(OH)_2 + 4H_2SO_4 \rightarrow 2Mn(SO_4)_2 + 6H_2O \tag{5.3}$$

$$2Mn(SO_4)_2 + 4KI \rightarrow 2MnSO_4 + 2K_2SO_4 + 2I_2. \tag{5.4}$$

The quantity of iodine is measured by titrating with sodium thiosulfate ($Na_2S_2O_3$) until the orange color from I_2 is no longer apparent:

$$4Na_2S_2O_3 + 2I_2 \rightarrow 2Na_2S_4O_6 + 4NaI. \tag{5.5}$$

Starch is added near the end of the titration because it turns deep purple in the presence of I_2, and gives a more obvious color endpoint for the test.

The quantity of $MnO(OH)_2$ formed in the first step is directly proportional to the available dissolved oxygen, and the amount of iodine formed in the second step is directly proportional to the $MnO(OH)_4$. Therefore, the titration measures a quantity of iodine directly related to the original dissolved oxygen concentration. Disadvantages of the Winkler test include chemical interferences and the inconvenience of performing a wet chemical test in the field. These two disadvantages can be overcome by using a dissolved oxygen electrode, or probe.

The simplest (and historically the first) type of oxygen probe is shown in Fig. 5-1. The principle of operation is that of a galvanic cell. If lead and silver electrodes are put in an electrolyte solution with a microammeter between, the reaction at the lead electrode is

$$Pb + 2OH^- \rightarrow PbO + H_2O + 2e^-. \tag{5.6}$$

At the lead electrode, electrons are liberated and travel through the microammeter to the silver electrode where the following reaction takes place:

$$2e^- + \tfrac{1}{2}O_2 + H_2O \rightarrow 2OH^-. \tag{5.7}$$

The reaction does not occur, and the microammeter does not register any current, unless free dissolved oxygen is available. The meter must be constructed and calibrated so that

84   ENVIRONMENTAL ENGINEERING

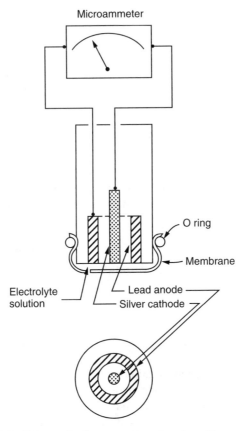

Figure 5-1. Schematic diagram of a galvanic cell oxygen probe.

the electricity recorded is proportional to the concentration of oxygen in the electrolyte solution.

In the commercial models, the electrodes are insulated from each other with nonconducting plastic and are covered with a permeable membrane with a few drops of electrolyte between the membrane and electrodes. The amount of oxygen that travels through the membrane is proportional to the dissolved oxygen concentration. Dissolved oxygen probes are convenient for fieldwork, but need careful maintenance and calibration (usually against Winkler results). Most oxygen probes are sensitive to changes in temperature and have thermisters attached to the probe so that temperature adjustments can be made in the field.

BIOCHEMICAL OXYGEN DEMAND

The rate of oxygen use is commonly referred to as biochemical oxygen demand (BOD). Biochemical oxygen demand is not a specific pollutant, but rather a measure of the amount of oxygen required by bacteria and other microorganisms engaged in stabilizing decomposable organic matter over a specified period of time.

The BOD test is often used to estimate the impacts of effluents that contain large amounts of biodegradable organics such as that from food processing plants and feedlots, municipal wastewater treatment facilities, and pulp mills. A high oxygen demand indicates the potential for developing a dissolved oxygen sag (see previous chapter) as the microbiota oxidize the organic matter in the effluent. A very low oxygen demand indicates either clean water or the presence of a toxic or nondegradable pollutant.

The BOD test was first used in the late 1800s by the Royal Commission on Sewage Disposal as a measure of the amount of organic pollution in British rivers. At that time, the test was standardized to run for 5 days at 18.3°C. These numbers were chosen because none of the British rivers had headwater-to-sea travel times greater than 5 days, and the average summer temperature for the rivers was 18.3°C. Accordingly, this should reveal the "worst case" oxygen demand in any British river. The BOD incubation temperature was later rounded to 20°C, but the 5-day test period remains the current, if somewhat arbitrary, standard.

In its simplest version, the 5-day BOD test (BOD_5) begins by placing water or effluent samples into two standard 60- or 300-mL BOD bottles (Fig. 5-2). One sample is analyzed immediately to measure the initial dissolved oxygen concentration in the effluent, often using a Winkler titration. The second BOD bottle is sealed and stored at 20°C in the dark. (The samples are stored in the dark to avoid photosynthetic oxygen generation.) After 5 days the amount of dissolved oxygen remaining in the sample is measured. The difference between the initial and ending oxygen concentrations is the BOD_5.

The oxidation of organic matter follows an exponential decay curve, as in Fig. 4-6. If the dissolved oxygen concentrations were measured daily, the results would produce curves like those shown in Fig. 5-3. In this example, sample A had an initial dissolved oxygen concentration of 8 mg/L, which dropped to 2 mg/L in 5 days. The BOD therefore is $8 - 2 = 6$ mg/L.

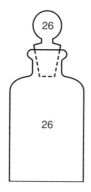

Figure 5-2. A biochemical oxygen demand (BOD) bottle.

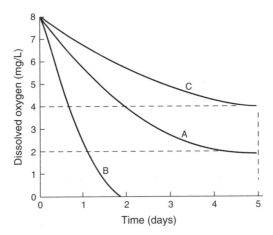

Figure 5-3. Typical oxygen uptake curves in a BOD test.

Sample B also had an initial dissolved oxygen concentration of 8 mg/L, but the oxygen was used so fast that it dropped to 0 by the second day. Since there is no measurable dissolved oxygen left after 5 days, the BOD of sample B must be more than $8 - 0 = 8$ mg/L, but we do not know how much more because the organisms in the sample might have used more dissolved oxygen if it had been available. Samples like this require diluting the sample. Typically, five $\frac{1}{10}$ dilutions are recommended for wastewaters of unknown origin. Suppose sample C in Fig. 5-3 is sample B diluted by $\frac{1}{10}$. The BOD$_5$ for sample B would be

$$\frac{8-4}{0.1} = 40 \text{ mg/L}.$$

It is possible to measure the BOD of any organic material (e.g., sugar) and thus estimate its influence on a stream, even though the material in its original state might not contain the microorganisms necessary to break down organic matter. *Seeding* is a process in which the microorganisms that oxidize organic matter are added to the BOD bottle. Seeding also facilitates measurement of very low BOD concentrations. The seed source can be obtained from unchlorinated domestic wastewater or surface water that receives degradable wastewater effluents.

Suppose we use the water previously described in curve A as seed water since it obviously contains microorganisms (it has a 5-day BOD of 6 mg/L). We now put 100 mL of an unknown solution into a bottle and add 200 mL of seed water, thus filling the 300-mL bottle. Assuming that the initial dissolved oxygen of this mixture is 8 mg/L and the final dissolved oxygen is 1 mg/L, the total oxygen consumed is 7 mg/L. Some of this is due to the seed water, because it also has a BOD, and only a portion is due to the decomposition of the unknown material. The oxygen consumed due to the seed water is

$$6 \times \frac{2}{3} = 4 \text{ mg/L},$$

because only two-thirds of the bottle is seed water, and only the seed water has a BOD of 6 mg/L. The remaining oxygen consumed (7 − 4 = 3 mg/L) must be due to the unknown material. Equation (5.8) shows how to calculate the BOD$_5$ for a diluted, seeded effluent sample,

$$\text{BOD (mg/L)} = \frac{(I - F) - (I' - F')(X/Y)}{D}, \tag{5.8}$$

where

I = initial dissolved oxygen in the bottle containing both effluent sample and seeded dilution water,
F = final dissolved oxygen in the bottle containing the effluent and seeded dilution water,
I' = initial dissolved oxygen of the seeded dilution water,
F' = final dissolved oxygen of the seeded dilution water,
X = mL of seeded dilution water in sample bottle,
Y = total mL in the bottle, and
D = dilution of the sample.

EXAMPLE 5.1. Calculate the BOD$_5$ of a water sample, given the following data:

— Temperature of sample = 20°C,
— Initial dissolved oxygen is saturation,
— Dilution is 1:30, with seeded dilution water,
— Final dissolved oxygen of seeded dilution water is 8 mg/L,
— Final dissolved oxygen bottle with sample and seeded dilution water is 2 mg/L, and
— Volume of BOD bottle is 300 mL.

From Table 5-1, dissolved oxygen saturation at 20°C is 9.2 mg/L; hence, this is the initial dissolved oxygen. Since the BOD bottle contains 300 mL, a 1:30 dilution with seeded water would contain 10 mL of sample and 290 mL of seeded dilution water, and, by Eq. (5.8)

$$\text{BOD}_5 \text{ (mg/L)} = \frac{(9.2 - 2) - (9.2 - 8)(290/300)}{0.033} = 183 \text{ mg/L}$$

BOD is a measure of oxygen use, or potential oxygen use. An effluent with a high BOD may be harmful to a stream if the oxygen consumption is great enough to cause anaerobic conditions. Obviously, a small trickle going into a great river will have negligible effect, regardless of the BOD concentration involved. Conversely, a large flow into a small stream may seriously affect the stream even though the

BOD concentration might be low. Engineers often talk of "pounds of BOD," a value calculated by multiplying the concentration by the flow rate, with a conversion factor, so that

$$\text{lb BOD/day} = [\text{mg/L BOD}] \times \begin{bmatrix} \text{flow in million} \\ \text{gallons per day} \end{bmatrix} \times 8.34. \quad (5.9)$$

The BOD of most domestic sewage is about 250 mg/L, while many industrial wastes run as high as 30,000 mg/L. The potential detrimental effect of untreated dairy waste that might have a BOD of 20,000 mg/L is quite obvious.

As discussed in Chapter 4, the BOD curve can be modeled using Eq. (4.8):

$$\text{BOD}(t) = L_0(1 - e^{-k_1't}),$$

where

$\text{BOD}(t)$ = amount of oxygen required by the microorganisms at any time t (mg/L),
L_0 = ultimate carbonaceous oxygen demand (mg/L),
k_1' = deoxygenation rate constant (days^{-1}), and
t = time (days).

When it is necessary to know both k_1' and L_0, as when modeling the dissolved oxygen profile in a stream, both are measured using laboratory BOD tests.

There are a number of techniques for calculating k_1' and L_0. One of the simplest is a method devised by Thomas (1950). Rewriting Eq. (4.8) using common logarithms results in

$$\text{BOD}(t) = L_0(1 - 10^{-k_1't}),$$

which can be rearranged to read

$$\left(\frac{t}{\text{BOD}(t)}\right)^{1/3} = (2.3\,k_1'L_0)^{-1/3} + \left(\frac{k_1'^{2/3}}{3.43 L_0^{1/3}}\right)t. \quad (5.10)$$

This equation is in the form of a straight line

$$x = a + bt,$$

where x is $(t/\text{BOD}(t))^{1/3}$, the intercept (a) is $(2.3 k_1' L_0)^{-1/3}$, and the slope (b) is $k_1'^{2/3}/(3.43 L_0^{1/3})$.

By plotting BOD versus t, the slope (b) and intercept (a) can be used to solve for k_1' and L_0:

$$k_1' = 2.61\,(b/a)$$

$$L_0 = 1/(2.3 k_1' a^3).$$

EXAMPLE 5.2. The BOD versus time data for the first 5 days of a BOD test are obtained as follows:

Time (days)	BOD (mg/L)
2	10
4	16
6	20

Calculate k_1' and L_0.

The $(t/BOD(t))^{1/3}$ values 0.585, 0.630, and 0.669 are plotted as shown in Fig. 5-4. The intercept $(a) = 0.545$ and the slope $(b) = 0.021$. Thus:

$$k_1' = 2.61 \left(\frac{0.021}{0.545}\right) = 0.10 \text{ day}^{-1}$$

$$L_0 = \frac{1}{2.3(0.10)(1.545)^3} = 26.8 \text{ mg/L}.$$

If, instead of stopping the BOD test after 5 days, we allowed the test to continue and measured the dissolved oxygen each day, we might get a curve like that shown in Fig. 5-5. Note that after about 5 days the curve turns sharply upward. This discontinuity is due to the demand for oxygen by the microorganisms that decompose nitrogenous organic compounds to inorganic nitrogen. In the following example, microorganisms decompose a simple organic nitrogen compound, urea ($NH_2 \cdot CO \cdot NH_2$), releasing

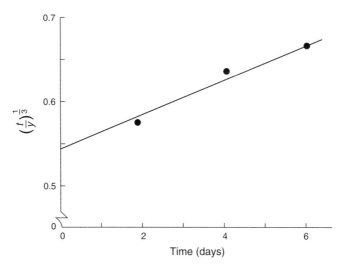

Figure 5-4. Plot of k_1' and L_0, for Example 5.2.

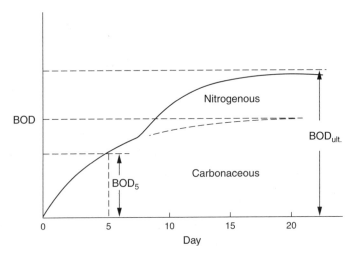

Figure 5-5. Long-term BOD. Note that BOD_{ult} here includes both ultimate carbonaceous BOD (L_0) and ultimate nitrogenous BOD.

ammonia (NH_3; NH_4^+ in ionized form), which is further decomposed into nitrite (NO_2^-) and nitrate (NO_3^-):

$$NH_2 \cdot CO \cdot NH_2 + H_2O \rightarrow 2NH_3 + CO_2 \qquad \text{ammonification} \qquad (5.11)$$

$$NH_4^+ + 1\tfrac{1}{2}O_2 \rightarrow NO_2^- + 2H^+ + H_2O \qquad \text{nitrification, step 1} \qquad (5.12)$$

$$NO_2^- + \tfrac{1}{2}O_2 \rightarrow NO_3^-. \qquad \text{nitrification, step 2} \qquad (5.13)$$

Note that the first step, ammonification, does not require oxygen; it can be done by a wide variety of aerobic and anaerobic plants, animals, and microbes.

The BOD curve is thus divided into nitrogenous and carbonaceous BOD areas. The ultimate BOD, as shown in Fig. 5-5, includes both nitrogenous and carbonaceous BOD. For streams and rivers with travel times greater than about 5 days, the ultimate demand for oxygen must include the nitrogenous demand.

Although the use of BOD_{ult} (carbonaceous plus nitrogenous) in dissolved oxygen sag calculations is not strictly accurate, the ultimate BOD may be estimated as

$$BOD_{ult} = a(BOD_5) + b(TKN), \qquad (5.14)$$

where TKN is the total Kjeldahl nitrogen (organic nitrogen plus ammonia, in mg/L), and a and b are constants.

The state of North Carolina, for example, has used $a = 1.2$ and $b = 4.0$ for calculating the ultimate BOD, which was then substituted for the ultimate carbonaceous BOD (L_0) in the dissolved oxygen sag equation.

CHEMICAL OXYGEN DEMAND

One problem with the BOD test is that it takes 5 days to run. If the organic compounds were oxidized chemically instead of biologically, the test could be shortened considerably. Such oxidation can be accomplished with the chemical oxygen demand (COD) test. Because nearly all organic compounds are oxidized in the COD test, while only some are decomposed during the BOD test, COD results are always higher than BOD results. One example of this is wood pulping waste, in which compounds such as cellulose are easily oxidized chemically (high COD) but are very slow to decompose biologically (low BOD).

The standard COD test uses a mixture of potassium dichromate and sulfuric acid to oxidize the organic matter (HCOH), with silver (Ag^+) added as a catalyst. A simplified example of this reaction is illustrated below, using dichromate ($Cr_2O_7^{2-}$) and hydrogen ions (H^+):

$$2Cr_2O_7^{2-} + 3HCOH + 16H^+ \xrightarrow{heat + Ag^+} 3CO_2 + 11H_2O + 4Cr^{3+}. \quad (5.15)$$

A known amount of a solution of $K_2Cr_2O_7$ in moderately concentrated sulfuric acid is added to a measured amount of sample, and the mixture is boiled in air. In this reaction, the oxidizing agent, hexavalent chromium (Cr^{VI}), is reduced to trivalent chromium (Cr^{III}). After boiling, the remaining Cr^{VI} is titrated against a reducing agent, usually ferrous ammonium sulfate. The difference between the initial amount of Cr^{VI} added to the sample and the Cr^{VI} remaining after the organic matter has been oxidized is proportional to the chemical oxygen demand.

TOTAL ORGANIC CARBON

Since the ultimate oxidation of organic carbon is to CO_2, the total combustion of a sample yields some information about the potential oxygen demand in an effluent sample. A far more common application of total organic carbon testing is to assess the potential for creating *disinfection by-products*. Disinfection by-products are the result of halogens (e.g., bromine, chlorine) or ozone interacting with naturally occurring organic carbon compounds during the drinking water disinfection process. For example, trihalomethane, a carcinogen, is created when halogens displace three hydrogen ions on methane. Water that is high in total organic carbon has a greater potential to develop disinfection by-products. Some of the organics can be removed by adding levels of treatment specific for organic carbon absorption; however, it is usually not economically feasible to remove all naturally occurring organics from finished drinking water.

Total organic carbon is measured by oxidizing the organic carbon to CO_2 and H_2O and measuring the CO_2 gas using an infrared carbon analyzer. The oxidation is done by direct injection of the sample into a high-temperature (680–950°C) combustion chamber or by placing a sample into a vial containing an oxidizing agent such as potassium persulfate, sealing and heating the sample to complete the oxidation, then measuring the CO_2 using the carbon analyzer.

TURBIDITY

Water that is not clear but is "dirty," in the sense that light transmission is inhibited, is known as turbid water. Many materials can cause turbidity, including clays and other tiny inorganic particles, algae, and organic matter. In the drinking water treatment process, turbidity is of great importance, partly because turbid water is aesthetically displeasing, and also because the presence of tiny colloidal particles makes it more difficult to remove or inactivate pathogenic organisms.

Turbidity is measured using a turbidimeter. Turbidimeters are photometers that measure the intensity of scattered light. Opaque particles scatter light, so scattered light measured at right angles to a beam of incident light is proportional to the turbidity. Formazin polymer is currently used as the primary standard for calibrating turbidimeters, and the results are reported as nephelometric turbidity units (NTU).

COLOR, TASTE, AND ODOR

Color, taste, and odor are important measurements for determining drinking water quality. Along with turbidity, color, taste, and odor are important from the standpoint of aesthetics. If water looks colored, smells bad, or tastes swampy, people will instinctively avoid using it, even though it might be perfectly safe from the public health aspect. Color, taste, and odor problems in drinking water are often caused by organic substances such as algae or humic compounds, or by dissolved compounds such as iron.

Color can be measured visually by comparison with potassium chloroplatinate standards or by scanning at different spectrophotometric wavelengths. Turbidity interferes with color determinations, so the samples are filtered or centrifuged to remove suspended material. Odor is measured by successive dilutions of the sample with odor-free water until the odor is no longer detectable. (Odor-free water is prepared by passing distilled, deionized water through an activated charcoal filter.) This test is obviously subjective and depends entirely on the olfactory senses of the tester. Panels of testers are used to compensate for variations in individual perceptions of odor.

Taste is evaluated using three methods: the flavor threshold test (FTT), the flavor rating assessment (FRA), and the flavor profile analysis (FPA). For the FTT, water samples are diluted with increasing amounts of reference water until a panel of taste testers concludes that there is no perceptible flavor. In the FRA, a panel of testers is asked to rate the flavor from very favorable to very unfavorable. The oldest, and most useful, of the taste tests is the FPA, which measures both taste and odor of a water sample in comparison to taste and odor reference standards. The intensity of specific tastes and odors are described on a 12-point, ranging from no taste or odor (0) to taste or odor (12).

pH

The pH of a solution is a measure of hydrogen (H^+) ion concentration, which is, in turn, a measure of acidity. Pure water dissociates slightly into equal concentrations of hydrogen and hydroxyl (OH$^-$) ions:

$$H_2O \leftrightarrow H^+ + OH^-. \tag{5.16}$$

An excess of hydrogen ions makes a solution acidic, whereas a dearth of H$^+$ ions, or excess of hydroxyl ions, makes it basic. The equilibrium constant for this reaction, K_w, is the product of H$^+$ and OH$^-$ concentrations and is equal to 10^{-14}. This relationship may be expressed as

$$[H^+][OH^-] = K_w = 10^{-14}, \tag{5.17}$$

where [H$^+$] and [OH$^-$] are the concentrations of hydrogen and hydroxyl ions, respectively, in moles per liter. Considering Eq. (5.16) and solving Eq. (5.17), in pure water, [H$^+$] and [OH$^-$] are in equal concentrations:

$$[H^+] = [OH^-] = 10^{-7} \text{ moles/L}.$$

The hydrogen ion concentration is so important in aqueous solutions that an easier method of expressing it has been devised. Instead of speaking in terms of moles per liter, we define a quantity pH as the negative logarithm of [H$^+$] so that

$$\text{pH} = -\log_{10}[H^+] = \log_{10}\frac{1}{[H^+]} \tag{5.18}$$

or

$$[H^+] = 10^{-\text{pH}}. \tag{5.19}$$

In a neutral solution the H$^+$ concentration is 10^{-7}, so the pH is 7. As the H$^+$ concentration increases the pH decreases. For example, if the H$^+$ concentration is 10^{-4}, the pH is 4, and the solution is acidic. In this solution, we see that the OH$^-$ concentration is $10^{-14}/10^{-4}$, or 10^{-10}. Since 10^{-4} is much greater than 10^{-10}, the solution contains a large excess of H$^+$ ions, confirming that it is indeed acidic. Any solution where the H$^+$ concentration is less than 10^{-7}, or the pH is greater than 7, would be basic. The pH range in dilute samples is from 0 (very acidic) to 14 (very alkaline), and in water samples is rarely below 4 or above 10.

The measurement of pH is now almost universally done using electronic pH meters. A typical pH meter consists of a potentiometer, a glass electrode and a reference electrode (or a single, "combination" electrode), and a temperature-compensating device. The glass electrode is sensitive to H$^+$ activity and converts the signal to electric current, which can be read as electrode potential (mV) or pH.

The pH of an effluent or water sample is important in almost all phases of drinking water and wastewater treatment. In water treatment as well as in disinfection and corrosion control, pH is important in ensuring proper chemical treatment. Aquatic organisms are sensitive to pH changes, as well as to the actual pH of the water. Few aquatic organisms tolerate waters with a pH less than 4 or greater than 10. Acid mine drainage, unregulated acids or bases in industrial effluents, or atmospheric acid deposition may alter the pH of a water body substantially and have detrimental effects on aquatic life.

ALKALINITY

Alkalinity measures the buffering capacity of the water against changes in pH. Water that has a high alkalinity can accept large doses of acids or bases without altering the pH significantly. Waters with low alkalinity, such as rainwater or distilled water, can experience a drop in the pH with only a minor addition of an acid or base.

In natural waters much of the alkalinity is provided by the carbonate/bicarbonate buffering system. Carbon dioxide (CO_2) dissolves in water to form carbonic acid (H_2CO_3), which dissociates and is in equilibrium with bicarbonate (HCO_3^-) and carbonate (CO_3^{2-}) ions:

$$CO_2 \text{ (gas)} \leftrightarrow CO_2 \text{ (dissolved)} \tag{5.20}$$

$$CO_2 \text{ (dissolved)} + H_2O \leftrightarrow H_2CO_3 \leftrightarrow H^+ + HCO_3^- \leftrightarrow 2H^+ + CO_3^{2-}. \tag{5.21}$$

If an acid is added to the water, the hydrogen ion concentration is increased, and this combines with both the carbonate and bicarbonate ions, driving the equilibrium to the left, releasing carbon dioxide into the atmosphere. As long as there is bicarbonate and carbonate, the added hydrogen ions will be absorbed by readjustment of the equilibrium equation. Only when all of the carbonate and bicarbonate ions are depleted will the addition of acid cause a drop in pH.

The amount of bicarbonate in water is supplemented by naturally occurring carbonates such as $CaCO_3$ (limestone) that dissolve as acidic rain water comes into contact with watershed soils or the stream bed. The $CaCO_3$ dissolves to form calcium bicarbonate ($Ca(HCO_3)_2$), which dissociates and increases the bicarbonate concentration in the water:

$$CaCO_3 + H_2CO_3 \rightarrow Ca(HCO_3)_2 \leftrightarrow Ca^{2+} + 2HCO_3^-. \tag{5.22}$$

The effect of alkalinity on the pH of a water sample is shown in Fig. 5-6.

Alkalinity is determined by measuring the amount of acid needed to lower the pH in a water sample to a specific endpoint; the results are usually reported in standardized units as milligrams $CaCO_3$ per liter. Poorly buffered water may have alkalinities lower than 40 mg $CaCO_3$/L while water sampled from a stream flowing through a limestone or "karst" region may have alkalinities greater than 200 mg $CaCO_3$/L.

SOLIDS

Wastewater treatment is complicated by the dissolved and suspended inorganic material it contains. In discussion of water treatment, both dissolved and suspended materials are called solids. The separation of these solids from the water is one of the primary objectives of treatment.

Total solids include any material left in a container after the water is removed by evaporation, usually at 103–105°C. Total solids can be separated into *total suspended solids* (solids that are retained on a 2.0-μm filter) and *total dissolved solids*

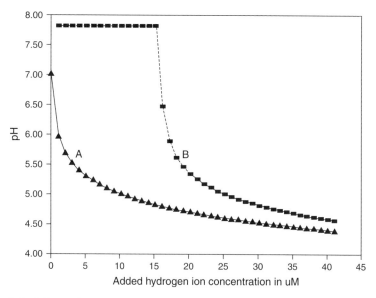

Figure 5-6. Effect of alkalinity in buffering against pH changes. (A) acid is added to deionized water (very low alkalinity); (B) acid is added to monobasic phosphate buffer solution (high alkalinity).

(dissolved and colloidal material that passes through the filter). The difference between total suspended solids and total dissolved solids is illustrated in the following example:

A teaspoonful of table salt dissolves in a glass of water, forming a water-clear solution. However, the salt will remain behind if the water evaporates. Sand, however, will not dissolve and will remain as sand grains in the water and form a turbid mixture. The sand will also remain behind if the water evaporates. The salt is an example of a dissolved solid, whereas the sand is a suspended solid.

Suspended solids are separated from dissolved solids using a special crucible, called a *Gooch crucible*. The Gooch crucible has holes on the bottom on which a glass fiber filter is placed (Fig. 5-7). The water sample is drawn through the crucible with the aid of a vacuum. The suspended material is retained on the filter, while the dissolved fraction passes through. If the initial dry weight of the crucible and filter is known, the subtraction of this from the total weight of the crucible, filter, and the dried solids caught in the filter yields the weight of suspended solids, expressed in milligrams per liter.

Solids may be classified in another way: those that are volatilized at a high temperature (550°C) and those that are not. The former are known as *volatile solids*, the later as *fixed solids*. Volatile solids are usually organic compounds. At 550°C some inorganics are also decomposed and volatilized, but this is not considered as a serious drawback. Example 5.3 illustrates the relationship between total solids and total volatile solids.

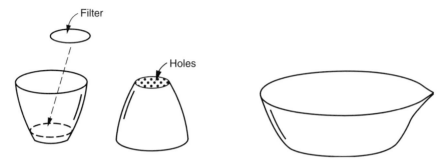

Figure 5-7. The Gooch crucible, with filter, for determining suspended solids, and the evaporating dish used for determining total solids.

EXAMPLE 5.3. Given the following data:

—Weight of a dish (such as shown in Fig. 5-7) = 48.6212 g,
—100 mL of sample is placed in the dish and evaporated. New weight of dish and dry solids = 48.6432 g.
—The dish is placed in a 550°C furnace, then cooled. New weight = 48.6300 g.

Find the total, volatile, and fixed solids.

$$\text{Total solids} = \frac{(\text{dish} + \text{dry solids}) - (\text{dish})}{\text{sample volume}}$$

$$= \frac{48.6432 - 48.6212}{100}$$

$$= (220)10^{-6} \text{ g/mL}$$

$$= (220)10^{-3} \text{ mg/mL}$$

$$= 220 \text{ mg/L}.$$

$$\text{Fixed solids} = \frac{(\text{dish} + \text{unburned solids}) - (\text{dish})}{\text{sample volume}}$$

$$= \frac{48.6300 - 48.6212}{100}$$

$$= 88 \text{ mg/L}$$

$$\text{Total volatile solids} = \text{Total solids} - \text{Total fixed solids}$$

$$= 220 - 88$$

$$= 132 \text{ mg/L}$$

NITROGEN AND PHOSPHORUS

Recall from Chap. 4 that nitrogen and phosphorus are important nutrients for biological growth. Nitrogen occurs in five major forms in aquatic environments: organic nitrogen, ammonia, nitrite, nitrate, and dissolved nitrogen gas; phosphorus occurs almost entirely as organic phosphate and inorganic orthophosphate or polyphosphates.

Ammonia is one of the intermediate compounds formed during biological metabolism and, together with organic nitrogen, is considered an indicator of recent pollution. Aerobic decomposition of organic nitrogen and ammonia eventually produces nitrite (NO_2^-) and finally nitrate (NO_3^-). High nitrate concentrations, therefore, may indicate that organic nitrogen pollution occurred far enough upstream that the organics have had time to oxidize completely. Similarly, nitrate may be high in groundwater after land application of organic fertilizers if there is sufficient residence time (and available oxygen) in the soils to allow oxidation of the organic nitrogen in the fertilizer.

Because ammonia and organic nitrogen are pollution indicators, these two forms of nitrogen are often combined in one measure, called *Kjeldahl nitrogen*, after the scientist who first suggested the analytical procedure. A popular alternative to the technically difficult Kjeldahl test is to measure total nitrogen and nitrate + nitrite separately. The difference between the two concentrations equals organic nitrogen plus ammonia.

Phosphorus is usually measured as total phosphorus (all forms combined) or dissolved phosphorus (portion that passes through a 0.45-μm membrane filter). Dissolved orthophosphate (PO_4^-) is an important indicator of water pollution because it is easily and rapidly taken up by biota, and therefore is almost never found in high concentrations in unpolluted waters.

The various forms of nitrogen and phosphorus can all be measured analytically by colorimetric techniques. In colorimetry, the ion in question combines with a reagent to form a colored compound; the color intensity is proportional to the original concentration of the ion. For example, in the "Phenate Method" for ammonia analysis, an intensely blue compound (indophenol) is created from the reaction between ammonia, hypochlorite, and phenol, with sodium nitroprusside as a catalyst (Clesceri *et al.* 1999). The color is measured photometrically, or occasionally by visual comparison to color standards.

A photometer, illustrated in Fig. 5-8, consists of a light source, a filter, the sample, and a photocell. The filter allows only those wavelengths of light to pass through that the compounds being measured will absorb. Light passes through the sample to the photocell, which converts light energy into electric current. An intensely colored sample will absorb a considerable amount of light and allow only a limited amount of light to pass through and thus create little current. On the other hand, a sample containing very little of the chemical in question will be lighter in color and allow almost all of the light to pass through, and set up a substantial current.

The intensity of light transmitted by the colored solution obeys the *Beer–Lambert Law*

$$\log_{10} \frac{P_0}{P} = ebc = A, \tag{5.23}$$

98   ENVIRONMENTAL ENGINEERING

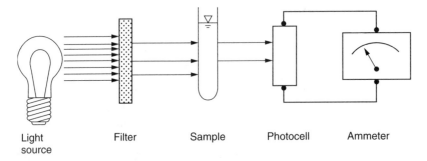

Figure 5-8. Elements of a filter photometer.

where

P_0 = radiant power of incident light,
P = radiant power of light after it passes through the sample,
e = absorbtivity (L mol^{-1} cm^{-1}),
b = path length (cm),
c = concentration of absorbing substance (moles L^{-1}), and
A = absorbance (no units).

A photometer, as shown in Fig. 5-8, measures the difference between the intensity of light passing through the sample (P in Eq. (5.23)) and the intensity of light passing through clear distilled water or a reference sample (P_0) and reports *absorbance* (A) as well as *percent transmission* (%T):

$$A = \log_{10} \frac{1}{T} = \log_{10} \frac{100}{\%T}. \tag{5.24}$$

Typically, in colorimetric analyses, a standard dilution series is used to estimate the concentration of an unknown sample, as illustrated in Example 5.4.

EXAMPLE 5.4. Several known concentrations of ammonia and an unknown sample were analyzed using the phenate method, and the color was measured with a photometer. Find the ammonia concentration of the unknown sample.

Standards	Absorbance
0 μg/L of ammonia	0.050
5 μg/L of ammonia	0.085
10 μg/L of ammonia	0.121
50 μg/L of ammonia	0.402
100 μg/L of ammonia	0.747
350 μg/L of ammonia	2.450
Unknown sample	1.082

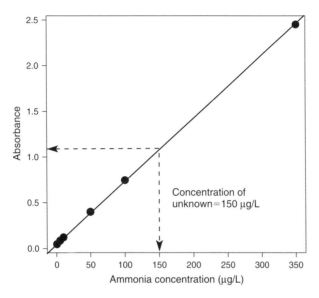

Figure 5-9. Calculation using colorimetric standards.

From Fig. 5-9 where ammonia concentration of the standards vs absorbance results in a straight line, we see that an absorbance of 1.082 (the unknown) corresponds to an ammonia concentration of 150 μg/L.

Although most nitrogen and phosphorus analyses are done using a spectrophotometer, other techniques are growing in acceptance. Selective ion electrodes are available for measuring ammonia, nitrite, and nitrate (the pH meters described earlier are selective ion electrodes that measure H^+). Ion chromatography (ICP) can be used to measure nitrite, nitrate, and phosphate, as well as total nitrogen and total phosphorus if the samples are first digested (oxidized) to convert all forms of nitrogen or phosphorus to nitrate and phosphate. Ion chromatography involves passing a water sample through a series of ion exchange columns that separate the anions so that they are released to a detector at different times. For simple (i.e., not particularly accurate) measurements, field kits that provide premeasured packets of chemicals for testing nitrogen and phosphorus in water and soil samples are now available. These kits usually use colorimetric techniques similar to the more sophisticated versions used in analytical labs, but rely on color reference cards rather than a spectrophotometer for determining chemical concentrations.

PATHOGENS

From the public health standpoint, the bacteriological quality of water is as important as the chemical quality. A large number of infectious diseases may be transmitted by water,

Table 5-2. Examples of Common Waterborne Pathogens

Microorganism	Effects on humans
Bacteria	
Campylobacter	Gastroenteritis
Clostridium botulinum	Gastroenteritis (botulism)
Clostridium perfringens	Gastroenteritus
E. coli O157:H7	Gastroenteritis
Legionella	Pneumonia-like pulmonary disease
Salmonella paratyphi	Paratyphoid
Salmonella typhi	Typhoid fever
Shigella (several species)	Shigellosis (dysentery)
Staphylococcus aureus	Gastroenteritis
Vibrio comma (*V. cholerae*)	Cholera
Yersinia enterocolitica	Gastroenteritis
Protozoans	
Cryptosporidium	Cryptosporidiasis
Entamoeba histolytica	Amoebic dysentery
Giardia lamblia	Giardiasis
Viruses	
Hepatitis A virus	Hepatitis
Poliovirus	Poliomyelitis

among them typhoid and cholera. Although we clearly desire drinking water that is not contaminated by *pathogens* (disease-causing organisms), determining whether the organisms are present in water, and whether they represent a health threat, is relatively complicated. First, there are many pathogens. Table 5-2 lists just a few of the most common waterborne microbial pathogens. Each has a specific detection procedure and must be screened individually. Second, the concentration of these organisms, although large enough to spread disease, may be so small as to make their detection impossible, like the proverbial needle in a haystack.

How can we measure for bacteriological quality? The answer lies in the concept of *indicator organisms* that, while not necessarily directly harmful, indicate the possible presence of other pathogens.

The indicator most often used is *Escherichia coli* (*E. coli*), a member of the *coliform* bacteria group (coliform bacteria are nonspore forming, rod-shaped bacteria capable of fermenting lactose within 48 h at 35°C). Although many coliforms occur naturally in aquatic environments, *E. coli*, often called fecal coliforms, are associated with the digestive tracts of warm-blooded animals. Fecal coliforms are particularly good indicator organisms because they are easily detected with a simple test, generally harmless (some strains are very pathogenic, but most are not), and do not survive long outside their host. The presence of fecal coliforms in a water sample does not prove the presence of pathogens, nor does the absence of fecal coliforms ensure the absence of pathogens. However, if a large number of fecal coliforms are present, there is a good chance of recent pollution by wastes from warm-blooded animals.

This last point should be emphasized. The presence of coliforms does not prove that there are pathogenic organisms in the water, but indicates that such organisms *might* be present. A high coliform count is thus suspicious, and the water should not be consumed, even though it may be safe.

There are several ways to measure fecal coliforms. One of the most widely used methods is the membrane filter (MF) technique. A water sample is filtered through a sterile micropore filter by suction, thereby capturing any coliforms. The filter is placed in a Petri dish containing a sterile culture medium that promotes the growth of the fecal coliforms while inhibiting other organisms. After 24 h of incubation at 35°C, the number of shiny metallic red dots (fecal coliform colonies) is counted. The concentration of coliforms is typically expressed as coliforms/100 mL of sample. The equipment used for such tests is shown in Fig. 5-10.

The second method of measuring for coliforms is called the most probable number (MPN) test. This test is based on the observation that in lactose broth, coliforms will produce gas and make the broth cloudy. The production of gas is detected by placing a small tube upside down inside a larger tube (Fig. 5-11) so as not to have air bubbles in the smaller tube. After incubation, if gas is produced, some of the gas will become trapped in the smaller tube and this, along with a cloudy broth, will indicate that tube has been inoculated with at least one coliform. The MPN test is often used when the sample is very turbid, brackish, or from a mud or sediment sample, all of which are conditions that interfere with the MF technique.

A third way of measuring the coliforms is by a proprietary device called a Coli-Count. A sterile pad with all the necessary nutrients is dipped into the water sample and incubated, and the colonies are counted. The pad is designed to absorb exactly 1 mL of sample water so that the colonies counted give a coliform concentration per milliliter. Although fast and simple, Coli-Count results are not accepted for testing drinking water.

A growing concern in pathogen testing is detecting the presence of virulent strains of *E. coli* (e.g., *E. coli* O157:H7) in food and drinking water supplies. The standard MF and MPN tests do not distinguish between pathogenic and harmless strains of *E. coli*; genetic testing is normally used to determine which strains of the bacteria are present.

Over the past two decades we have seen an increasing emphasis on using other indicator microorganisms to supplement or replace the *E. coli* test. For example, the enterococcus subgroup of fecal streptococcus bacteria (*Streptococcus faecalis*, *S. faecium*, *S. gallinarum*, and *S. avium*) has been found to be excellent indicators of the quality of recreational waters (e.g., swimming beaches). As with *E. coli*, enterococcus bacteria are normal inhabitants in the gastrointestinal tract of warm-blooded animals and are easily enumerated using membrane filtration followed by incubation on selective growth medium.

Pathogenic viruses constitute a particularly difficult group of organisms to identify and enumerate. Because of this, routine testing for viruses is rarely done unless there is an outbreak of disease or you are testing the safety of reclaimed wastewater. (Low coliform counts are not a reliable measure of pathogen inactivation in reclaimed

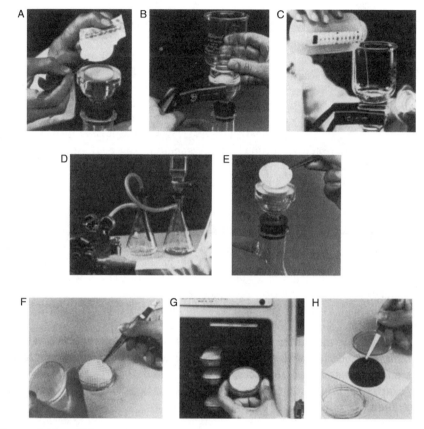

Figure 5-10. Millipore filter apparatus for coliform measurement. The procedure is illustrated in the sequence of photos starting at top left: (A) Millipore filter is put on filter support; (B) funnel is replaced; (C) measured sample is poured into funnel; (D) suction is applied using laboratory vacuum; (E) suction is released and filter removed; (F) filter is centered on growth medium in Petri dish; (G) Petri dish is incubated at 35°C for 24 h; (H) coliform colonies are counted using a microscope.

wastewater because some pathogens are more resistant to disinfection methods than coliforms.)

HEAVY METALS

Heavy metals such as arsenic, copper, and mercury can harm aquatic organisms, or bioaccumulate in the food chain, even if the metal concentration in water is relatively low. Consequently, the method of measuring metals in water must be very sensitive. There are a large variety of methods available to measure metals in water samples, and the choice of method often depends on the desired sensitivity as well as cost. Heavy metals are usually measured using flame, electrothermal (graphite furnace), or

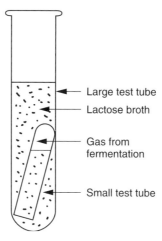

Figure 5-11. Test tubes used for most probable number (MPN) coliform test.

cold-vapor atomic absorption (AA), inductively coupled plasma (ICP) and inductively coupled plasma/mass spectrometry (ICP/MS), and colorimetric techniques. Samples can be filtered and analyzed for dissolved metals or digested using strong acids to measure total metals.

In flame AA a solution of lanthanum chloride is added to the sample, and the treated sample is sprayed into a flame using an atomizer. Each metallic element in the sample imparts a characteristic color to the flame, whose intensity is then measured spectrophotometrically. Graphite furnace AA methods use an electrically heated device to atomize metal elements, and can measure much lower concentrations of metals than flame AA, but often have "matrix" interference problems caused by salts and other compounds in the sample. Cold vapor AA is used primarily to measure arsenic and mercury. Inductively coupled plasma and ICP/MS are less sensitive to matrix problems and cover a wide range of concentrations.

OTHER ORGANIC COMPOUNDS

One of the most diverse (and difficult) areas of pollution assessment is the measurement of toxic, carcinogenic, or other potentially harmful organic compounds in water. These organics encompass the disinfection by-products introduced earlier, as well as pesticides, detergents, industrial chemicals, petroleum hydrocarbons, and degradation products that these chemicals become as they are altered chemically or biologically in the environment (e.g., DDT biodegrades to hazardous DDD and DDE).

Some of the methods described earlier in this chapter can be used to assess the overall content of organics in water (e.g., total organic carbon analysis). *Gas chromatography* (GC) and *high-performance liquid chromatography* (HPLC) are effective methods for measuring minute quantities of specific organics. Gas chromatography uses a mobile phase (carrier gas) and a stationary phase (column packed with an inert

104   ENVIRONMENTAL ENGINEERING

granular solid) to separate organic chemicals. The organics are vaporized, and then allowed to move through the column at different rates that are unique to each organic chemical. After separation in the column, the amount of each organic is determined using a detector that is sensitive to the type of organic being measured. Peak height and travel time are used to identify and quantify each organic. High-performance liquid chromatography is similar to gas chromatography except that the mobile phase is a high-pressure liquid solvent.

CONCLUSION

Only a few of the most important tests used in water pollution control have been discussed in this chapter. Hundreds of analytical procedures have been documented, many of which can be performed only with special equipment and skilled technicians. Understanding this, and realizing the complexity, variations, and objectives of some of the measurements of water pollution, how would you answer someone who brings a jug of water to your office, sets it on your desk, and asks, "Can you tell me if this water is polluted?"

PROBLEMS

5.1 Given the following BOD test results:

— Initial dissolved oxygen = 8 mg/L
— Final dissolved oxygen = 0 mg/L
— Dilution = 1/10

what can you say about

a. BOD_5?
b. BOD ultimate?
c. COD?

5.2 If you had two bottles full of lake water and kept one dark and the other in daylight, which would have a higher dissolved oxygen after a few days? Why?

5.3 Name three types of samples you would need to seed if you wanted to measure their BOD.

5.4 The following data were obtained for a sample:

— total solids = 4000 mg/L
— suspended solids = 5000 mg/L
— volatile suspended solids = 2000 mg/L
— fixed suspended solids = 1000 mg/L

Which of these numbers is questionable and why?

5.5 A sample of water has a BOD_5 of 10 mg/L. The initial dissolved oxygen in the BOD bottle was 8 mg/L, and the dilution was 1/10. What was the final dissolved oxygen in the BOD bottle?

5.6 If the BOD_5 of a waste is 100 mg/L, draw a curve showing the effect of adding progressively higher doses of chromium (a toxic chemical) on the BOD_5.

5.7 An industry discharges 10 million gallons a day of waste that has a BOD_5 of 2000 mg/L. How many pounds of BOD_5 are discharged?

5.8 If you dumped half a gallon of milk every day into a stream, what would be your discharge in pounds of BOD_5 per day?

5.9 Consider the following data from a BOD test:

Day	DO (mg/L)	Day	DO (mg/L)
0	9	5	6
1	9	10	6
2	9	15	4
3	8	20	3
4	7	25	3

What is the: (a) BOD_5, (b) ultimate carbonaceous BOD (L_0), and (c) ultimate BOD (BOD_{ult})? Why was there no oxygen used until the third day? If the sample had been seeded, would the final dissolved oxygen have been higher or lower? Why?

5.10 Given the same standard ammonia samples as in Example 5.4, what would the ammonia concentration be if your unknown measured an absorbance of 0.050?

5.11 Suppose you ran a multiple-tube coliform test and got the following results: 10-mL samples, all 5 positive; 1-mL samples, all 5 positive; 0.1-mL samples, all 5 negative. Use the table in *Standard Methods* to estimate the concentration of coliforms.

5.12 If fecal coliform bacteria were to be used as an indicator of viral pollution as well as an indicator of bacterial pollution, what attributes must *E. coli* have relative to viruses?

5.13 Draw a typical BOD curve. Label the (a) ultimate carbonaceous BOD (L_0), (b) ultimate nitrogenous BOD, (c) ultimate BOD (BOD_{ult}), and (d) 5-day BOD. On the same graph, plot the BOD curve if the test had been run at 30°C instead of at the usual temperature. Also plot the BOD curve if a substantial amount of toxic materials were added to the sample.

5.14 Consider the BOD data below (no dilution, no seed).

a. What is the ultimate carbonaceous BOD?
b. What might have caused the lag at the beginning of the test?
c. Calculate k_1' (the reaction rate constant).

Day	DO (mg/L)
0	8
1	8
3	7
5	6.5
7	6
15	4
20	4

5.15 A sample of wastewater is estimated to have a BOD of 200 mg/L.

a. What dilution is necessary for this BOD to be measured by the usual technique?
b. If the initial and final dissolved oxygen of the test thus conducted is 9.0 and 4.0 mg/L, and the dilution water has a BOD of 1.0 mg/L, what is the dilution?

Chapter 6

Water Supply

A supply of water is critical to the survival of life as we know it. People need water to drink, animals need water to drink, and plants need water to drink. The basic functions of society require water: cleaning for public health, consumption for industrial processes, and cooling for electrical generation. In this chapter, we discuss water supply in terms of:

- the hydrologic cycle and water availability,
- groundwater supplies,
- surface water supplies, and
- water transmission.

The direction of our discussion is that sufficient water supplies exist for the world, and for the nation as a whole, but many areas are water poor while others are water rich. Adequate water supply requires engineering the supply and its transmission from one area to another, keeping in mind the environmental effects of water transmission systems. In many cases, moving the population to the water may be less environmentally damaging than moving the water. This chapter concentrates on measurement of water supply, and the following chapter discusses treatment methods available to clean up the water once it reaches areas of demand.

THE HYDROLOGIC CYCLE AND WATER AVAILABILITY

The hydrologic cycle is a useful starting point for the study of water supply. This cycle, illustrated in Fig. 6-1, includes precipitation of water from clouds, infiltration into the ground or runoff into surface water, followed by evaporation and transpiration of the water back into the atmosphere. The rates of precipitation and evaporation/transpiration help define the baseline quantity of water available for human consumption. *Precipitation* is the term applied to all forms of moisture falling to the ground, and a range of instruments and techniques for measuring the amount and intensity of rain, snow, sleet, and hail have been developed. The average depth of precipitation over a given region, on a storm, seasonal, or annual basis, is required in many water availability studies. Any open receptacle with vertical sides is a common rain gauge, but varying wind and splash effects must be considered if amounts collected by different gauges are to be compared.

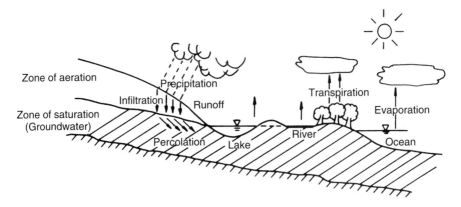

Figure 6-1. The hydrologic cycle.

Evaporation and *transpiration* are the movement of water back to the atmosphere from open water surfaces and from plant respiration. The same meteorological factors that influence evaporation are at work in the transpiration process: solar radiation, ambient air temperature, humidity, and wind speed. The amount of soil moisture available to plants also affects the transpiration rate. Evaporation is measured by measuring water loss from a pan. Transpiration can be measured with a *phytometer*, a large vessel filled with soil and potted with selected plants. The soil surface is hermetically sealed to prevent evaporation; thus moisture can escape only through transpiration. Rate of moisture escape is determined by weighing the entire system at intervals up to the life of the plant. Phytometers cannot simulate natural conditions, so results have limited value. However, they can be used as an index of water demand by a crop under field conditions, and thus relate to calculations that help an engineer determine water supply requirements for that crop. Because it is often not necessary to distinguish between evaporation and transpiration, the two processes are often linked as *evapotranspiration*, or the total water loss to the atmosphere.

GROUNDWATER SUPPLIES

Groundwater is both an important direct source of supply that is tapped by wells and a significant indirect source, since surface streams are often supplied by subterranean water.

Near the surface of the earth, in the *zone of aeration*, soil pore spaces contain both air and water. This zone, which may have zero thickness in swamplands and be several hundred feet thick in mountainous regions, contains three types of moisture. After a storm, *gravity water* is in transit through the larger soil pore spaces. *Capillary water* is drawn through small pore spaces by capillary action and is available for plant uptake. *Hygroscopic moisture* is water held in place by molecular forces during all except the driest climatic conditions. Moisture from the zone of aeration cannot be tapped as a water supply source.

In the *zone of saturation*, located below the zone of aeration, the soil pores are filled with water, and this is what we call *groundwater*. A stratum that contains a substantial amount of groundwater is called an *aquifer*. At the surface between the two zones, called the *water table* or *phreatic surface*, the hydrostatic pressure in the groundwater is equal to the atmospheric pressure. An aquifer may extend to great depths, but because the weight of overburden material generally closes pore spaces, little water is found at depths greater than 600 m (2000 ft).

The amount of water that can be stored in the aquifer is the volume of the void spaces between the soil grains. The fraction of voids volume to total volume of the soil is termed *porosity*, so that

$$\text{Porosity} = (\text{Volume of voids})/(\text{Total volume}). \tag{6.1}$$

However, not all of this water is available because it is so tightly tied to the soil particles. The amount of water that can be extracted is known as *specific yield*, defined as the percent of total volume of water in the aquifer that will drain freely from the aquifer.

The flow of water out of a soil is illustrated in Fig. 6-2 and analyzed using the continuity equation, as

$$Q = Av, \tag{6.2}$$

where

Q = flow rate (m³/s),
A = area of porous material through which flow occurs (m²), and
v = superficial velocity (m/s).

The superficial velocity is of course not the actual velocity of the water in the soil, since the volume occupied by the solid particles greatly reduces the available area for flow. If a is the area available for flow, then

$$Q = Av = av',$$

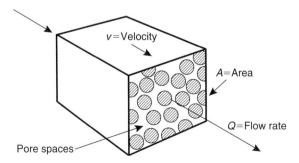

Figure 6-2. The flow of water through a soil sampler.

where v' = the actual velocity in the soil, and a = the area available for flow. Solving for v',

$$v' = (Av)/a. \tag{6.3}$$

If a sample of soil is of some length L, then

$$v' = (Av)/a = (AvL)/(aL) = v/\text{porosity}. \tag{6.4}$$

Water flowing through the soil at a velocity v' loses energy, just as water flowing through a pipeline or an open channel. The head loss is defined as

$$\frac{\Delta h}{\Delta L}, \tag{6.5}$$

where h is the pressure head. The flow through a porous medium such as soil is related to the head loss using the Darcy equation,

$$Q = KA \frac{\Delta h}{\Delta L}, \tag{6.6}$$

where

K = coefficient of permeability (m³/day-m²), and
A = cross-sectional area (m²).

The *coefficient of permeability* varies dramatically for different soils ranging from about 0.04 m³/day-m² for clay to over 200 m³/day-m² for gravel.

Typical values of porosity specific yield and the coefficient of permeability are shown in Table 6-1. The coefficient of permeability is measured commonly in the laboratory using a permeameter, which consists of a soil sample through which a fluid like water is forced. The flow rate is measured for a given driving force, and the permeability calculated.

Table 6-1. Estimate of Average Permeability and Porosity for Selected Materials[a]

Material	% Porosity	% Specific yield	Permeability (gal/day-ft²)	K_p (m³/day-m²)
Clay	45	3	1	0.04
Sand	35	25	800	32
Gravel	25	22	5000	200
Sandstone	15	8	700	28
Granite	1	0.5	0.1	0.004

[a] *Source*: R. K. Linsley and J. B. Franzini, *Elements of Hydrology*, McGraw–Hill, New York. Copyright © 1958. Used with permission of the McGraw–Hill Book Company.

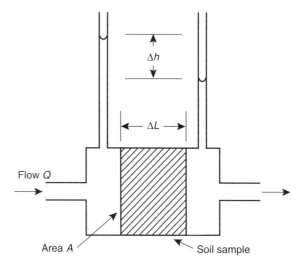

Figure 6-3. Permeameter for Example 6.1.

EXAMPLE 6.1. A soil sample is installed in a permeameter as shown in Fig. 6-3. The length of the sample is 0.1 m, and it has a cross-sectional area of 0.05 m². The water pressure placed on the sample is 2 m, and a flow rate of 2.0 m³/day is observed. What is the coefficient of permeability?

Using the above equation, and solving for K, we have

$$K = \frac{Q}{A(\Delta h/\Delta L)} = \frac{2.0}{0.05 \times (2/0.1)} = 2 \text{ m}^3/\text{m}^2\text{-day}.$$

An aquifer confined between two impermeable surfaces (*aquacludes*) is called a *confined aquifer* and can be thought of as a very large permeameter. The pressure loss due to the flow can be determined by measuring the water level in two wells, the second one being directly downstream of the other.

EXAMPLE 6.2. A confined aquifer is 6 m deep and the coefficient of permeability in the soil is 2 m³/day-m². The wells are 100 m apart, and the difference in the water elevation in the wells is 3.0 m. Find the flow rate and the superficial velocity through the aquifer.

The slope of the pressure gradient, $\Delta h/\Delta L = 3/100 = 0.03$, and the flow rate for a section of aquifer 1 m wide is

$$Q = KA\frac{\Delta h}{\Delta L} = 2 \times 6 \times 0.03 = 0.35 \text{ m}^3/\text{day}.$$

The superficial velocity is

$$v = \frac{Q}{A} = \frac{0.36}{1 \times 6} = 0.06 \text{ m/day}.$$

If a well is sunk into an unconfined aquifer, shown in Fig. 6-4, and water is pumped out, the water in the aquifer will begin to flow toward the well. As the water approaches the well, the area through which it flows gets progressively smaller, and therefore a higher superficial (and actual) velocity is required. The higher velocity of course results in an increasing loss of energy, and the pressure gradient must increase, forming a *cone of depression*. The reduction in the water table is known in groundwater terms as a *drawdown*. If the rate of water flowing toward the well is equal to the rate of water being pumped out of the well, the condition is at equilibrium, and the drawdown remains constant. If, however, the rate of water pumping is increased, the radial flow toward the well must compensate, and this results in a deeper cone or drawdown.

Consider a cylinder shown in Fig. 6-5 through which water flows toward the center. Using Eq. (6.6),

$$Q = KA\frac{\Delta h}{\Delta L} = K(2\pi r h)\frac{dh}{dr},$$

where r = radius of the cylinder and $2\pi r h$ = surface area of the cylinder. If water is pumped out of the center of the cylinder at the same rate as water is moving in through

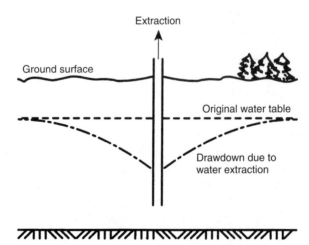

Figure 6-4. Drawdown in the groundwater table when water is pumped out of a well.

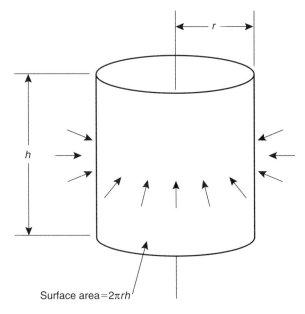

Figure 6-5. A cylinder with water flowing through its sides toward the center.

the cylinder surface area, the above equation can be integrated

$$\int_r^{r_1} Q\frac{dr}{r} = 2\pi K \int_h^{h_1} h\,dh$$

or

$$Q \ln \frac{r_2}{r_1} = \pi K(h_1^2 - h_2^2) \quad (6.7a)$$

$$Q = \frac{\pi K(h_1^2 - h_2^2)}{\ln(r_1/r_2)}. \quad (6.7b)$$

Note that the integration is between any two arbitrary values of r and h.

These equations can be used to estimate the pumping rate for a given drawdown any distance away from a well, using the water level measurements in two observation wells in an *unconfined aquifer* (where the water table is free to change), as shown in Fig. 6-6. Also, knowing the diameter of a well, it is possible to estimate the drawdown at the well, the critical point in the cone of depression. If the drawdown is depressed all the way to the bottom of the aquifer, the well "goes dry" — it cannot pump water at the desired rate. Although the derivation of the foregoing equations are for an unconfined aquifer, the same situation would occur for a confined aquifer, where the pressure would be measured by observation wells.

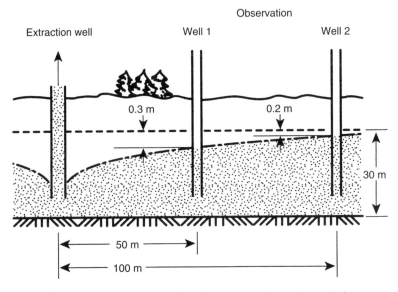

Figure 6-6. Two monitoring wells define the extent of drawdown during extraction.

EXAMPLE 6.3. A well is 0.2 m in diameter and pumps from an unconfined aquifer 30 m deep at an equilibrium (steady-state) rate of 1000 m³ per day. Two observation wells are located at distances 50 and 100 m, and they have been drawn down by 0.2 and 0.3 m, respectively. What is the coefficient of permeability and estimated drawdown at the well?

$$K = \frac{Q \ln(r_1/r_2)}{\pi \left(h_1^2 - h_2^2\right)} = \frac{1000 \ln(100/50)}{\pi \left[(29.8)^2 - (29.7)^2\right]} = 37.1 \text{ m}^3/\text{m}^2\text{-day}.$$

Now if the radius of the well is assumed to be 0.2/2 = 0.1 m, this can be plugged into the same equation, as

$$Q = \frac{\pi K \left(h_1^2 - h_2^2\right)}{\ln(r_1/r_2)} = \frac{\pi \times 1.97 \times \left[(27^2) - h_2^2\right]}{\ln(50/0.1)} = 1000$$

and solving for h_2,

$$h_2 = 28.8 \text{ m}.$$

Since the aquifer is 30 m deep, the drawdown at the well is 30 − 28.8 = 1.2 m.

Multiple wells in an aquifer can interfere with each other and cause excessive drawdown. Consider the situation in Fig. 6-7 where first a single well creates a cone of depression. If a second extraction well is installed, the cones will overlap, causing

Water Supply 115

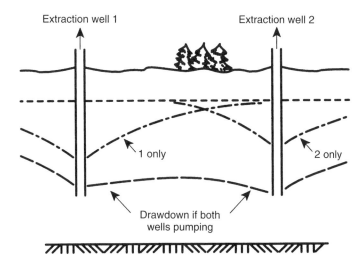

Figure 6-7. Interference between two extraction wells.

greater drawdown at each well. If many wells are sunk into an aquifer, the combined effect of the wells could deplete the groundwater resources and all wells would "go dry."

The reverse is also true, of course. Suppose one of the wells becomes an injection well, then the injected water flows from this well into the others, building up the groundwater table and reducing the drawdown. The judicious use of extraction and injection wells is one way that the flow of contaminants from hazardous waste or refuse dumps can be controlled.

Finally, a lot of assumptions are made in the above discussion. First, it is assumed that the aquifer is homogeneous and infinite; that is, it sits on a level aquaclude and that the permeability of the soil is the same at all places for an infinite distance in all directions. Second, steady-state and uniform radial flow is assumed. The well is assumed to penetrate the entire aquifer, and is open for the entire depth of the aquifer. Finally, the pumping rate is assumed to be constant. Clearly any of these assumptions may be unwarranted and cause the analysis to be faulty. This model of aquifer behavior is a relatively simple illustration. Modeling the behavior of groundwater is a complex and sophisticated science.

SURFACE WATER SUPPLIES

Surface water supplies are not as reliable as groundwater sources since quantities often fluctuate widely during the course of a year or even a week, and water quality is affected by pollution sources. If a river has an average flow of 10 cubic feet per second (cfs), this does not mean that a community using the water supply can depend on having 10 cfs available at all times.

The variation in flow may be so great that even a small demand cannot be met during dry periods, and storage facilities must be built to save water during wetter periods. Reservoirs should be large enough to provide dependable supplies. However, reservoirs are expensive and, if they are unnecessarily large, represent a waste of community resources.

One method of estimating the proper reservoir size is use of a *mass curve* to calculate historical storage requirements and then to calculate risk and cost using statistics. Historical storage requirements are determined by summing the total flow in a stream at the location of the proposed reservoir, and plotting the change of total flow with time. The change of water demand with time is then plotted on the same curve. The difference between the total water flowing in and the water demanded is the quantity that the reservoir must hold if the demand is to be met. The method is illustrated by Example 6.4.

EXAMPLE 6.4. A reservoir is needed to provide a constant flow of 15 cfs. The monthly stream flow records, in total cubic feet, are

Month	J	F	M	A	M	J	J	A	S	O	N	D
Million ft³ of water	50	60	70	40	32	20	50	80	10	50	60	80

The storage requirement is calculated by plotting the cumulative stream flow as in Fig. 6-8. Note that the graph shows 50 million ft³ for January, $60 + 50 = 110$ million ft³ for February, $70 + 110$ million ft³ for March, and so on.

The demand for water is constant at 15 cfs, or

$$15 \times 10^6 \frac{\text{ft}^3}{\text{s}} \times 3600 \frac{\text{s}}{\text{h}} \times 24 \frac{\text{h}}{\text{day}} \times 30 \frac{\text{days}}{\text{month}} = 38.8 \times 10^6 \frac{\text{ft}^3}{\text{month}}.$$

This constant demand is represented in Fig. 6-8 as a straight line with a slope of 38.8×10^6 ft³/month, and is plotted on the curved supply line. Note that the stream flow in May was lower than the demand, and this was the start of a drought lasting until June. In July the supply increased until the reservoir could be filled up again, late in August. During this period the reservoir had to make up the difference between demand and supply, and the capacity needed for this time was 60×10^6 ft³. A second drought, from September to November required 35×10^6 ft³ of capacity. The municipality therefore needs a reservoir with a capacity of 60×10^6 ft³ to draw water from throughout the year.

A mass curve like Fig. 6-8 is not very useful if only limited stream flow data are available. Data for one year yield very little information about long-term variations. The data in Example 6.4 do not indicate whether the 60 million cfs deficit was the worst drought in 20 years, or an average annual drought, or occurred during an unusually wet year.

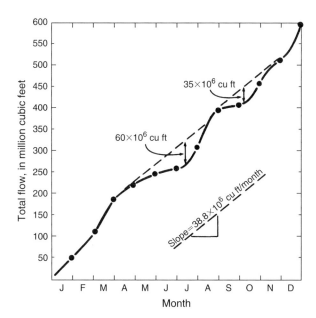

Figure 6-8. Mass curve for determining required reservoir capacity.

Long-term variations may be estimated statistically when actual data are not available. Water supplies are often designed to meet demands of 20-year cycles, and about once in 20 years the reservoir capacity will not be adequate to offset the drought. The community may choose to build a larger reservoir that will prove inadequate only every 50 years, for example. A calculation comparing the additional capital investment to the added benefit of increased water supply will assist in making such a decision. One calculation method requires first assembling required reservoir capacity data for a number of years, ranking these data according to the drought severity, and calculating the drought probability for each year. If the data are assembled for n years and the rank is designated by m, with $m = 1$ for the largest reservoir requirement during the most severe drought, the probability that the supply will be adequate for any year is given by $m/(n+1)$. For example, if storage capacity will be inadequate, on the average, one year out of every 20 years,

$$m/(n+1) = 1/20 = 0.05.$$

If storage capacity will be inadequate, on the average, one year out of every 100 years,

$$m/(n+1) = 1/100 = 0.01.$$

The calculation of storage is illustrated in Example 6.5.

EXAMPLE 6.5. A reservoir is needed to supply water demand for 9 out of 10 years. The required reservoir capacities, which were determined by the method of Example 6.4, are shown below:

Year	Required reservoir capacity (m^3 × 10^6)	Year	Required reservoir capacity (m^3 × 10^6)
1961	60	1971	53
1962	40	1972	62
1963	85	1973	73
1964	30	1974	80
1965	67	1975	50
1966	46	1976	38
1967	60	1977	34
1968	42	1978	28
1969	90	1979	40
1970	51	1980	45

These data must now be ranked, with the highest required capacity, or worst drought, getting rank 1, the next highest 2, and so on. Data were collected for 20 years, so that $n = 20$ and $n + 1 = 21$. Next, $m/(n+1)$ is calculated for each drought

Rank	Capacity (m^3 × 10^6)	$m/(n+1)$	Rank	Capacity (m^3 × 10^6)	$m/(n+1)$
1	90	0.05	11	50	0.52
2	85	0.1	12	46	0.57
3	80	0.14	13	45	0.62
4	73	0.19	14	42	0.67
5	67	0.24	15	40	0.71
6	62	0.29	16	40	0.76
7	60	0.33	17	38	0.81
8	60	0.38	18	34	0.86
9	53	0.43	19	30	0.90
10	51	0.48	20	28	0.95

These data are plotted in Fig. 6-9. A semilog plot often yields an acceptable straight line. If the reservoir capacity is required to be adequate 9 years out of 10, it may be inadequate 1 year out of 10. Entering Fig. 6-9 at $m/(n + 1) = 1/10 = 0.1$, we find that

$$m/n + 1 = 2/21 = 0.1.$$

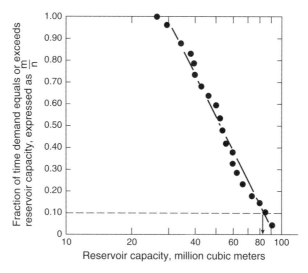

Figure 6-9. Frequency analysis of reservoir capacity.

The 10% probability of adequate capacity requires a reservoir capacity of 82 million m^3. Had the community only required adequate capacity 1 year out of 5, $m/(n+1) = 0.2$ and, from Fig. 6-9, a reservoir capacity of 71 million m^3 would have sufficed.

This procedure is a *frequency analysis* of a recurring natural event. The frequencies chosen for investigation were once in 10 years and once in 5 years, or a "10-year drought" and a "5-year drought," but droughts occurring 3 years in a row and then not again for 30 years still constitute "10-year droughts." Planning for a 10-year recurrence interval, though usually reliable, is not absolute.

WATER TRANSMISSION

Water can be transported from a ground or surface supply either directly to the water users in a community or initially to a water treatment facility. Water is transported by different types of conduits, including:

- Pressure conduits: tunnels, aqueducts, and pipelines
- Gravity-flow conduits: grade tunnels, grade aqueducts, and pipelines.

The location of the well field or river reservoir defines the length of the conduits, while the topography indicates whether the conduits are designed to carry the water in open-channel flow or under pressure. The profile of a water supply conduit should follow the hydraulic grade line to take advantage of gravity and minimize pumping costs.

Distribution reservoirs and water towers are also necessary in the transmission system to help level out peak demands. They are sized to meet three design constraints:

- hourly fluctuations in water consumption within the service area,
- short-term shutdown of the supply network for servicing, and
- back-up water requirements to control fires.

These reservoirs are most often constructed as open or covered basins, elevated tanks, or, as in the past, standpipes. Adequately designed distribution or service reservoirs require supply conduits feeding them to carry about 50% excess water over the average daily demand of the system or subsystem.

Elements of Closed-Conduit Hydraulics

Before discussing flow, some important properties of fluids are reviewed.

The *density* of a fluid is its mass per unit volume. Density is expressed in terms of g/cm^3 or kg/L or kg/m^3, and in slugs/ft^3 or lb-s^2/ft^4 in the English system. The density of water at 20°C is 1 g/cm^3 or 1.94 slugs/ft^3.

Specific weight is the force exerted by gravity on a unit volume of fluid. Specific weight is related to density as

$$w = \rho g,$$

where

$w =$ specific weight (in kg/m^3 or lb/ft^3),
$\rho =$ density (in kg/m^3 or lb-s^2/ft^4), and
$g =$ gravitational constant (in m/s^2 or ft/s^2).

Kilograms (or grams) are the units for both mass and weight in the metric system. The specific weight of water at 20°C is 1000 kg/m^3 in the metric system and 62.4 lb/ft^3 in the English system. *Specific gravity* of a liquid is the ratio of its density to that of pure water at a standard temperature. The specific gravity of water is 1.00. The *viscosity* of a fluid is a measure of its resistance to shear or angular deformation, and is defined as the proportionality constant relating the shear stress τ to the rate of deformation du/dy, and is usually written as

$$\tau = \mu \, du/dy.$$

This expression assumes that shear stress is directly proportional to the rate of deformation, and holds for fluids known as Newtonian fluids. For non-Newtonian fluids, like biological sludge, shear stress is not proportional to the deformation rate. Thus the term viscosity, when applied to non-Newtonian fluids such as biological sludge, is significant only if either the deformation rate or the shear stress is also specified. The unit of viscosity in the metric system is the poise, or gm/cm-s. The viscosity of water at 20°C (68.4°F) is 0.01 poise, or 1 centipoise. In the English system unit of viscosity is lb-s/ft^2. Therefore, 1 lb-s/ft^2 equals 479 poise.

Table 6-2. Viscosity of Water as a Function of Temperature

Temperature		Viscosity	
°F	°C	lb-s/ft²	Centipoise
40	4.4	3.1×10^{-5}	1.5
50	10.0	2.7×10^{-5}	1.3
60	15.5	2.3×10^{-5}	1.1
68.4	20.0	2.1×10^{-5}	1.0
70	21.0	2.0×10^{-5}	0.96
80	26.6	1.8×10^{-5}	0.86
90	32.2	1.6×10^{-5}	0.77

The *kinematic viscosity* v is the absolute viscosity divided by the fluid density, or

$$v = \mu/\rho.$$

The dimensions of kinematic viscosity are cm²/s.

Fluid viscosity is a function of temperature. Some representative values of the viscosity of water are shown in Table 6-2.

Closed-Conduit Flow

One of the most fundamental principles of hydraulics states that in a system, the total energy of a perfect liquid under ideal conditions does not change as it flows from point to point. The total energy is the sum of the position energy, the pressure energy, and the velocity energy. These are usually stated in terms of meters or feet of fluid, so that:

Z = static head = elevation (in m or ft),
P = pressure (in kg/m² or lb/ft²),
w = specific weight of the fluid (in kg/m³),
v = velocity (in m/s or ft/s),
g = gravitational acceleration (in m/s² or ft/s²),
P/w = pressure energy or pressure head, and
$v^2/2g$ = velocity energy or velocity head.

Consider a system like that shown in Fig. 6-10. Water is flowing through the pipe from a reservoir with constant surface elevation. Assuming no losses in the system, the total energy or total head of water remains constant, although energy or head may be converted from one form to another within the system. At Point 1, at the surface of the reservoir, all the energy is static head, while at Points 3, 4, and 5, the energy is distributed among static, pressure, and velocity head. At Point 6 the jet of water enters the atmosphere and the energy of the jet is the sum of the velocity head and the static head. The total energy, however, is constant at all points in the system.

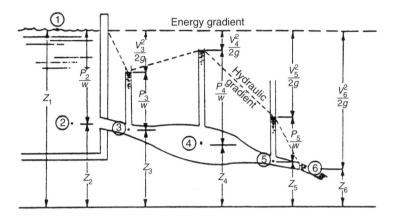

Figure 6-10. Hydraulics for ideal closed-conduit flow.

The principle of constant energy is known as the Bernoulli theorem when applied to fluid flow. Taking into account losses within the system, the Bernoulli theorem for any two points in the system may be written

$$Z_1 + \frac{P_1}{w} + \frac{v_1^2}{2g} = Z_2 + \frac{P_2}{w} + \frac{v_2^2}{2g} + h_L, \tag{6.8}$$

in which h_L represents the energy losses within the system. The energy loss may occur in many places in a system, such as valves, bends, and sudden changes in pipe diameter. One of the major losses of energy is in the friction between the moving fluid and the pipe wall.

Because of the practical problems involved in the application of more elegant and theoretically sound friction loss equations, engineers often use fitted or empirical exponential equations for flow calculations. Among these, the Hazen–Williams formula is most widely used in the United States for flow in pressure pipes and the Manning formula for flow in open channels or pipes not having full flow. These formulas are limited to turbulent flow of water and to common ambient temperatures. The Hazen–Williams formula is

$$v = 1.318 C r^{0.63} s^{0.54}, \tag{6.9a}$$

where

- v = the mean velocity of flow (in ft/s),
- r = the hydraulic radius (area divided by wetted perimeter, in ft),
- s = slope of the hydraulic grade line, and
- C = the Hazen–Williams friction coefficient.

When combined with the continuity equation, $Q = Av$, the discharge Q, in gallons per minute for a circular pipe of diameter D, in inches, is given by

$$Q = 0.285CD^{2.63}s^{0.54}. \tag{6.9b}$$

In metric units, these two equations are

$$v = 0.849Cr^{0.63}s^{0.54} \tag{6.10a}$$

$$Q = 0.278CD^{2.63}s^{0.54}, \tag{6.10b}$$

where D and r are in meters and v and Q are in meters per second (m/s) and cubic meters per second (m³/s), respectively. The nomographs shown in Fig. 6-11 are solutions to the above equations in both English and metric units. Table 6-3 summarizes Hazen–Williams coefficients for various pipe materials.

EXAMPLE 6.6. The pressure drop through a 6-in. asbestos-cement pipe, 3000 ft in length is 20 psi. What is the flow rate?

$$h_L = (20 \text{ pei})(2.31 \text{ ft water/psi}) = 46.2 \text{ ft}$$

$$s = \frac{46.2 \text{ ft}}{3000 \text{ ft}} = \frac{15.4 \text{ ft}}{1000 \text{ ft}}.$$

From Table 6-3, C is 140 and from Fig. 6-11, the discharge is 460 gpm.

Energy or head is lost at the entrance to a pipe or conduit; at valves, meters, fittings, and other irregular features; at enlargements; and at flow contractions. These losses, called *minor losses*, are in excess of friction losses over the same length of straight pipe or conduit and may be expressed as

$$h_L = K\frac{v^2}{2g}, \tag{6.11}$$

where K = the energy loss coefficient of minor losses in closed conduits, and for which values of K may be estimated by using Table 6-4.

EXAMPLE 6.7. The loss for a flow of 1.0 cfs through a given 6-in. main with a gate valve wide open is 20 ft. Find the head loss with the gate valve 75% closed.

From Table 6-4, the increase in K value when the valve is partially closed is $K = 24.0$:

$$v = \frac{Q}{A} = \frac{1.0}{0.2} = 5 \text{ ft/s}$$

$$h_L = h_0 + K\frac{v^2}{2g} = 20 + 24\left(\frac{5^2}{64.4}\right) = 29.2 \text{ ft.}$$

Two pipes, two systems of pipes, or a single pipe and a system of pipes are said to be equivalent when their losses of head for equal rates of flow are equal (or flow is equal for equal loss of head). Compound pipes, whether in parallel or in series (Fig. 6-12),

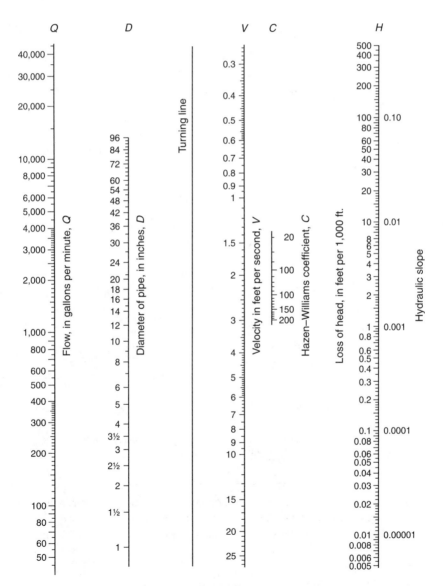

Figure 6-11. Hazen–Williams equation nomograph, English and metric units.

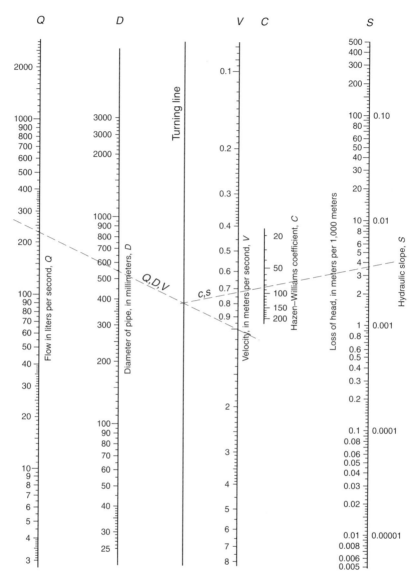

Figure 6-11. (*continued*)

may be reduced to single, equivalent pipes. The following examples are illustrative:

EXAMPLE 6.8. For the parallel pipes as shown in Fig. 6-12a, find the diameter of equivalent pipe (length is assumed to be 1000 ft).

1. Loss of head through pipe 1 must always equal loss of head through pipe 2 between points A and B.

2. Assume any arbitrary head loss, say 10 ft.
3. Calculate head loss in feet per 1000 ft for pipes 1 and 2.
   Pipe 1: $(10/1300) \times (1000) = 7.7$ ft/1000 ft
   Pipe 2: $(10/1400) \times (1000) = 7.1$ ft/1000 ft.
4. Use Fig. 6-11 to find flow in gallons per minute (gpm).
   Pipe 1: $D = 8$ in., $s = 0.0077$, $Q = 495$ gpm
   Pipe 2: $D = 6$ in., $s = 0.0071$, $Q = 220$ gpm
   Total Q through both pipes = 715 gpm.
5. Using Fig. 6-11 with $s = 0.010$ and $Q = 715$ gpm, equivalent pipe size is found to be 8.8 in. in diameter.

Table 6-3. Values of C for the Hazen–Williams Equation

Type of pipe	C
Asbestos cement	140
Brass	130–140
Brick sewer	100
Cast iron	
New, unlined	130
Old, unlined	40–120
Cement-lined	130–150
Bitumastic enamel-lined	140–150
Tar-coated	115–135
Concrete or concrete-lined	
Steel forms	140
Wooden forms	120
Centrifugally spun	135
Copper	130–140
Fire hose (rubber-lined)	135
Galvanized iron	120
Glass	140
Lead	130–140
Masonry conduit	120–140
Plastic	140–150
Steel	
Coal-tar enamel-lined	145–150
New, unlined	140–150
Riveted	110
Tin	130
Vitrified	100–140
Wood stove	120

Table 6-4. Minor Losses of Head

Nature of resistance	Loss (K)	Nature of resistance	Loss (K)
45° elbow	Use 3/4 loss for 90° bend of same radius	22.5° elbow	Use 1/2 loss for 90° bend of same radius
Angle valve wide open	2–5		
		Entrance losses	
Butterfly valve		Pipe projects into tank	0.8–1.0
$\theta = 10°$	1	Pipe end flush with tank	0.5
$\theta = 40°$	10	Slightly rounded	0.23
$\theta = 70°$	320	Bell-mouthed	0.04
Check valves		Outlet losses from pipe into still water or atmosphere	
Horizontal lifts	8–12		
Ball	65–70		
Swing	0.6–2.5		1.0
Gate valves		Sudden contraction	
Wide open		$d/D = 0.25$	0.42
0.25 closed	1.2	$d/D = 0.5$	0.33
0.5 closed	5.6	$d/D = 0.75$	0.19
0.75 closed	24.0		
		Sudden enlargement	
Globe valves wide open	10	$d/D = 0.25$	0.92
		$d/D = 0.5$	0.56
90° elbow		$d/D = 0.75$	0.91
Regular flanged	0.21–0.30		
Long-radius flanged	0.14–0.23		
Short-radius screwed	0.9		
Medium-radius screwed	0.75		
Long-radius screwed	0.60		

EXAMPLE 6.9. For the pipes in series as shown in Fig. 6-12b, find the diameter of equivalent pipe (length is assumed to be 1000 ft).

1. Quantity of water flowing through pipes 3 and 4 is the same.
2. Assume any arbitrary flow through pipes 3 and 4, say 500 gpm.

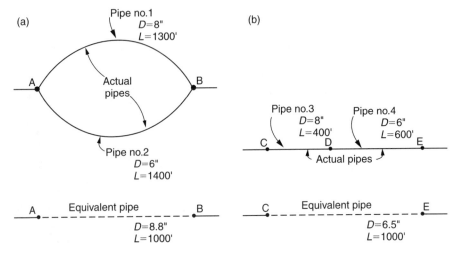

Figure 6-12. (a) Equivalent pipes, parallel. (b) Equivalent pipes, series.

3. Using Fig. 6-11, find head loss for pipes 3 and 4.
   Pipe 3: $D = 8$ in., $L = 400$ ft, $Q = 500$ gpm, $h^1 = s_1 \times L_1 = 0.008 \times 400 = 3.2$ ft
   Pipe 4: $D = 6$ in., $L = 600$ ft, $Q = 500$ gpm, $h^1 = s_1 \times L_1 = 0.028 \times 600 = 16.8$ ft
   Total head loss in both pipes = 20 ft.
4. Using Fig. 6-11 with head loss = 20 ft, $s = 20/1000$, and $Q = 500$ gpm, equivalent pipe size is found to be 6.5 in. in diameter.

More complex systems may be reduced to a single equivalent pipe by piecework conversion of real and equivalent pipes. Although the calculations become tedious, solutions of network flow problems are dependent on the same basic physical principles as those for single pipes; that is, the principles of energy conservation and continuity must be satisfied throughout the network.

Pumps and Pumping[1]

Pumps are mechanical devices for converting other forms of energy to hydraulic energy. When interposed in a pipe, they add energy to the liquid passing through the pipes. The added energy is almost always pressure energy. Pumps, like motor vehicles, are not individually designed for public works projects, except for very large and unusual

---

[1] This discussion on pumps and pumping is adapted from F. E. McJunkin and P. A. Vesilind, *Practical Hydraulics for Public Works Engineers*, published as a separate issue by *Public Works Magazine* (1968).

installations. Rather, they are selected from predesigned and manufactured units readily available for a wide range of applications.

Economical selection requires that attention be given to: (1) the normal pumping rate and the minimum and maximum rates that the pump will ever be called on to deliver; (2) the total head capacity to meet flow requirements; (3) suction head, or lift; (4) pump characteristics, including speed, number of pumps, power source, and other spatial and environmental requirements; and (5) the nature of the liquid to be pumped.

Rotodynamic and *displacement* pumps are the two types most often encountered in environmental engineering. Rotodynamic pumps impart kinetic energy into the liquid using a rotating element or *impeller*, shaped to force water outward at right angles to the axis of rotation in *radial flow*, to force the liquid in the axial direction in *axial flow*, or to give the liquid both radial and axial velocity in *mixed flow*. *Centrifugal pumps* are radial flow and mixed flow devices, and *propeller pumps* are axial flow pumps. Displacement pumps include the *reciprocating* type, in which a piston draws water into a cylinder on one stroke and forces it out on the next, and the *rotary* type, in which two cams or gears mesh together and rotate in opposite directions to force water continuously past them. There are also *jet pumps (ejectors), airlift pumps, hydraulic rams, diaphragm pumps*, and others that may be useful in special circumstances.

Pump hydraulics are shown in Fig. 6-13. The *static suction head* on a pump is the vertical distance from the free liquid surface on the intake to the pump centerline. If there is no free surface, the gauge pressure at the pump flange (suction or discharge)

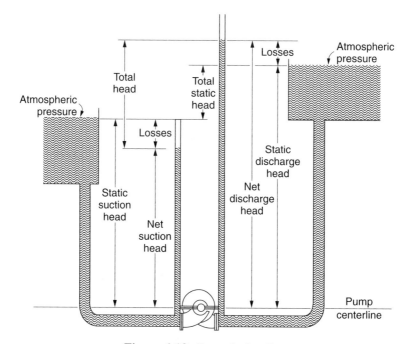

Figure 6-13. Pump hydraulics.

for zero discharge, corrected to the centerline pump elevation, should be used in lieu of the static head. The static suction head may be positive or negative; negative static suction head is sometimes called *suction lift*. The net suction head is the difference between the static suction head and friction head losses, including entrance loss, for the capacity under consideration. The *static discharge head* is the vertical distance from the pump centerline to the free surface on the discharge line. Thus, the *net discharge head* is the sum of the static discharge head and the friction head losses. The total head H is given by

$$H = h_d - h_a, \tag{6.12}$$

where h_d is the net discharge head and h_s is the net static head.

In pumping liquids, the pressure anywhere in the suction line should remain higher than the vapor pressure of the liquid in order to avoid air binding, priming loss, and cavitation. The energy available for moving liquid through the suction line to the impeller, known as the *net positive suction head (NPSH)*, is the sum of the net suction head and any pressure existing in the suction supply line, less the vapor pressure of the liquid at the pumping temperature. Any vacuum is to be treated as a negative pressure. The useful work done by a pump is the product of the weight of liquid pumped and the head developed by the pump. The power, or work/time, required is the *water horsepower* (WHP). Thus

$$\text{WHP} = QHw, \tag{6.13}$$

where Q = pump discharge, H = total head, and w = specific weight. For water at 68°F, Q in gallons per minute, and H in feet,

$$\text{WHP} = \frac{QH}{3960}.$$

The *brake horsepower* (BHP) is the total power required to drive a pump. The pump efficiency η is the ratio of the water horsepower to the brake horsepower, or

$$\eta = \frac{\text{WHP}}{\text{BHP}} \times 100. \tag{6.14}$$

For water at 68°F

$$\text{BHP} = \frac{100QH}{3960\eta}.$$

BHP for other water temperatures and other liquids can be determined by correcting for the change in specific weight. The temperature correction for natural water is usually negligible.

Pump Characteristic Curves

The total head developed by a pump, the power required to drive it, and the resulting efficiency vary with the capacity. The interrelations of head, power, efficiency, and capacity are shown in Fig. 6-14 and are called *pump characteristic curves*:

- The *head capacity curve H–Q* shows the relationship between capacity and total head. Pumps are often classified according to the shape of the head capacity curve.
- The *efficiency curve η–Q* shows the relationship between capacity and efficiency.
- The *power curve P–Q* shows the relationship between capacity and power input.

System Head Curves

Head loss in a pumping system increases with increasing flow through the system, and can be shown graphically as a *system head curve* like that in Fig. 6-15. The system head loss for any flow rate is the sum of friction head loss and the total static head in the system. Static head is present whether the pump is operating or not, and is plotted as the lower portion of the system head curve.

Friction losses, including minor losses, are determined as outlined in the previous section of this chapter. The system piping and fittings may be converted to one equivalent pipe, and head losses for several flow rates may be determined readily from the Hazen–Williams nomograph. Alternatively, the head loss through all pipes and fittings may be computed for a single flow rate, and losses for other flow rates may be determined from the relationship

$$\frac{Q_1}{Q_2} = \left[\frac{h_{L_1}}{h_{L_2}}\right]^{0.54}. \tag{6.15}$$

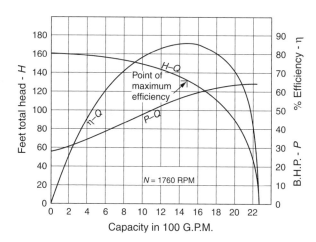

Figure 6-14. Pump characteristic curves.

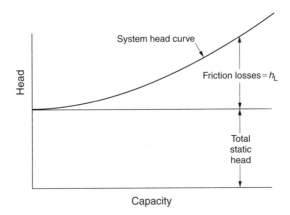

Figure 6-15. System head curve.

For zero discharge, total head is equal to the total static head. This point plus several computed points will suffice to plot the curve. Static head in a system will vary as tanks and reservoirs are filled or drawn down. In such cases, system curves may readily be constructed for minimum and maximum heads, thereby enabling prediction of system pumping capacity for the entire range of possible static head conditions. For elaborate installations, an economic analysis of the tradeoff between pumping and piping costs may be justified. For comparatively short pipelines, however, friction losses should not be more than about 20% of the static head.

Operating Head and Discharge

The usual design condition is that a system will be given and the proper pump must be selected. The intersection of a pump head capacity curve with a system curve on which it is superimposed is the *operating point*. Figure 6-16 is an example. The operating point is the discharge and head at which a given system and given pump will operate. Operating efficiency and power requirements will also be located by this superposition. A pump that has an operating point at or near its peak efficiency should be selected.

Placing two identical pumps in series doubles the pumping head. Conversely, two identical pumps in parallel will double the pumping capacity (see Fig. 6-17). Doubling the pumping head or capacity will not double the *system* capacity, however. Figure 6-18 shows that the added capacity for two pumps in parallel results in a greater friction head loss and that the system capacity is not doubled. Similarly, pumping in series will double neither the system head nor the discharge.

CONCLUSION

As the hydrologic cycle indicates, water is a renewable resource because of the driving force of energy from the sun. The earth is not running out of water, though enough

Water Supply 133

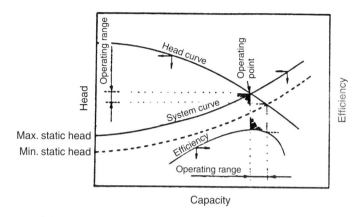

Figure 6-16. Determination of pump operating point.

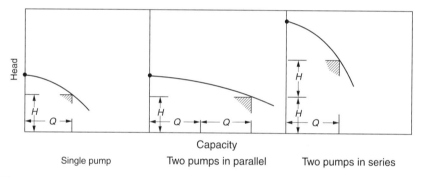

Figure 6-17. Pump characteristic curves with two pumps in parallel and in series.

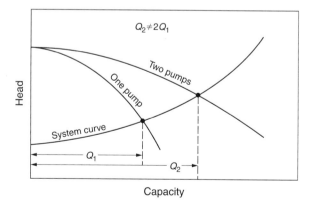

Figure 6-18. System head curve with two pumps in parallel.

134 ENVIRONMENTAL ENGINEERING

water or enough clean water may not be available in some areas because of climate and water use. Both groundwater and surface water supplies are available to varying degrees over the entire earth's surface, and can be protected by sound engineering and environmental judgment.

The next chapter addresses methods of preparing and treating water for distribution and consumption once the supply has been provided.

PROBLEMS

6.1 A storage reservoir is needed to ensure a constant flow of 20 cfs to a city. The monthly stream flow records are

Month	J	F	M	A	M	J	J	A	S	O	N	D
Million ft^3 of water	60	70	85	50	40	25	55	85	20	55	70	90

Calculate the storage requirements.

6.2 An unconfined aquifer is 10 m thick and is being pumped so that one observation well placed at a distance of 76 m shows a drawdown of 0.5 m. On the opposite side of the extraction well is another observation well, 100 m from the extraction well, and this well shows a drawdown of 0.3 m. Assume the coefficient of permeability is 50 m^3/day-m^2.

 a. What is the discharge of the extraction well?
 b. Suppose the well at 100 m from the extraction well is now pumped. Show with a sketch what this will do to the drawdown.
 c. Suppose the aquifer sits on an aquaclude that has a slope of 1/100. Show with a sketch how this would change the drawdown.

6.3 If a faucet drips at a rate of two drops per second, and it takes 25,000 drops to make one gallon of water, how much water is lost each day? Each year? If water costs $1.60 per 1000 gallons, how long would the water leak until its cost equalled the 75 cents in parts needed to fix the leak (if you fix it yourself)? How long if you called a plumber for $40.00 minimum rate plus 75 cents in parts?

6.4 A 24-in. pipe with centrifugally spin concrete lining carries a flow of 10 mgd. Find the head loss per unit length and the velocity.

6.5 A 6-in. main is carrying a flow of 1.0 cfs. What head loss is produced by two regular flanged 45° elbows and a gate valve one-half closed?

6.6 Three pipes, each 1500 ft long, are arranged in parallel. Two have diameters of 6 in. each, and the diameter of the third is 10 in. All three have a Hazen–Williams C value of 100. Find the diameter of the equivalent pipe that is also 1500 ft long.

6.7 A pump delivers 900 gpm against a total head of 145 ft. Assuming a pump efficiency of 90%, what is the BHP?

6.8 Given two identical pumps with the characteristics shown in Fig. 6-14, find the discharge for a head of 100 ft if (a) both pumps are operated in series and (b) both pumps are operated in parallel.

Chapter 7

Water Treatment

Many aquifers and isolated surface waters are high in water quality and may be pumped from the supply and transmission network directly to any number of end uses, including human consumption, irrigation, industrial processes, or fire control. However, clean water sources are the exception in many parts of the world, particularly regions where the population is dense or where there is heavy agricultural use. In these places, the water supply must receive varying degrees of treatment before distribution.

Impurities enter water as it moves through the atmosphere, across the earth's surface, and between soil particles in the ground. These background levels of impurities are often supplemented by human activities. Chemicals from industrial discharges and pathogenic organisms of human origin, if allowed to enter the water distribution system, may cause health problems. Excessive silt and other solids may make water aesthetically unpleasant and unsightly. Heavy metal pollution, including lead, zinc, and copper, may be caused by corrosion of the very pipes that carry water from its source to the consumer.

The method and degree of water treatment are important considerations for environmental engineers. Generally speaking, the characteristics of raw water determine the treatment method. Most public water systems are relied on for drinking water as well as for industrial consumption and fire fighting, so that human consumption, the highest use of the water, defines the degree of treatment. Thus, we focus on treatment techniques that produce potable water.

A typical water treatment plant is diagrammed in Fig. 7-1. It is designed to remove odors, color, and turbidity as well as bacteria and other contaminants. Raw water entering a treatment plant usually has significant turbidity caused by colloidal clay and silt particles. These particles carry an electrostatic charge that keeps them in continual motion and prevents them from colliding and sticking together. Chemicals like alum (aluminum sulfate) are added to the water both to neutralize the particles electrically and to aid in making them "sticky" so that they can coalesce and form large particles called flocs. This process is called coagulation and flocculation and is represented in stages 1 and 2 in Fig. 7-1.

COAGULATION AND FLOCCULATION

Naturally occurring silt particles suspended in water are difficult to remove because they are very small, often colloidal in size, and possess negative charges, and are thus

136   ENVIRONMENTAL ENGINEERING

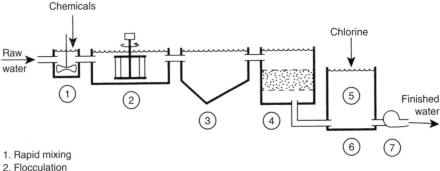

1. Rapid mixing
2. Flocculation
3. Settling
4. Filtration
5. Chlorination
6. Clear well storage
7. Pumping to distribution system

Figure 7-1. Diagram of a typical water treatment facility.

prevented from coming together to form large particles that could more readily be settled out. The removal of these particles by settling requires first that their charges be neutralized and second that the particles be encouraged to collide with each other. The charge neutralization is called coagulation, and the building of larger flocs from smaller particles is called flocculation.

A fairly simple but not altogether satisfactory explanation of coagulation is available in the double-layer model. Figure 7-2 is a representation of the static electric field surrounding the particle. The solid particle is negatively charged, and attracts positively charged ions — counterions — from the surrounding fluid. Some of these negative ions are so strongly attracted that they are virtually attached to the particle and travel with it, thereby forming a slippage plane. Around this inner layer is an outer layer of ions consisting mostly of positive ions, but they are less strongly attracted, are loosely attached, and can slip off. The charge on the particle as it moves through the fluid is the negative charge, diminished in part by the positive ions in the inner layer. The latter is called the zeta potential.

If the net negative charge is considered a repulsive charge, since the neighboring particles are also so charged, the charge may be pictured as in Fig. 7-3A. In addition to this repulsive charge, however, all particles carry an attractive electrostatic charge, van der Waals force, that is a function of the molecular structure of the particle. This attractive charge is also shown in Fig. 7-3A. The combination of these forces results in a net repulsive charge, an energy barrier, or "energy hill," that prevents the particles from coming together. The objective of coagulation is to reduce this energy barrier to 0 so that the particles no longer repel each other. Adding trivalent cations to the water is one way to reduce the energy barrier. These ions are electrostatically attracted to the negatively charged particle and, because they are more positively charged, they displace the monovalent cations. The net negative charge, and thus the net repulsive force, is thereby reduced, as shown in Fig. 7-3B. Under this condition,

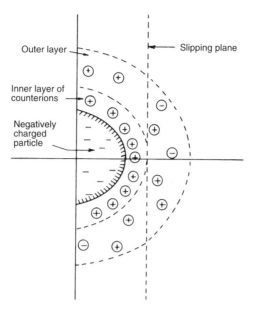

Figure 7-2. Charges on a suspended particle, as explained by the double-layer theory.

the particles do not repel each other and, on colliding, stick together. A stable colloidal suspension can be destabilized in this way, and the larger particles will not remain suspended.

Alum (aluminum sulfate) is the usual source of trivalent cations in water treatment. Alum has an advantage in addition to its high positive charge: some fraction of the aluminum ions may form aluminum oxide and hydroxide by the reaction

$$Al^{3+} + 3OH^- \rightarrow Al(OH)_3 \downarrow.$$

These complexes are sticky and heavy and will greatly assist in the clarification of the water in the settling tank if the unstable colloidal particles can be made to come in contact with the floc. This process is enhanced through an operation known as flocculation.

A flocculator introduces velocity gradients into the water so that the particles in a fast-moving stream can catch up and collide with slow-moving particles. Such velocity gradients are usually introduced by rotating paddles, as shown in Fig. 7-4. The power required for moving a paddle through the water is

$$P = \frac{C_D A \rho v^2}{2}, \qquad (7.1)$$

where

P = power (N/s or ft-lb/s),
A = paddle area (m² or ft²),

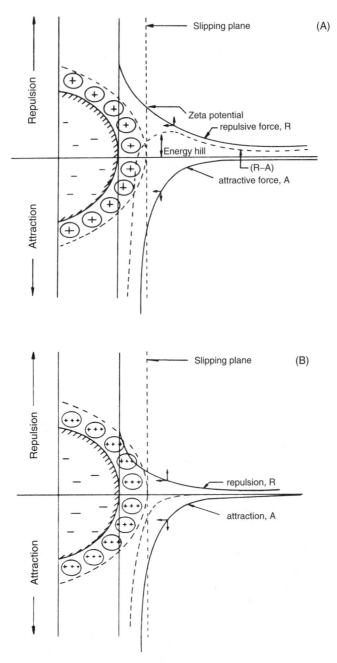

Figure 7-3. Reduction of the net charge on a particle as a result of the addition of trivalent counterions. (A) Particle carries net negative charge and van der Waals positive charge; energy barrier prevents coagulation. (B) Addition of trivalent cations reduces energy barrier, and coagulation is possible.

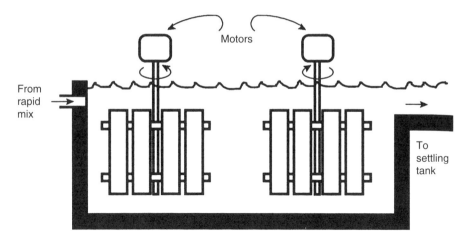

Figure 7-4. Flocculator used in water treatment.

ρ = fluid density (kg/m³ or lb-s²/ft³), and
C_D = drag coefficient.

The velocity gradient produced as a result of a power input in a given volume of water V is

$$G = \left(\frac{P}{V\mu}\right)^{1/2}, \qquad (7.2)$$

where

G = velocity gradient (in s^{-1}),
μ = viscosity (in dyne-s/cm² or lb-s/ft²), and
V = tank volume (in m³ or ft³).

Generally accepted design standards require G to be between 30 and 60 s^{-1}.

Time is also an important variable in flocculation, and the term $G\bar{t}$ is often used in design, where $\bar{t}$ is the hydraulic retention time in the flocculation basin. $G\bar{t}$ values are typically between 10^4 and 10^5.

EXAMPLE 7.1. A water treatment plant is designed for 30 million gallons per day (mgd). The flocculator dimensions are length = 100 ft, width = 50 ft, depth = 16 ft. Revolving paddles attached to four horizontal shafts rotate at 1.7 rpm. Each shaft supports four paddles that are 6 in. wide and 48 in. long. Paddles are centered 6 ft from the shaft. Assume $C_D = 1.9$, and the mean velocity of water is 35% of the paddle velocity. Find the velocity differential between the paddles and the water. At 50°F, the density of water is 1.94 lb-s²/ft³ and the viscosity is 2.73×10^{-5} lb-s/ft². Calculate the value of G and the time of flocculation (hydraulic retention time).

The rotational velocity is

$$v_t = \frac{2\pi r n}{60},$$

where r = radius in feet and n = rpm, so that

$$v_t = \frac{(2\pi)(6)(1.7) m}{60} = 1.07 \text{ ft/s}.$$

The velocity differential between paddles and fluid is assumed to be 65% of v_t, so that

$$v = 0.65 v_t = (0.65)(1.7) = 0.70 \text{ ft/s}.$$

Total power input from Eq. (7.1) is

$$P = \frac{(1.9)(16)(0.5 \text{ ft})(48 \text{ ft})(1.94 \text{ lb-s}^2/\text{ft}^3)(0.70 \text{ ft/s})^3}{2} = 243 \text{ ft-lb/s},$$

and, from Eq. (7.2), the velocity gradient is

$$G = \sqrt{\left(\frac{243}{(100)(50)(16)(2.73 \times 10^{-5})}\right)} = 10.5 \,\frac{\text{ft/s}}{\text{ft}},$$

which is a little low. The time of flocculation is

$$\bar{t} = \frac{V}{Q} = \frac{(100)(50)(16)(7.48)(24)(60)}{(30)10^5} = 28.7 \text{ min},$$

so that the $G\bar{t}$ value is 1.8×10^4. This is within the accepted range.

SETTLING

When the flocs have been formed they must be separated from the water. This is invariably done in gravity settling tanks that allow the heavier-than-water particles to settle to the bottom. Settling tanks are designed to approximate uniform flow and to minimize turbulence. Hence, the two critical elements of a settling tank are the entrance and exit configurations. Figure 7-5 shows one type of entrance and exit configuration used for distributing the flow entering and leaving the water treatment settling tank.

Alum sludge is not very biodegradable and will not decompose at the bottom of the tank. After some time, usually several weeks, the accumulation of alum sludge at the bottom of the tank is such that it must be removed. Typically, the sludge exits through a mud valve at the bottom and is wasted either into a sewer or to a sludge holding and drying pond. In contrast to water treatment sludges, sludges collected in wastewater treatment plants can remain in the bottom of the settling tanks only a matter of hours before starting to produce odoriferous gases and floating some of the solids. Settling tanks used in wastewater treatment are discussed in Chap. 9. The water leaving a settling tank is essentially clear. Polishing is performed with a rapid sand filter.

Water Treatment 141

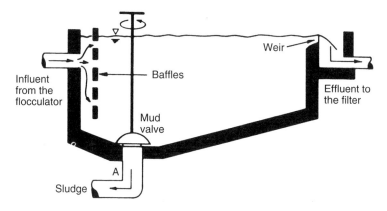

Figure 7-5. Settling tank used in water treatment.

FILTRATION

The movement of water into the ground and through soil particles and the cleansing action the particles have on contaminants in the water are discussed in Chap. 6. Picture the extremely clear water that bubbles up from "underground streams" as spring water. Soil particles help filter the ground water, and through the years environmental engineers have learned to apply this natural process in water treatment and supply systems, and have developed what we now know as the rapid sand filter. The actual process of separating impurities from carrying liquid by rapid sand filtration involves two processes: filtration and backwashing.

Figure 7-6 shows a cutaway of a slightly simplified version of a rapid sand filter. Water from the settling basins enters the filter and seeps through the sand and gravel bed, through a false floor, and out into a clear well that stores the finished water. Valves A and C are open during filtration.

The rapid sand filter eventually becomes clogged and must be cleaned. Cleaning is performed hydraulically. The operator first shuts off the flow of water to the filter, closing valves A and C, then opens valves D and B, which allow wash water (clean water stored in an elevated tank or pumped from the clear well) to enter below the filter bed. This rush of water forces the sand and gravel bed to expand and jolts individual sand particles into motion, rubbing against their neighbors. The light colloidal material trapped within the filter is released and escapes with the wash water. After a few minutes, the wash water is shut off and filtration is resumed.

The solid impurities in the water are removed by many processes, the most important of which are straining, sedimentation, interception, and diffusion (see Fig. 7-7). Straining, possibly the most important mechanism, takes place exclusively in the first few centimeters of the filter medium. As the filtering process begins, straining removes only particles in the water large enough to get caught in the pores (A in Fig. 7-7). After a time, these trapped particles themselves begin to form a screen that has smaller openings than the original filter medium. Smaller particles suspended in the water are

142   ENVIRONMENTAL ENGINEERING

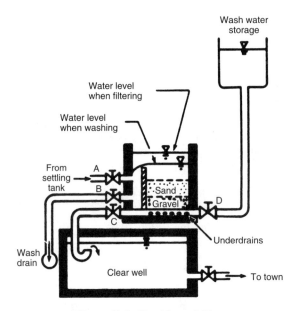

Figure 7-6. Rapid sand filter.

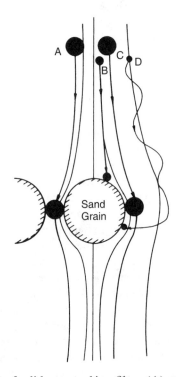

Figure 7-7. Mechanisms of solids removal in a filter: (A) straining, (B) sedimentation, (C) interception, and (D) diffusion.

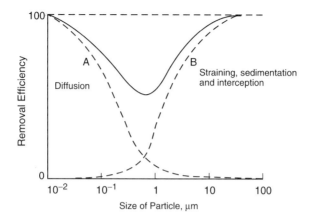

Figure 7-8. Removal of various-sized particles in a filter.

trapped by this mat and immediately begin acting as part of the screen. Thus, removal efficiency owing to screening tends to increase in some proportion to the time of the filtration phase.

In sedimentation, larger and heavier particles do not follow the fluid streamline around the sand grain, and settle on the grain (B in Fig. 7-7). Interception occurs with particles that do follow the streamline, but are too large and are caught because they brush up against the sand grain (C in Fig. 7-7). Finally, very small particles are experiencing Brownian motion and may collide with the sand grain by chance. This process is called diffusion (D in Fig. 7-7).

The first three mechanisms are most effective for larger particles, while diffusion can occur only for colloidal particles. A typical removal efficiency curve for different-sized particles is shown in Fig. 7-8. Efficiency removal is high for both large and small particles, and substantially reduced for midsized (about 1 μm) particles. Unfortunately, many viruses, bacteria, and fine clay particles are about 1 μm in size, and thus the filter is less effective in the removal of these particles.

Filter beds are often classified as single medium, dual media, or trimedia. The latter two are often utilized in wastewater treatment because they permit solids to penetrate into the bed, have more storage capacity, and thus increase the required time between backwashings. Also, multimedia filters tend to spread head loss buildup over time and further permit longer filter runs.

Head loss through the sand is a primary condition in filter design. As sand gets progressively dirtier the head loss increases. Figure 7-9 shows a simplified representation of head loss in a filter. Although the head losses experienced in a particular application cannot be predicted, the head loss in clean sand may be estimated by several different equations. One of the oldest and most widely used methods is the Carman–Kozeny equation. Head loss in clean sand while filtering may be estimated by first considering the filter to be a mass of pipes, in which case the Darcy–Weisbach head loss equation applies,

$$h_L = f \frac{L}{D} \frac{v^2}{2g}, \tag{7.3a}$$

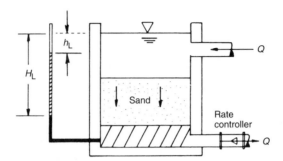

Figure 7-9. Head loss in a filter as measured by a manometer. Note that h_L is head loss in a clean filter and H_L is head loss in a dirty filter.

where

L = depth of filter,
D = "pipe" diameter,
v = velocity in "pipes," and
f = friction factor.

However, the "pipes" or channels through the sand are not straight and D varies, so we can substitute $D = 4R$, where

$$R = \frac{\text{area}}{\text{wetted\_perimeter}} = \frac{\pi D^2/4}{\pi D} = \frac{D}{4} = \text{hydraulic\_ratio.}$$

We thus have

$$h_L = f\frac{L}{4R}\frac{v^2}{2g}. \tag{7.3b}$$

The velocity of the water approaching the sand is

$$v_a = \frac{Q}{A'}, \tag{7.4a}$$

and the velocity through the bed is

$$v = \frac{v_a}{e}, \tag{7.4b}$$

where

e = porosity, or fraction of open spaces in the sand,
Q = flow rate, and
A' = surface area of the sand bed.

The total channel volume is the porosity of the bed multiplied by the total volume, or eV. The total solids volume is $(1 - e)$ times the total volume, which is also equal to the number of particles times the volume of the particles. The total volume is thus

$$\frac{NV_P}{1-e},$$

where N = number of particles and V_P = volume occupied by each particle. The total channel volume is

$$e\frac{NV_P}{1-e},$$

and since the total wetted surface is NA_P, where A_P is the surface area of each particle, the hydraulic radius

$$R = \frac{\text{area}}{\text{wetted\_perimeter}} = \frac{\text{volume}}{\text{area}} = \frac{e\frac{NV_P}{1-e}}{NA_P} = \left(\frac{e}{1-e}\right)\left(\frac{V_P}{A_P}\right).$$

For spherical particles, $V_P/A_P = d/6$, but for particles that are not true spheres, $V_P/A_P = \phi(d/6)$, where ϕ is a shape factor. For example, $\phi = 0.95$ for Ottawa sand, a common filter sand. We thus have

$$R = \left(\frac{e}{1-e}\right)\left(\phi\frac{d}{6}\right). \tag{7.5}$$

Substituting Eqs. (7.4a), (7.4b), and (7.5) into Eq. (7.3b),

$$h_L = f\left(\frac{L}{\phi d}\right)\left(\frac{1-e}{e^3}\right)\left(\frac{v_a^2}{g}\right). \tag{7.6}$$

The friction factor may be approximated as

$$f = 150\left(\frac{1-e}{R}\right) + 1.75, \tag{7.7}$$

where R is the Reynolds number and

$$R = \phi\left(\frac{\rho v_a d}{\mu}\right).$$

The preceding discussion is applicable to a filter bed made of only one size particle. For a bed made of nonuniform sand grains,

$$d = \frac{6}{\phi}\left(\frac{V}{A}\right)_{\text{avg}} = \frac{6}{\phi}\left(\frac{V_{\text{avg}}}{A_{\text{avg}}}\right), \tag{7.8}$$

146 ENVIRONMENTAL ENGINEERING

where V_{avg} = average volume of all particles and A_{avg} = average area of all particles. Substituting Eq. (7.8) into Eq. (7.6) yields

$$h_L = f\left(\frac{L}{6}\right)\left(\frac{1-e}{e^3}\right)\left(\frac{v_a^2}{g}\right)\left(\frac{V}{A}\right)_{\text{avg}} \tag{7.9}$$

and

$$\left(\frac{A}{V}\right)_{\text{avg}} = \frac{6}{\phi}\sum \frac{x}{d'}, \tag{7.10}$$

where x = weight fraction of particles retained between any two sieves and d' = geometric mean diameter between those sieves

$$h_L = f\left(\frac{L}{6}\right)\left(\frac{1-e}{e^3}\right)\left(\frac{v_a^2}{g}\right)\sum \frac{x}{d'}. \tag{7.11}$$

This equation holds for nonstratified beds, such as are found in a slow sand filter, since the fraction factor does not vary with depth. When particles are stratified, as in a rapid sand filter, we can write Eq. (7.11) as

$$\frac{dh_L}{dL} = \left[\left(\frac{1}{\phi}\right)\left(\frac{1-e}{e^3}\right)\left(\frac{v_a^2}{g}\right)\right] f\frac{1}{d'}.$$

The total head loss is thus

$$h_L = \int_0^{h_L} dh_L = \left[\left(\frac{1}{\phi}\right)\left(\frac{1-e}{e^3}\right)\left(\frac{v_a^2}{g}\right)\right]\int_0^L f\frac{1}{d'}\,dL. \tag{7.12}$$

Since $dL = L\,dx$, where dx is the proportion of particles of size d

$$h_L = \left[\left(\frac{1}{\phi}\right)\left(\frac{1-e}{e^3}\right)\left(\frac{v_a^2}{g}\right)\right]L\int_{x=0}^{x=1} f\frac{dx}{d}, \tag{7.13}$$

and if the particles between adjacent sieve sizes are assumed uniform,

$$h_L = \left[\left(\frac{L}{\phi}\right)\left(\frac{1-e}{e^3}\right)\left(\frac{v_a^2}{g}\right)\right]\sum \frac{fx}{d'}. \tag{7.14}$$

EXAMPLE 7.2. A sand consisting of the following sizes is used:[1]

Sieve number	% of sand retained on sieve, $\times 10^2$	Geometric mean sand size, ft $\times 10^{-3}$
14–20	1.10	3.28
20–28	6.60	2.29
28–32	15.94	1.77
32–35	18.60	1.51
35–42	19.10	1.25
42–48	17.60	1.05
48–60	14.30	0.88
60–65	5.10	0.75
65–100	1.66	0.59

The filter bed measures 20×20 ft^2 and is 2 ft deep. The sand has a porosity of 0.40 and a shape factor of 0.95. The filtration rate is 4 gal/min-ft^2. Assume the viscosity is 3×10^{-5} lb-s/ft^2. Find the head loss through the clean sand.

The solution is shown in tabular form:

Reynolds number, R	Friction factor, f	x/d	$f(x/d)$
1.80	51.7	3.4	174
1.37	67.4	28.8	1,941
1.06	86.6	90.1	7,802
0.91	100.6	123.2	12,394
0.75	121.7	152.8	18,595
0.63	144.6	167.6	24,235
0.53	171.5	162.5	27,868
0.45	201.7	68.0	13,715
0.35	258.8	28.1	7,272

Column 1: The approach velocity is

$$v_a = 4 \left(\frac{\text{gal}}{\text{min-ft}^2}\right) \left(\frac{1 \text{ft}^3}{7.481 \text{ gal}}\right) \left(\frac{1 \text{ min}}{60 \text{ s}}\right) = 8.9 \times 10^{-3} \text{ ft/s}.$$

For the first particle size, $d = 3.28 \times 10^{-3}$ ft, and

$$R = \frac{(0.95)(1.94)(89)(10^{-3})(3.28)(10^{-3})}{3 \times 10^{-5}} = 1.80.$$

---

[1] Adapted from J. W. Clark, W. Viessman, and M. J. Hammer, *Water Supply and Pollution Control*, 3rd ed., Thomas Crowell, New York, 1977.

Column 2: From Eq. (7.7)

$$f = 150\left(\frac{1-0.4}{1.8}\right) + 1.75 = 51.75.$$

Columns 3 and 4: For the first size, $x = 1.10\%$ and $d = 3.28 \times 10^{-3}$

$$f\frac{x}{d} = \frac{(51.75)(0.011)}{3.28 \times 10^{-3}} = 174.$$

The last column is summed: $\sum f(x/d) = 113{,}977$, and from Eq. (7.14), we have

$$h_L = \frac{2}{0.95}\left(\frac{1-0.4}{(0.4)^3}\right)\left(\frac{(8.9 \times 10^{-3})^2}{32.2}\right)(113{,}977) = 5.78 \text{ ft.}$$

The deposition of material during the filtering process increases the head loss through the filter. A method of predicting head loss takes advantage of this limitation, predicting head loss as

$$H_L = h_L + \sum_{i=1}^{n}(h_i)_t. \tag{7.15}$$

Here, H_L is the total head loss through the filter, h_L is the clear water head loss at time zero, and $(h_L)_t$ is the head loss in the ith layer of the medium in the filter at time t. All values are in meters. Head loss within an individual layer, $(h_L)_t$, is related to the amount of materials caught by the layer

$$(h_L)_t = x(q_i)_t^y. \tag{7.16}$$

Here $(q_i)_t$ is the amount of material collected in the ith layer at time t (in mg/cm^3), and x and y are experimental constants. Head loss data for sand and anthracite are summarized in Fig. 7-10.

A filter run is the time a filter operates before it must be cleaned. The end of a filter run is indicated by excessive head loss or excessive turbidity of the filtered water. If either of these occurs, the filter must be washed.

EXAMPLE 7.3. The filter described in Example 7.2 is run so that it reduces the suspended solids from 55 to 3 mg/L during a filter run of 4 h. Assume this material is captured in the first 6 in. of sand, where the sand grain size is adequately described as 0.8 mm. What is the head loss of the dirty filter?

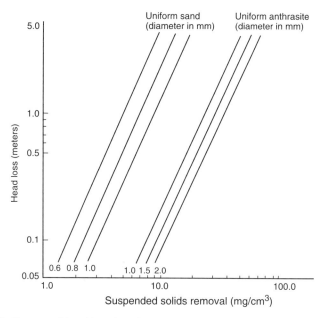

Figure 7-10. Increased head loss in a filter owing to materials deposits. (Adapted from Metcalf and Eddy, Inc., *Wastewater Engineering*, McGraw-Hill, New York, 1979.)

The total water through the filter is

$$4\frac{\text{gal}}{\text{min-ft}^2} \times (20 \text{ ft} \times 20 \text{ ft}) \times 4 \text{ h} \times 60 \frac{\text{min}}{\text{h}} = 384{,}000 \text{ gal}$$

$$384{,}000 \text{ gal} \times 3.785 \frac{\text{L}}{\text{gal}} = 1.45 \times 10^6 \text{ L}.$$

Solids removal is $55 - 3 = 52$ mg/L

$$52 \text{ mg/L} \times 1.45 \times 10^6 \text{ L} = 75.4 \times 10^6 \text{ mg}.$$

Effective filter volume $= 20 \text{ ft} \times 20 \text{ ft} \times 0.5 \text{ ft} = 200 \text{ ft}^3$

$$200 \text{ ft}^3 \times 0.02832 \text{ m}^3/\text{ft}^3 \times 10^6 \text{ cm}^3/\text{m}^3 = 5.65 \times 10^6 \text{ cm}^3.$$

Suspended solids removal per unit filter volume is

$$\frac{75.4 \times 10^6 \text{ mg}}{5.65 \times 10^6 \text{ cm}^3} = 13.3 \text{ mg/cm}^3.$$

From Fig. 7-9, $h_L \approx 5 \text{ m} = 16.45 \text{ ft}$, so the total head loss of the dirty filter is

$$5.78 \text{ ft} + 16.45 \text{ ft} = 22.2 \text{ ft}.$$

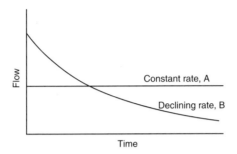

Figure 7-11. Flow through a filter operated at a constant rate (A) or declining rate (B).

Note that in the preceding example, the head loss of the dirty filter exceeds the total head available, if we assume that the filter is perhaps 10 ft deep. Obviously, a very dirty filter, as in the preceding example, has a region of negative pressure in the filter bed. This is not recommended since the negative pressure may cause gases to come out of solution and thus further impede the effectiveness of filtration.

The flow rate in most filters is controlled by a value that allows only a given volume of water to pass through, regardless of the pressure. Such rate controllers allow the filter to operate at a constant rate, as shown by curve A in Fig. 7-11. An alternative method of filter operation is to allow the water to flow through at a rate governed by the head loss, as shown by curve B in Fig. 7-11. The relative advantages of these two methods of filter operation are still being debated.

DISINFECTION

After filtration, the finished water is often disinfected with chlorine (Step 5 in Fig. 7-1). Disinfection kills the remaining microorganisms in the water, some of which may be pathogenic. Chlorine from bottles or drums is fed in correct proportions to the water to obtain a desired level of chlorine in the finished water. When chlorine comes in contact with organic matter, including microorganisms, it oxidizes this material and is in turn itself reduced. Chlorine gas is rapidly hydrolyzed in water to form hydrochloric acid and hypochlorous acid, by the reaction

$$Cl_2 + H_2O \rightarrow HOCl + H^+ + Cl^-.$$

The hypochlorous acid itself ionizes further:

$$HOCl \rightarrow OCl^- + H^+.$$

At the temperatures usually found in water supply systems, the hydrolysis of chlorine is usually complete in a matter of seconds, while the ionization of HOCl is instantaneous. Both HOCl and OCl$^-$ are effective disinfectants and are called free available chlorine in water. Free available chlorine kills pathogenic bacteria and thus disinfects the water.

Many water plant operators prefer to maintain a residual of chlorine in the water. Then if organic matter, like bacteria, enters the distribution system, there is sufficient chlorine present to eliminate this potential health hazard.

Chlorine may have adverse secondary effects. Chlorine is thought to combine with trace amounts of organic compounds in the water to produce chlorinated organic compounds that may be carcinogenic or have other adverse health effects. A recent risk analysis (Morris 1992) has found a weak positive association between bladder and rectal cancer and consumption of chlorinated drinking water, indicating that there may be some risk of carcinogenesis. The authors suggest use of both chlorine and ammonia as disinfectants in order to reduce the chlorine concentration used. Disinfection by ozonation, bubbling ozone through the water, also avoids the risk of side effects from chlorination.

A number of municipalities also add fluorine to drinking water, since fluorine has been shown to prevent tooth decay in children and young adults. The amount of fluorine added is so small that it does not participate in the disinfection process.

From the clear well (step 6 in Fig. 7-1) the water is pumped into the distribution system. This is a closed network of pipes, all under pressure. In most cases, water is pumped to an elevated storage tank that not only serves to equalize pressures but provides storage for fires and other emergencies as well.

CONCLUSION

Water treatment is often necessary if surface water supplies, and sometimes groundwater supplies, are to be available for human use. Because the vast majority of cities use one water distribution system for households, industries, and fire control, large quantities of water often must be made available to satisfy the highest use, which is usually drinking water.

However, does it make sense to produce drinkable water and then use it to irrigate lawns? Growing demand for water has prompted serious consideration of dual water supplies: one high-quality supply for drinking and other personal use, and one of lower quality, perhaps reclaimed from wastewater, for urban irrigation, fire fighting, and similar applications. Many engineers are convinced that the next major environmental engineering concern will be the availability and production of water to meet an ever-increasing demand. The job, therefore, is far from done.

PROBLEMS

7.1 A water treatment system must be able to deliver 15 mgd of water to a city of 150,000 people. Estimate: (1) the diameter of three equally sized mixing basins 10 ft deep, with a detention time of 2 min, (2) the length, width, and corresponding surface area of three flocculator basins that are 10 ft deep, (3) the surface area of three settling basins that are 10 ft deep with 2-h detention periods, and (4) the required area of each of 15 rapid sand filters rated at 2 gal/min-ft^2.

7.2 An engineer suggests the following design parameters for a city's proposed rapid sand filter: flow rate = 0.6 m³/s, and loading rate to filter = 125.0 m³/day-m². How much surface area is required for the filter? Select the number of equally sized filters, and size these filters assuming a width-to-length ratio of 1.0 to 2.5 with a maximum surface area of each filter tank being 75 m².

7.3 A flocculator is designed to treat 25 mgd. The flocculator basin is 100 ft long, 50 ft wide, and 20 ft deep and is equipped with 12-in. paddles 48 ft long. The paddles are attached to four horizontal shafts, two paddles per shaft, their center line is 8.0 ft from the shaft, and they rotate at 2.5 rpm. Assume the velocity of the water is 25% of the velocity of the paddles, the water temperature is 50°F, and the drag coefficient is 1.7.

Calculate (1) the time of flocculation, (2) the velocity differential between the water and the paddle, (3) the hydraulic power (and subsequent energy consumption), and (4) the values of G and $G\bar{t}$.

7.4 Propose some use and disposal options for water treatment sludges collected in the settling tanks following flocculation basins.

7.5 For the house or dormitory where you live, suggest which water uses require potable water and for which a lower water quality would be adequate.

Chapter 8

Collection of Wastewater

"The Shambles" is a street or area in many medieval English cities, like London and York. During the eighteenth and nineteenth centuries, the Shambles were commercialized areas, with meat packing as a major industry. The butchers of the Shambles would throw all of their waste into the street where it was washed away by rainwater into drainage ditches. The condition of the street was so bad that it contributes its name to the English language originally as a synonym for butchery or a bloody battlefield.

In old cities, drainage ditches like those at the Shambles were constructed for the sole purpose of moving stormwater out of the cities. In fact, discarding human excrement into these ditches was illegal in London. Eventually, the ditches were covered over and became what we now know as *storm sewers*. As water supplies developed and the use of the indoor water closet increased, the need for transporting domestic wastewater, called sanitary waste, became obvious. Sanitary wastes were first discharged into the storm sewers, which then carried both sanitary waste and stormwater and were known as *combined sewers*. Eventually a new system of underground pipes, known as *sanitary sewers,* was constructed for removing the sanitary wastes.

Cities and parts of cities built in the twentieth century almost all built separate sewers for sanitary waste and stormwater. In the United States, separation was mandated by federal water quality legislation in 1972. The design of sewers for stormwater is discussed in Chap. 11. Emphasis in this chapter is on estimating quantities of domestic and industrial wastewaters, and in the design of sanitary sewerage systems to handle these flows.

ESTIMATING WASTEWATER QUANTITIES

The term *sewage* is used here to mean only domestic wastewater. Domestic wastewater flows vary with the season, day of the week, and hour of the day. Figure 8-1 shows typical daily flow for a residential area. In addition to sewage, however, sewers also must carry industrial wastes, infiltration, and inflow, and the amount of flow contributed by each of these sources must be estimated for design purposes.

The quantity of industrial wastes may usually be established by water use records, or the flows may be measured in manholes that serve only a specific industry, using a small flow meter, like a Parshall flume, in a manhole. The flow is calculated as a direct

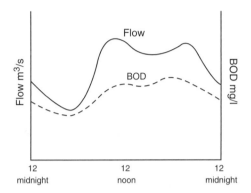

Figure 8-1. Typical dry-weather wastewater flow for a residential area.

proportion of the flow depth. Industrial flows often vary considerably throughout the day and continuous recording is necessary.

Infiltration is the flow of groundwater into sanitary sewers. Sewers are often placed below the groundwater table, and any cracks in the pipes will allow water to seep in. Infiltration is least for new, well-constructed sewers, but can be as high as 500 m^3/km–day (200,000 gal/mile–day). For older systems, 700 m^3/km–day (300,000 gal/mile–day) is the commonly estimated infiltration. Infiltration flow is detrimental since the extra volume of water must go through the sewers and the wastewater treatment plant. It should be reduced as much as possible by maintaining and repairing sewers and keeping sewerage easements clear of large trees whose roots could severely damage the sewers.

Inflow is storm water that is collected unintentionally by the sanitary sewers. A common source of inflow is a perforated manhole cover placed in a depression, so that stormwater flows into the manhole. Sewers laid next to creeks and drainageways that rise up higher than the manhole elevation, or where the manhole is broken, are also a major source. Illegal connections to sanitary sewers, such as roof drains, can substantially increase the wet weather flow over the dry weather flow. The ratio of dry weather flow to wet weather flow is usually between 1:1.2 and 1:4.

The three flows of concern when designing sewers are the average flow, the peak or maximum flow, and the extreme minimum. The ratio of the average flow to both the maximum and minimum flows is a function of the total flow, since a higher average daily discharge implies a larger community in which the extremes are evened out. Figure 8-2 shows commonly experienced ratios of average to extremes as a function of the average daily discharge.

SYSTEM LAYOUT

Sewers that collect wastewater from residences and industrial establishments almost always operate as open channels or gravity flow conduits. Pressure sewers are used in

Collection of Wastewater 155

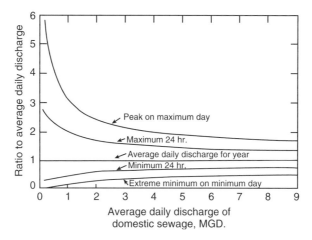

Figure 8-2. Relation of the average daily flow of domestic wastewater to the extremes of flow.

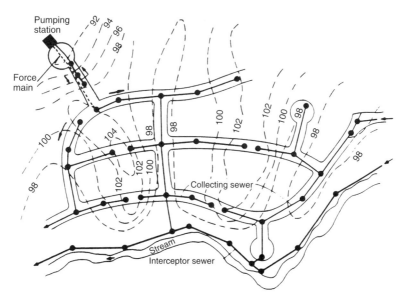

Figure 8-3. Typical wastewater collection system layout. (Adapted from J. Clark, W. Viessman, and M. Hammer, *Water Supply and Sewerage*, IEP, New York, 1977.)

a few places, but these are expensive to maintain and are useful only when there are severe restrictions on water use or when the terrain is such that gravity flow conduits cannot be efficiently maintained.

A typical system for a residential area is shown in Fig. 8-3. Building connections are usually made with clay or plastic pipe, 6 in. in diameter, to the *collecting sewers*

156 ENVIRONMENTAL ENGINEERING

that run under the street. Collecting sewers are sized to carry the maximum anticipated peak flows without surcharging (filling up) and are ordinarily made of clay, cement, concrete, or cast iron pipe. They discharge in turn into *intercepting sewers*, or *interceptors*, that collect from large ares and discharge finally into the wastewater treatment plant.

Collecting and intercepting sewers must be placed at sufficient grade to allow for adequate flow velocity during periods of low flow, but not so steep as to promote excessively high velocities when flows are at their maximum. In addition, sewers must have manholes, usually every 120 to 180 m (400 to 600 ft) to facilitate cleaning and repair. Manholes are necessary whenever the sewer changes grade (slope), size, or direction. Typical manholes are shown in Fig. 8-4.

Gravity flow may be impossible in some locations, or may be uneconomical, so that the wastewater must be pumped. A typical packaged pumping station is shown in Fig. 8-5, and its use is indicated on the system layout in Fig. 8-3.

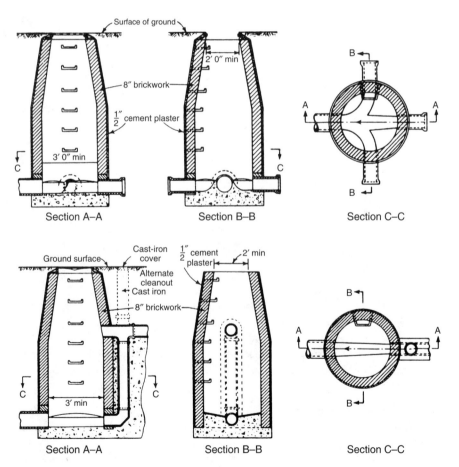

Figure 8-4. Typical manholes used for collecting sewers. (Courtesy of ASCE.)

Collection of Wastewater   157

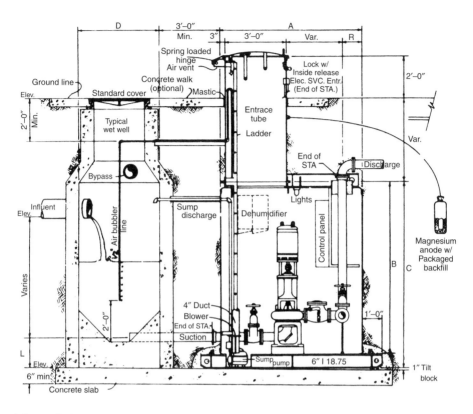

Figure 8-5. Diagram of typical pumping station for wastewater. (Courtesy of Gorman-Rupp.)

SEWER HYDRAULICS

Sewer design begins by selecting a reasonable layout and establishing the expected flows within each pipe linking the manholes. For large systems, this often involves the use of economic analysis to determine exactly what routing provides the optimal system. For most smaller systems, this is an unnecessary refinement, and a reasonable system may be estimated and sketched by hand.

The average discharge is estimated on the basis of the population served in the drainage area, and the maximum and minimum flows are calculated as shown in Fig. 8-2. Once this is done, the design is a search for the right pipe diameter and grade (slope) that will allow the minimum flow to exceed a velocity necessary for conveyance of solids, while keeping the velocity at maximum flow less than a limit at which undue erosion and structural damage can occur to the pipes. The velocity is usually held between the following limits:

- minimum = 0.6 m/s (∼2 ft/s)
- maximum = 3.0 m/s (∼10 ft/s).

158 ENVIRONMENTAL ENGINEERING

The velocity in sewers is usually calculated by using the Manning equation, based on the Chezy open channel flow equation

$$v = c\sqrt{Rs}, \tag{8.1}$$

where

v = velocity of the flow (m/s),
r = pipe hydraulic radius, or the area divided by the wetted perimeter (m),
s = slope of the pipe, and
c = Chezy coefficient.

For circular pipes, the Manning equation is developed by setting

$$c = \frac{k}{n} R^{1/6}, \tag{8.2}$$

where n = roughness factor and k = constant, so that

$$V = \frac{k}{n} R^{2/3} s^{1/2}. \tag{8.3}$$

If v is in feet per second and r is in feet, $k = 1.486$. In metric units with v in meters per second and r in meters, $k = 1.0$. The slope s is dimensionless, calculated as the fall over distance. The term n is the *roughness coefficient*, and its value depends on the pipe material, increasing for rougher pipe. Table 8-1 gives values of n for some types of sewer pipe.

EXAMPLE 8.1. An 8-in.-diameter cast iron sewer is to be set at a grade of 1-m fall per 500-m length. What will be the flow in this sewer when it is flowing full?

Using English units and noting that $n = 0.013$ from Table 8-1, by Eq. (8.3)

$$v = \frac{1.486}{0.013} \left[\frac{\frac{\pi(8/12)^2}{4}}{\pi(8/12)} \right]^{2/3} \left[\frac{1}{500} \right]^{1/2}$$

$$v = 1.54 \text{ ft/s},$$

and since the area is

$$A = \frac{\pi(8/12)^2}{4} = 0.35 \text{ ft}^2$$

$$Q = Av = (0.35)(1.54) = 0.54 \text{ cfs}.$$

Table 8-1. Manning Roughness Coefficient n

Type of channel, closed conduits	Roughness coefficient n
Cast iron	0.013
Concrete, straight	0.011
Concrete, with bends	0.013
Concrete, unfinished	0.014
Clay, vitrified	0.012
Corrugated metal	0.024
Brickwork	0.013
Sanitary sewers coated with slime	0.013

For circular conduits flowing full, the Manning formula may be rearranged as

$$Q = \frac{0.00061}{n} D^{8/3} s^{1/2}, \qquad (8.4)$$

where Q = discharge (cfs) and D = pipe diameter (inside, inches). Instead of making this calculation, a nomograph that represents a graphical solution of the Manning equation may be used. Figures 8-6 and 8-7 are such nomographs in English and metric units, respectively.

Note that the Manning equation (Eq. (8.3)) calculates velocities in sewers flowing full, and thus uses the maximum expected flow in the sewer. For flows less than the maximum, the velocity is calculated using the *hydraulic elements* graph shown as Fig. 8-8. In Fig. 8-8, the hydraulic radius (area per wetted perimeter) is calculated for various ratios of the depth when partially full (d) to the depth when completely full (D, the inside diameter of the pipe). The velocity v is calculated using Eq. (8.3) for any depth d, and the discharge is obtained from multiplying the velocity times the area. Experimental evidence has shown that the friction factor also varies slightly with the depth, as shown in Fig. 8-8, but this is often ignored in the calculations.

The maximum flow in a given sewer does not occur when the sewer is full but when d/D is about 0.95, because the additional flow area gained by increasing the d/D from 0.95 to 1.00 is very small compared with the large additional pipe surface area that the flow experiences.

EXAMPLE 8.2. From Example 8.1, it was found that the flow carried by an 8-in. cast iron pipe flowing full at a grade of 1/500 is 0.54 cfs with a velocity of 1.54 ft/s. What will be the velocity in the pipe when the flow is only 0.1 cfs?

$$\frac{q}{Q} = \frac{0.1}{0.54} = 0.185.$$

From the hydraulic elements chart

$$\frac{d}{D} = 0.33$$

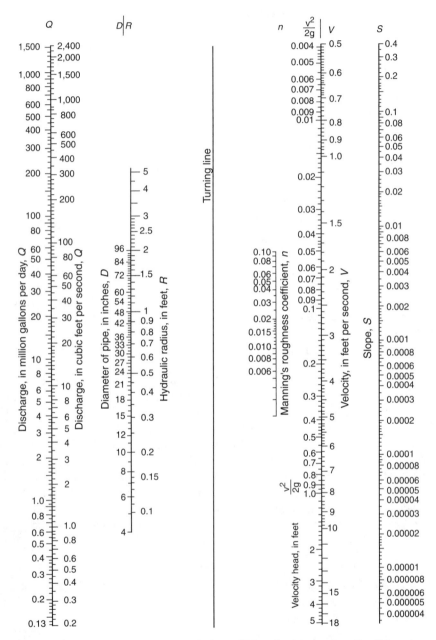

Figure 8-6. Nomograph for the solution of Manning's equation, English units.

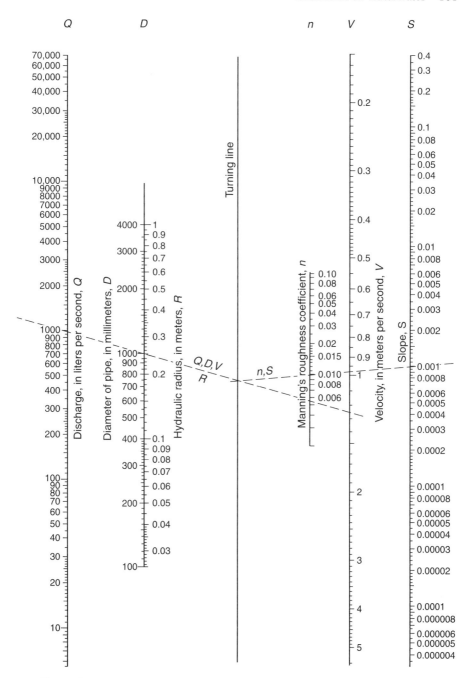

Figure 8-7. Nomograph for the solution of Manning's equation, metric units.

162   ENVIRONMENTAL ENGINEERING

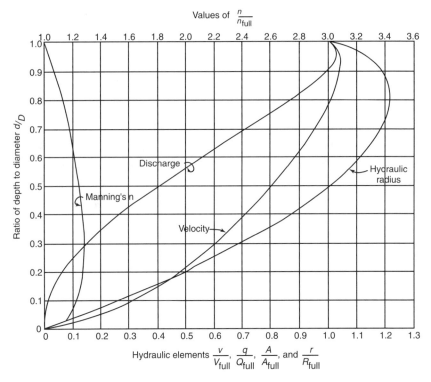

Figure 8-8. Hydraulic elements chart for open channel flow.

and, since $D = 8$ in.,

$$d = (0.33)(8 \text{ in.}) = 2.64 \text{ in.}$$

$$\frac{v}{V} = 0.64,$$

and since v, the velocity when flowing full, is 1.54 ft/s,

$$v = (0.64)(1.54 \text{ ft/s}) = 0.98 \text{ ft/s}.$$

If the desired slope between manholes is not sufficient to maintain adequate velocities at minimum flow, additional slope must be provided or a larger pipe diameter must be used. On the other hand, if the slope of the pipe is too steep, it must be reduced. Often a drop manhole is used, which is, in effect, a method of wasting energy (see Fig. 8-4).

Standard American convention requires that a head loss of 0.1 ft (0.06 m) be allowed for the head loss through each manhole. The invert elevation of the pipe leaving the manhole is thus at least 0.1 ft lower than the invert elevation of the pipe coming in.

Collection of Wastewater 163

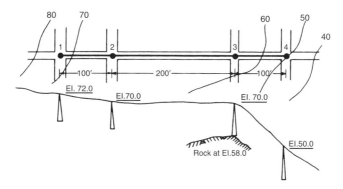

Figure 8-9. Sewer layout for Example 8.3.

EXAMPLE 8.3. The system shown in Fig. 8-9 is to be designed given the following flows: maximum flow = 3.2 mgd, minimum flow = 0.2 mgd, minimum allowable velocity = 2 ft/s, and maximum allowable velocity = 12 ft/s.

All manholes should be about 10 ft deep, and there is no additional flow between Manhole 1 and Manhole 4. Design acceptable invert elevations for this system. (The nomograph of Fig. 8-6 is used in this solution.)

From Manhole 1 to Manhole 2:

1. The street slopes at 2/100, so choose the slope of the sewer $s = 0.02$.
2. Assume $n = 0.013$ and try $D = 12$ in. From the nomograph of Fig. 8-6, connect $n = 0.013$ and $s = 0.02$, and extend that straight line to the turning line.
3. Connect the point on the turning line with $D = 12$ in.
4. Read $v = 6.1$ ft/s for $Q = 3.2$ mgd from intersection of the line drawn in Step 3. This is acceptable for the sewer flowing full.
5. To check for minimum velocity,

$$q/Q = 0.2/3.2 = 0.063,$$

and from the hydraulic elements chart, Fig. 8-8, $q/Q = 0.063$ intersects the discharge curve at $d/D = 0.2$, which intersects the velocity curve at $v_p/v = 0.48$, and

$$v_p = 0.48(6.1 \text{ ft/s}) = 2.9 \text{ ft/s}.$$

6. The downstream invert elevation of Manhole 1 is ground elevation minus 10 ft, or 62.0 ft. The upstream invert elevation of Manhole 2 is thus $62.0 - 2.0 = 60.0$ ft. Allowing 0.1 ft for head loss in the manhole, the downstream invert elevation is 59.9 ft.

From Manhole 2 to Manhole 3, the slope will be a problem because of rock. Try a larger pipe, $D = 18$ in. Repeating steps 1 and 2,

7. Connect the point on the turning line with $D = 18$ in.

164   ENVIRONMENTAL ENGINEERING

8. From the nomograph, $v = 2.75$ ft/s for $Q = 3.2$ mgd, $v_p/v = 0.48$ and

$$v_p = 0.48(2.75 \text{ ft/s}) = 1.32 \text{ ft/s}.$$

Even with a slope of 0.002 (from the nomograph), the resulting velocity at minimum flow is too low. Try $s = 0.005$ with $D = 18$ in. for Steps 1–3. Then $Q = 4.7$ mgd and $v = 4.1$ ft/s for Step 4. From the hydraulic elements chart, $v_p = 1.9$ ft/s, which is close enough. Thus the upstream invert elevation of Manhole 3 is

$$59.9 - (0.005)(200) = 58.9 \text{ ft},$$

and the downstream is at 58.8 ft, still well above the rock. From Manhole 3 to Manhole 4, the street obviously has too much slope. Try using $D = 12$ in. and

$$s = (58.8 - 40)/100 = 0.188,$$

since 40 ft is the desired invert elevation of Manhole 4. Read $Q = 8.5$ mgd and $v = 19$ ft/s, but only 3.2 mgd is required at maximum flow; hence

$$q/Q = 3.2/8.5 = 0.38, \text{ and}$$
$$v_p = (0.78)(18 \text{ ft/s}) = 14.8 \text{ ft/s}.$$

This is too high. Use a drop manhole, with the invert at, say, elevation 45.0. In this case $s = 0.138$ and $Q = 7.3$ mgd, $v = 14.5$ s,

$$q/Q = 3.2/7.3 = 0.44, \text{ and}$$
$$v_p = (0.85)(14.5 \text{ ft/s}) = 12.3 \text{ ft/s},$$

which is close enough. The upstream invert elevation of Manhole 4 is thus at 45.0 ft and the downstream invert elevation can be at, say, 40.0 ft. The minimum velocity for this last step need not be checked. The final design is shown in Fig. 8-10.

It is often convenient to organize sewer calculations according to a table such as Fig. 8-11, a slightly simplified version of a table suggested by the American Society for Civil Engineers (American Society for Civil Engineers 1969). This reference should be consulted for details of accepted engineering practice in sewer design.

CONCLUSION

The methods discussed in this chapter are hand calculations, and few modern consulting engineers who design sewerage systems will use such cumbersome methods. Computer programs that not only solve for the hydraulic parameters, but evaluate the economic alternatives (e.g., a deeper cut or a bigger pipe) are generally used. Elegant programs are only useful, however, if the design process is well understood.

Collection of Wastewater 165

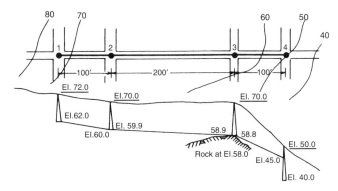

Figure 8-10. Final design for Example 8.3.

Figure 8-11. Work sheet for the design of sewers.

PROBLEMS

8.1 A vitrified clay pipe sewer is to discharge 4 cfs when laid on a grade of 0.0016 and flowing half full. What is the required diameter?

8.2 An 18-in. sewer, $n = 0.013$, is 15,000 ft long and laid on a uniform grade. Difference in elevation of the two ends is 4.8 ft. Find velocity and discharge when the sewer is flowing 0.6 full.

8.3 A 48-in. circular sewer is required to discharge 100 cfs when full. What is the required grade, according to the Manning formula, if $n = 0.015$?

8.4 A 12-in. concrete sewer, flowing just full, is laid on a grade of 0.00405 ft/ft. Find the velocity and rate of discharge.

8.5 A 12-in. sewer, $n = 0.013$, is laid on a grade of 3.0 ft/1000 ft. Find the velocity and rate of discharge when the sewer is flowing 0.4 full.

Chapter 9

Wastewater Treatment

As civilization developed and cities grew, domestic sewage and industrial waste were eventually discharged into drainage ditches and sewers, and the entire contents emptied into the nearest watercourse. For major cities, this discharge was often enough to destroy even a large body of water. As Samuel Taylor Coleridge described the city of Cologne (Köln) in Germany:

> In Köln, a town of monks and bones,
> And pavements fang'd with murderous stones,
> And rags and hags, and hideous wenches,
> I counted two-and-seventy stenches
> All well defined, and several stinks!
> Ye Nymphs that reign o'er sewers and sinks,
> The river Rhine, it is well known,
> Doth wash your city of Cologne;
> But tell me Nymphs! what power divine
> Shall henceforth wash the river Rhine?

During the nineteenth century, the River Thames was so grossly polluted that the House of Commons stuffed rags soaked with lye into cracks in the windows of Parliament to reduce the stench.

Sanitary engineering technology for treating wastewater to reduce its impact on watercourses, pioneered in the United States and England, eventually became economically, socially, and politically feasible. This chapter reviews these systems from the earliest simple treatment systems to the advanced systems used today. The discussion begins by reviewing those wastewater characteristics that make disposal difficult, showing why wastewater cannot always be disposed of on-site, and demonstrating the necessity of sewers and centralized treatment plants.

WASTEWATER CHARACTERISTICS

Discharges into a sanitary sewerage system consist of domestic wastewater, industrial discharge, and infiltration. The last adds to the total wastewater volume but is not itself a concern in wastewater disposal; infiltration will even dilute *municipal* sewage to some extent. These discharges vary widely with the size and type of industry and the amount of treatment applied before discharge into sewers. In the United States, the trend has been to mandate increasing pretreatment of wastewater in response to both regulations limiting discharges and the imposition of local sewer surcharges. Surcharges are levied by communities to cover the cost of unusual treatment or treatment of unusually large quantities of wastewater, as in Example 9.1.

EXAMPLE 9.1. A community normally levies a sewer charge of 20 cents/in.^{3}. For discharges in which the BOD > 250 mg/L and suspended solids (SS) > 300 mg/L, an additional \$0.50/kg BOD and \$1.00/kg SS are levied.

A chicken processing plant uses 2000 m^3 water per day and discharges wastewater with BOD = 480 mg/L and SS = 1530 mg/L. What is the plant's daily wastewater disposal bill?

The excess BOD and SS are, respectively,
(480 − 250) mg/L × 2000 in.^{3} × 1000 L/m^3 × 10^{-6} kg/mg = 460 kg excess BOD
(1530 − 300) mg/L × 2000 m^3 × 1000 L/m^3 × 10^{-6} kg/mg = 2460 kg excess SS.

The daily bill is thus
(2000 m^3)(\$0.20/m^3) + (460 kg$_{BOD}$)(\$0.50/kg$_{BOD}$) +
(2460 kg$_{SS}$)(\$1.00/kg$_{SS}$) = \$3090.00.

BOD is reduced and SS are removed by wastewater treatment, but heavy metals, motor oil, refractory organic compounds, radioactive materials, and similar exotic pollutants are not readily handled this way. Communities usually severely restrict the discharge of such substances by requiring pretreatment of wastewater.

Domestic sewage varies substantially in quantity and quality over time and from one community to the next. Typical variation for a small community is shown in Fig. 9-1. Table 9-1 shows typical values for the most important parameters of domestic wastewater.

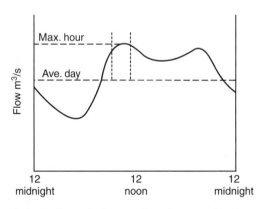

Figure 9-1. Daily variations in flow for a small community.

Table 9-1. Characteristics of Typical Domestic Wastewater

Parameter	Typical value for domestic sewage
BOD	250 mg/L
SS	220 mg/L
Phosphorus	8 mg/L
Organic and ammonia nitrogen	40 mg/L
pH	6.8
Chemical oxygen demand	500 mg/L
total solids	270 mg/L

ON-SITE WASTEWATER TREATMENT

The original on-site system, of course, is the *pit privy,* glorified in song and fable.[1] The privy, still used in camps and temporary residences and in many less industrialized countries, consists of a pit about 2 m (6 ft) deep into which human excrement is deposited. When a pit fills up, it is covered and a new one is dug. The composting *toilet,* which accepts both human excrement and food waste, and produces a useful compost, is a logical extension of the pit privy. In a dwelling with a composting toilet, wastewater from other sources, like washing, is discharged separately.

By far the greatest number of households with on-site disposal systems use a form of the *septic tank* and *tile field.* As shown in Fig. 9-2, a septic tank consists of a concrete box that removes the solids in the waste and promotes partial decomposition. The solid particles settle out and eventually fill the tank, thus necessitating periodic cleaning. The water overflows into a tile drain field that promotes the seepage of discharged water.

A tile field consists of a plastic pipe with holes in it laid in a 3-ft-deep trench. The effluent from the septic tank flows into the tile field pipes and seeps into the ground through these holes. Alternatively, seepage pits consisting of gravel and sand may be used for promoting absorption of effluent into the ground. The most important consideration in designing a septic tank and tile field system is the ability of the ground to absorb the effluent. Septic tank design includes the classification of soils that have been found to "perc," or allow the treated wastewater to percolate into the soil.

---

[1] A literary work on this theme is "The Passing of the Backhouse" by James Whitcomb Riley:

> But *when* the crust was on the snow and the sullen skies were gray,
> In sooth the building was no place where one would wish to stay,
> We did our duties promptly, there one purpose swayed the mind,
> We tarried not nor lingered long on what we left behind.
>
> The torture of that icy seat would make a Spartan sob;
>
> (lines 25–29)

170 ENVIRONMENTAL ENGINEERING

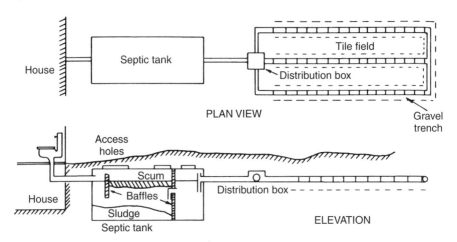

Figure 9-2. Septic tank and tile field used for on-site wastewater disposal.

Table 9-2. Adsorption Area Requirements for Private Residents

Percolation rate (in./min)	Required adsorption field area, per bedroom (ft^2)
Greater than 1	70
Between 1 and 0.5	85
Between 0.5 and 0.2	125
Between 0.2 and 0.07	190
Between 0.07 and 0.03	250
Less than 0.03	Unsuitable ground

The U.S. Public Health Service and all county and local departments of health have established guidelines for sizing the tile fields or seepage pits. Typical standards are shown in Table 9-2. Many areas in the United States have soils that percolate poorly, and septic tank/tile fields are not appropriate. Several other options are available for on-site wastewater disposal, one of which is shown in Fig. 9-3.

In urbanized areas, it has been several centuries since there was enough land available for on-site treatment and percolation. Up until the nineteenth century, this problem was solved by constructing large cesspools or holding basins for the wastewater; these holding basins needed periodic pumping as they filled up. The holding basins also created considerable public health problems, as in the Broad Street pump incident discussed in Chap. 4. A better way to move human waste out of a congested community was to use water as a carrier.

The *water closet*, as it is still known in Europe, has thus become a standard fixture of modern urban society. Some authors (Kirby *et al.* 1956) credit John Bramah with

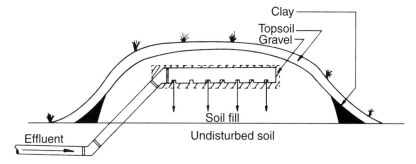

Figure 9-3. Alternative on-site disposal system.

its invention in 1778; others (Reyburn 1969) recognize it as the brainchild of Sir John Harrington in 1596.[2] The latter argument is strengthened by Sir John's original description of the device, although there is no record of his donating his name to the invention. The first recorded use of that euphemism is found in the regulation at Harvard University, where in 1735 it was decreed that "No Freshman shall go to the Fellows' John."

The wide use of waterborne wastewater disposal, however, concentrated all of the wastes of a community in one place. Cleanup then requires a major effort, and this requirement fostered what is known as central wastewater treatment.

CENTRAL WASTEWATER TREATMENT

The objective of wastewater treatment is to reduce the concentrations of specific pollutants to the level at which the discharge of the effluent will not adversely affect the environment or pose a health threat. Moreover, reduction of these constituents need only be to some required level. Although water can technically be completely purified by distillation and deionization, this is unnecessary and may actually be detrimental to the receiving water. Fish and other organisms cannot survive in deionized or distilled water.

For any given wastewater in a specific location, the *degree* and *type* of treatment are variables that require engineering decisions. Often the degree of treatment depends on the assimilative capacity of the receiving water. DO sag curves can indicate how much BOD must be removed from wastewater so that the DO of the receiving water is not depressed too far. The amount of BOD that must be removed is an effluent standard

---

[2] Sir John, a courtier and poet, installed his invention in his country house at Kelson, near Bath. Queen Elizabeth had one fitted soon afterward at Richmond Palace. The two books that were written about this innovation bear the strange titles, *A New Discourse on a Stale Subject, Called the Metamorphosis of Ajax,* and *An Anatomie of the Metamorphosed Ajax.* "Ajax" is a play on the words "a jakes," a synonym for water closet.

(discussed more fully in Chap. 11) and dictates in large part the type of wastewater treatment required.

To facilitate the discussion of wastewater, a "typical wastewater" (Table 9-1) is assumed, and it is further assumed that the effluent from this wastewater treatment must meet the following effluent standards:

BOD $\leq$ 15 mg/L
SS $\leq$ 15 mg/L
P $\leq$ 1 mg/L

Additional effluent standards could have been established but, for illustrative purposes, we consider only these three. The treatment system selected to achieve these effluent standards includes:

- Primary treatment: *physical processes* that remove nonhomogenizable solids and homogenize the remaining effluent.
- Secondary treatment: *biological processes* that remove most of the biochemical demand for oxygen.
- Tertiary treatment: *physical, biological, and chemical processes* that remove nutrients like phosphorus, remove inorganic pollutants, deodorize and decolorize effluent water, and carry out further oxidation.

PRIMARY TREATMENT

The most objectionable aspect of discharging raw sewage into watercourses is the floating material. Thus *screens* were the first form of wastewater treatment used by communities, and are used today as the first step in treatment plants. Typical screens, shown in Fig. 9-4, consist of a series of steel bars that might be about 2.5 cm apart. A screen in a modern treatment plant removes materials that might damage equipment or hinder further treatment. In some older treatment plants screens are cleaned by hand, but mechanical cleaning equipment is used in almost all new plants. The cleaning rakes are activated when screens get sufficiently clogged to raise the water level in front of the bars.

In many plants, the second treatment step is a *comminutor*, a circular grinder designed to grind the solids coming through the screen into pieces about 0.3 cm or less in diameter. A typical comminutor design is shown in Fig. 9-5.

The third treatment step is the removal of grit or sand from the wastewater. Grit and sand can damage equipment like pumps and flow meters and, therefore, must be removed. The most common *grit chamber* is a wide place in the channel where the flow is slowed enough to allow the dense grit to settle out. Sand is about 2.5 times denser than most organic solids and thus settles much faster. The objective of a grit chamber is to remove sand and grit without removing organic material. Organic material must be treated further in the plant, but the separated sand may be used as fill without additional treatment.

Wastewater Treatment   173

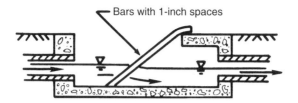

Figure 9-4. Bar screen used in wastewater treatment. The top picture shows a manually cleaned screen; the bottom picture shows a mechanically cleaned screen. (Photo courtesy of Envirex.)

Most wastewater treatment plants have a *settling tank* (Figs. 9-6 and 9-7) after the grit chamber, to settle out as much solid material as possible. Accordingly, the *retention tune* is long and turbulence is kept to a minimum. Retention time is the total time an average slug of water spends in the tank and is calculated as the time required

174   ENVIRONMENTAL ENGINEERING

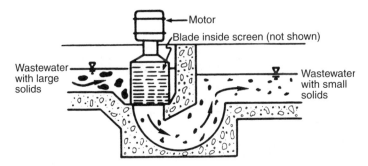

Figure 9-5. A comminutor used to grind up large solids.

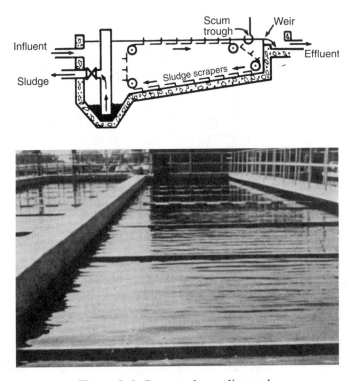

Figure 9-6. Rectangular settling tank.

to fill the tank. For example, if the tank volume is 100 m³ and the flow rate is 2 m³/min, the retention time is $100/2 = 50$ min.

The solids settle to the bottom of the tank and are removed through a pipe, while the clarified liquid escapes over a *V-notch weir*, which distributes the liquid discharge equally all the way around a tank. Settling tanks are also called *sedimentation tanks*

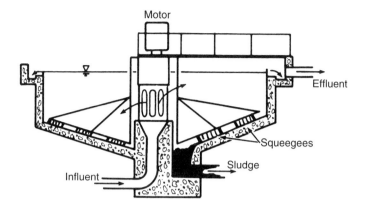

Figure 9-7. Circular settling tank.

or *clarifiers*. The settling tank that immediately follows screening and grit removal is called the *primary clarifier*. The solids that drop to the bottom of a primary clarifier are removed as *raw sludge*.

Raw sludge generally has a powerfully unpleasant odor and is full of water, two characteristics that make its disposal difficult. It must be stabilized to reduce further decomposition and dewatered for ease of disposal. Solids from processes other than the primary clarifier must be treated similarly before disposal. Treatment and disposal of wastewater sludge are discussed further in Chap. 10.

Settling Tank Design and Operation

Gravitational settling is, as has been noted, one of the most efficient means of separating solids from surrounding liquid. Solid/liquid separation can be achieved this way as long as the solids are more dense than the liquid.[3] Simple settling is, however, affected by other forces in certain circumstances. Particles settle in one of three general ways:

- Class I: *Discrete particle settling* occurs when particles settle unhindered by container walls or neighboring particles.
- Class II: *Flocculent settling* occurs when neighboring particles come into contact with each other, changing particle size and hindering settling.
- Class III: *Thickening* occurs when the entire mass of particles settle with the same velocity, and there is no interparticle movement.

The simplest case is Class I, discrete particle settling. A single particle, having attained terminal velocity, has no net force acting on it. The three forces acting on the particle — drag, buoyancy, and gravity — balance as in

$$F_g = F_D + F_B, \tag{9.1}$$

which becomes

$$\rho_s g V = \rho g V + C_D \frac{A v^2 \rho}{2}, \tag{9.2}$$

where

F_g = force due to gravity = mg,
F_D = drag force = $(C_D \rho v^2 A)/2$,
F_B = buoyancy = $\rho V g$,
m = mass of particle = $\rho_s V$,
g = gravitational acceleration,
ρ_s = particle density (in kg/m^2),
ρ = density of the medium,
C_D = drag coefficient,
v = velocity of particle in (m/s),
A = projected surface area of particle (in m^2), and
V = volume of particle (in m^3).

Solving for the velocity,

$$v = \sqrt{2V \frac{g(\rho_s - \rho)}{C_D A \rho}}. \tag{9.3}$$

If we now assume that the particle is a sphere,

$$v = \sqrt{\frac{4dg(\rho_s - \rho)}{3 C_D \rho}}, \tag{9.4}$$

---

[3] Separation of *less* dense solids may be achieved by flotation, or allowing the solids to float to the top. This method is used, for example, in ore separation, but is not applicable to solids separation in primary clarifiers.

which is the well-known Newtonian equation, where d is the particle diameter. If, as is the case in most settling in wastewater treatment, the Reynolds number (R) is sufficiently low and a laminar boundary layer is maintained, the drag coefficient can be expressed as

$$C_D = \frac{24}{R},$$

where $R = v d\rho/\mu$ and μ = fluid viscosity. If $C_D > 1$, this no longer holds, and the drag coefficient may be approximated as

$$C_D = \frac{24}{R} + \frac{3}{\sqrt{R}} + 0.34.$$

For laminar flow, substituting $C_D = 24/R$ into the Newtonian equation yields the Stokes equation

$$v = \frac{g(\rho_s - \rho)d^2}{18\mu}. \tag{9.5}$$

The velocity of particles may be related to the expected settling tank performance by idealizing a settling tank. As shown in Fig. 9-8, a rectangular tank is first divided into four zones: inlet zone, outlet zone, sludge zone, and settling zone. The first two zones are designated for the damping of currents caused by the influent and effluent, respectively. The sludge zone is storage space for the settled solids. Settling itself takes place only in the fourth (settling) zone.

Several initial assumptions are required in this analysis:

- Uniform flow occurs within the settling zone.
- All particles entering the sludge zone are removed.
- Particles are evenly distributed in the flow as they enter the settling zone.
- All particles entering the effluent zone escape the tank.

Thus, a particle entering the settling zone at the water surface has a settling velocity v_g and a horizontal velocity v_h such that the component of the two defines a trajectory

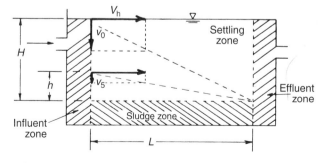

Figure 9-8. Schematic diagram of an ideal settling tank.

as shown in Fig. 9-8. In other words, the particle is just barely removed. Had the particle entered the settling zone at any other height, its trajectory would always have carried it into the sludge zone. Particles having this velocity are termed *critical particles* because particles with lower settling velocities are not all removed. For example, the particle having velocity v_s, entering the settling zone at the surface, will end up in the effluent zone and escape. If this same particle had entered at height h, it would have just been removed. Any of these particles that happen to enter the settling zone at height h or lower would be removed, and those entering above h would not. Since the particles entering the settling zone are equally distributed, the proportion of those particles with a velocity v that are removed is

$$v_s = h/H, \tag{9.6}$$

where H is the height of the settling zone. With reference to Fig. 9-8, similar triangles yield

$$\frac{v_0}{H} = \frac{v_s}{h} = \frac{v_b}{L}. \tag{9.7}$$

The time that the critical particle spends in the settling zone is

$$\bar{t} = \frac{L}{v_h} = \frac{h}{v_s} = \frac{H}{v_0}. \tag{9.8a}$$

Hence

$$v_0 = \frac{H}{\bar{t}}. \tag{9.8b}$$

The time t is also equal to the hydraulic retention time or V/Q, where Q is the flow rate and V is the volume of the settling zone. Moreover

$$V = AH, \tag{9.9a}$$

where A is the surface area of the settling zone. Thus

$$v_0 = \frac{H}{\bar{t}} = \frac{H}{AH/Q} = \frac{Q}{A}. \tag{9.9b}$$

Equation (9.9b) gives the *overflow rate*, an important design parameter for settling tanks. The units of the overflow rate are

$$v_0 = \frac{m}{s} = \frac{Q}{A} = \frac{m^3/s}{m^2}.$$

Although overflow rate is usually expressed as gallons/day–foot^2, it implies a velocity and is equal to the velocity of the critical particle. Thus, when the design of a clarifier is specified by the overflow rate, the critical particle is thereby defined.

Where any two of the parameters — overflow rate, retention time, and depth — are specified, the remaining parameter is also fixed.

EXAMPLE 9.2. A primary clarifier has an overflow rate of 600 gal/day–ft² and a depth of 6 ft. What is its hydraulic retention time?

$$v_0 = 600 \frac{\text{gal}}{\text{day–ft}^2} \times \frac{1 \text{ft}^3}{7.48 \text{ gal}} = 80.2 \text{ ft/day}$$

$$\bar{t} = \frac{H}{v_0} = \frac{6 \text{ ft}}{80.2 \text{ ft/day}} = 0.0748 \text{ day} = 1.8 \text{ h}.$$

We may obtain a better understanding of settling by considering the individual variables. Increasing the flow rate Q in a given tank increases the critical velocity v_0. Since fewer particles then have $v > v_0$, fewer particles are removed from the effluent. More particles could be removed if v_0 is *decreased*. This may be done by either reducing Q or increasing A. The latter term may be increased by changing the dimensions of the tank so that the depth is very shallow, and the length and width are large. If a tank 3 m deep is sliced in half and the two 1.5-m-deep slices are placed alongside each other, the horizontal velocity will be the same as in the 3-m tank but the surface area will be doubled and v_0 will have half the original value. Very shallow tanks would thus appear to be optimum primary clarifiers, except that they take up a great deal of land area, do not ensure even distribution of flow, and incur great expense in steel and concrete.[4]

The foregoing discussion incorporates a significant assumption that is not really true for wastewater treatment. Solids in wastewater do *not* settle as discrete particles but tend to form larger particles by clumping together in *flocculent settling*, which may be demonstrated by setting up a cylinder of dirty water and allowing the particles in it to settle. If the water is sampled at various elevations, the clarification of water with time can be observed (Fig. 9-9).

If the dirty water contains particles that all have the same settling velocity, there should be perfect clarification as the highest particle drops to the bottom. If $v = 120$ cm/2 h $= 60$ cm/h, the curve in Fig. 9-9A should result. Everything below the settling curve would be at the suspended solids concentration (1000 mg/L) and above the line the water would be clean.

If the dirty water is a mixture of equal numbers of particles of two different sizes, with settling velocities $v_a = 30$ cm/h and $v_b = 60$ cm/h, the curves would look like Fig. 9-9B. A mixture of different-sized particles would still produce straight-line settling curves.

Real-life wastewater, however, produces settling curves like those in Fig. 9-9C. The fraction of solids removed or reduced at various sampling ports plot as curved lines whose slopes increase with time instead of remaining constant. The increased velocities are due to collisions and the subsequent building of larger particles.

---

[4]Remember the max-mm problem in calculus? "What is the shape of a six-sided prism with the smallest surface area per given volume?"

180 ENVIRONMENTAL ENGINEERING

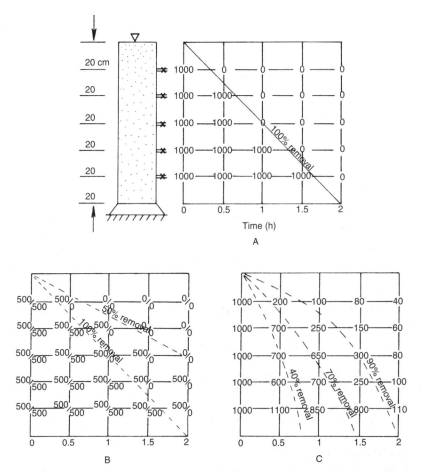

Figure 9-9. Settling test with a column of dirty water and periodic sampling at five ports. The initial suspended solids (SS) concentration ($t = 0$) is assumed to be 1000 mg/L.

Because of such flocculation, a shallow tank is not as efficient as ideal tank theory would suggest.

Removal efficiencies for slurries may be estimated from curves like those in Fig. 9-9C. The Figure shows that the entire tank experiences about 90% removal at a retention time of 2 h. The top section of the tank, however, is much clearer, because it contains only about 40 mg/L of SS, and thus experiences $[(1000 - 40)/1000] \times 100 = 96\%$ SS removal. The general equation for estimating removal, giving credit for cleaner water on top, is

$$R = P + \sum_{i-1}^{n-1} \left(\frac{h}{H}\right)(P_i - P), \tag{9.10}$$

where

R = % overall solids recovery,
P = solids recovery at the bottommost section,
P_i = solids recovery in section i,
n = number of sections,
h = height of each section, and
H = height of column ($H = nh$).

EXAMPLE 9.3. A chemical waste at an initial SS concentration of 1000 mg/L and flow rate of 200 m³/h is to be settled in a tank, $H = 1.2$ m deep, $W = 10$ m wide, and $L = 31.4$ m long. The results of a laboratory test are shown in Fig. 9-9C. Calculate the fraction of solids removed, the overflow rate, and the velocity of the critical particle.

The surface area of the tank is

$$A = WL = (31.4)(10) = 314 \, \text{m}^2.$$

The overflow rate is therefore

$$Q/A = 200/314 = 0.614 \, \text{m}^3/\text{h-m}^2.$$

The critical velocity is thus $v_0 = 0.614$ m/h. However, the waste in this instance undergoes flocculent settling rather than settling at the critical velocity. The hydraulic retention time is

$$\bar{t} = \frac{V}{Q} = \frac{AH}{Q} = \frac{(314)(1.2)}{200} = 1.88 \, \text{h}.$$

In Fig. 9-9C the 85% removal line approximately intersects the retention time of 1.88 h. Thus, 85% of the solids are removed. In addition to this, however, even better removal is indicated at the top of the water column. At the top 20 cm, assume the SS concentration is 40 mg/L, equal to $[(1000 - 4) \times 100]/1000 = 96\%$ removal, or 11% better than the entire column. The second shows $[(1000 - 60) \times 100]/1000 = 94\%$ removal and so on. The total amount removed, ignoring the bottommost section, is

$$R = P + \sum_{i=1}^{n-1}\left(\frac{h}{H}\right)(P_i - P) \tag{9.10}$$

$$R = 85 + (1/6)(11 + 9 + 5 + 4) = 90.9\%.$$

The solids capture efficiency of a primary clarifier for domestic waste is plotted as a function of retention time, as shown in Fig. 9-10.

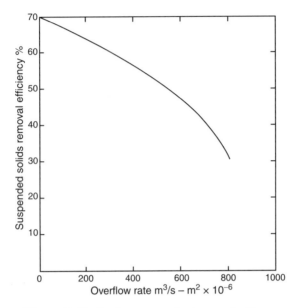

Figure 9-10. Performance of primary clarifiers.

Primary treatment is mainly removal of solids, although some BOD is removed as a consequence of the removal of decomposable solids. The wastewater described earlier might now have these characteristics:

	Raw wastewater	After primary treatment
BOD_5 (mg/L)	250	175
SS (mg/L)	220	60
P (mg/L)	8	7

A substantial fraction of the solids, as well as some BOD and a little P, has been removed as a consequence of the removal of raw sludge. After primary treatment the wastewater may move on to secondary treatment.

SECONDARY TREATMENT

Water leaving the primary clarifier has lost much of the solid organic matter but still contains high-energy molecules that decompose by microbial action, creating BOD. The demand for oxygen must be reduced (energy wasted) or else the discharge may create unacceptable conditions in the receiving waters. The objective of secondary treatment is to remove BOD, whereas the objective of primary treatment is to remove solids.

Wastewater Treatment   183

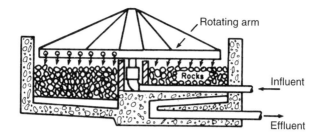

Figure 9-11. Trickling filter.

The *trickling filter*, shown in Fig. 9-11, consists of a filter bed of fist-sized rocks over which the waste is trickled. The name is something of a misnomer since no filtration takes place. A very active biological growth forms on the rocks, and these organisms obtain their food from the waste stream dripping through the rock bed. Air is either forced through the rocks or circulates automatically because of the temperature difference between the air in the bed and ambient temperatures. In older filters, the waste is sprayed onto the rocks from fixed nozzles. The newer designs use a rotating arm that moves under its own power, like a lawn sprinkler, distributing the waste evenly over the entire bed. Often the flow is recirculated and a higher degree of treatment attained.

Trickling filtration was a well-established treatment system at the beginning of the twentieth century. In 1914, a pilot plant was built for a different system that bubbled air through free-floating aerobic microorganisms. This process became established as the *activated sludge system*. Activated sludge differs from trickling filtration in that it recycles and reuses microorganisms, and the microorganisms are suspended in the liquid.

An activated sludge system, as shown in the block diagram in Fig. 9-12, includes a tank full of waste liquid from the primary clarifier and a mass of microorganisms. Air bubbled into this *aeration tank* provides the necessary oxygen for survival of the aerobic organisms. The microorganisms come in contact with dissolved organic matter in the wastewater, adsorb this material, and ultimately decompose the organic material

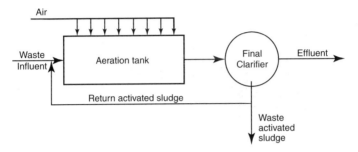

Figure 9-12. Block diagram of an activated sludge system.

to CO_2, H_2O, some stable compounds, and more microorganisms. The production of new organisms is relatively slow, and uses most of the aeration tank volume.

When most of the organic material, which is food for the microorganisms, has been used up, the microorganisms are separated from the liquid in a settling tank, sometimes called a *secondary* or *final clarifier*. The microorganisms remaining in the settling tank have no food available, become hungry, and are thus activated; hence the term *activated sludge*. The clarified liquid escapes over a weir and may be discharged into the receiving water. The settled microorganisms, now called *return activated sludge*, are pumped back to the head of the aeration tank where they find more food in the organic compounds in the liquid entering the aeration tank from the primary clarifier, and the process starts over again. Activated sludge treatment is a continuous process, with continuous sludge pumping and clean water discharge.

Activated sludge treatment produces more microorganisms than necessary and if the microorganisms are not removed, their concentration will soon increase and clog the system with solids. Some of the microorganisms must therefore be wasted. Disposal of such *waste activated sludge* is one of the most difficult aspects of wastewater treatment.

Activated sludge systems are designed on the basis of loading, or the amount of organic matter, or food, added relative to the microorganisms available. The food-to-microorganism (F/M) ratio is a major design parameter. Both F and M are difficult to measure accurately, but may be approximated by influent BOD and SS in the aeration tank, respectively. The combination of liquid and microorganisms undergoing aeration is known as *mixed liquor*, and the SS in the aeration tank are *mixed liquor suspended solids* (MLSS). The ratio of influent BOD to MLSS, the F/M ratio, is the *loading* on the system, calculated as pounds (or kg) of BOD per day per pound or kilogram of MLSS.

Relatively small F/M, or little food for many microorganisms, and a long aeration period (retention time in the tank) result in a high degree of treatment because the microorganisms can make maximum use of available food. Systems with these features are called *extended aeration* systems and are widely used to treat isolated wastewater sources, like small developments or resort hotels. Extended aeration systems create little excess biomass and little excess activated sludge to dispose of.

Table 9-3 compares extended aeration systems, conventional secondary treatment systems, and "high-rate" systems that have short aeration periods and high loading, and result in less efficient treatment.

Wastewater Treatment   185

Table 9-3. Loadings and Efficiencies of Activated Sludge Systems

Process	Loading: F/M (lb BOD/day/lb MLSS)	Aeration period (h)	BOD removal efficiency (%)
Extended aeration	0.05–0.2	30	95
Conventional	0.2–0.5	6	90
High rate	1–2	4	85

EXAMPLE 9.4. The BOD$_5$ of the liquid from the primary clarifier is 120 mg/L at a flow rate of 0.05 mgd. The dimensions of the aeration tank are $20 \times 10 \times 20$ ft^3 and the MLSS = 2000 mg/L. Calculate the F/M ratio:

$$\frac{\text{lb BOD}}{\text{day}} = \left(\frac{120 \text{ mg}}{\text{L}}\right)(0.05 \text{ mgd})\left(\frac{3.8 \text{ L}}{\text{gal}}\right)\left(\frac{1 \text{ lb}}{454 \text{ g}}\right)\left(\frac{1 \text{ g}}{1000 \text{ mg}}\right) = \left(\frac{50 \text{ lb}}{\text{day}}\right)$$

$$\text{lb MLSS} = (20 \times 10 \times 20) \text{ ft}^3 \left(\frac{2000 \text{ mg}}{\text{L}}\right)\left(\frac{3.8 \text{ L}}{\text{gal}}\right)\left(\frac{7.48 \text{ gal}}{\text{ft}^3}\right)\left(\frac{1 \text{ lb}}{454 \text{ g}}\right)$$

$$\times \left(\frac{1 \text{ g}}{1000 \text{ mg}}\right) = 229 \text{ lb}$$

$$\frac{F}{M} = \frac{50}{229} = 0.22 \frac{\text{lb BOD/day}}{\text{lb MLSS}}$$

When the microorganisms begin to metabolize the food, a great deal of oxygen is required. Accordingly, the DO in the aeration tank decreases suddenly after the point at which waste is introduced. The resulting low DO concentration at the influent end of the aeration tank may be detrimental to the microbial population. In order to sustain the microorganisms, the activated sludge process may incorporate a variation like *tapered aeration* or *step aeration* (see Fig. 9-13). In tapered aeration, air is blasted into the tank where needed, while in step aeration, waste is introduced at several locations in the tank, thus spreading out this initial oxygen demand.

Another process modification is *contact stabilization*, or *biosorption*, in which the sorption and bacterial growth phases are separated by a settling tank (see Fig. 9-14). Contact stabilization provides for growth at high solids concentrations, thus saving tank space. An activated sludge plant can often be converted into a contact stabilization plant when tank volume limits treatment efficiency.

The two principal means of introducing sufficient oxygen into the aeration tank are by bubbling compressed air through porous diffusers or by beating air in mechanically. These are shown in Figs. 9-15 and 9-16.

The success of the activated sludge system also depends on the separation of the microorganisms in the final clarifier. When the microorganisms do not settle out as anticipated, the sludge is said to be a *bulking sludge*. Bulking is often characterized by a biomass composed almost totally of filamentous organisms that form a kind of

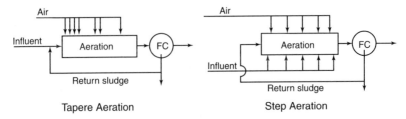

Figure 9-13. Diagrams of tapered aeration and step aeration.

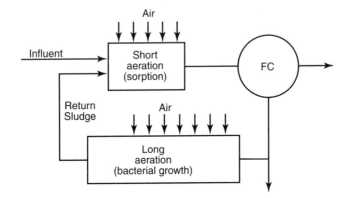

Figure 9-14. The biosorption modification of the activated sludge process.

lattice structure of their filaments that prevents settling.[5] A trend toward poor settling may be the forerunner of a badly upset and ineffective system. The settleability of activated sludge is most often described by the *sludge volume index* (SVI), which is determined by measuring the volume occupied by a sludge after settling for 30 min in a 1-L cylinder, and calculated as

$$\text{SVI} = \frac{(1000)(\text{volume of sludge after 30 min, mL})}{\text{mg/L of SS}}. \qquad (9.11)$$

EXAMPLE 9.5. A sample of sludge has an SS concentration of 4000 mg/L. After settling for 30 min in a 1-L cylinder, the sludge occupies 400 mL. Calculate the SVI:

$$\text{SVI} = \frac{(1000)(400\,\text{mL})}{4000\,\text{mg/L}} = 100.$$

SVI values below 100 are usually considered acceptable; SVI > 200 is a badly bulking sludge. Some common loadings, as a function of the SVI, for final clarifiers are

---

[5] You may picture this as a glass filled with cotton balls. When *water* is poured into the glass, the cotton filaments are not dense enough to settle to the bottom of the glass.

Wastewater Treatment   187

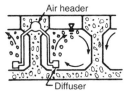

Diffused Aeration

Figure 9-15. Activated systems with diffused aeration. (Photo courtesy of Envirex.)

shown in Fig. 9-17. Some causes of poor settling are improper or varying F/M ratios, fluctuations in temperature, high concentrations of heavy metals, or deficiencies in nutrients. Cures include chlorination, changes in air supply, or dosing with hydrogen peroxide to kill the filamentous microorganisms. When sludge does not settle, the return activated sludge is "thin" because SS concentration is low, and the microorganism concentration in the aeration tank drops. A higher F/M ratio results, since there are fewer microorganisms to handle the same food input, and the BOD removal efficiency is reduced.

Biological Process Dynamics Applied to the Activated Sludge System

The organic materials in the influent meet two fates in an activated sludge system. Most are oxidized to CO_2 and water, but some high-energy compounds are used to build new microorganisms. The latter are known as *substrate* in biological process dynamics.

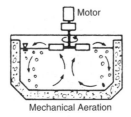

Mechanical Aeration

Figure 9-16. Activated systems with mechanical (surface) aeration. (Photo courtesy of Envirex.)

Substrate is usually measured indirectly as BOD, although total organic carbon or chemical oxygen demand (COD) may be more accurate measures of substrate concentration. The rate of new cell mass (microorganisms) produced as substrate is destroyed is expressed by

$$\frac{dX}{dt} = Y\frac{dS}{dT}, \qquad (9.12)$$

where

S = mass of substrate,
X = mass of microorganisms, or SS, and
Y = yield: mass of microorganisms produced per mass of substrate used.

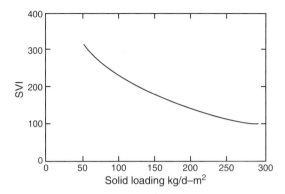

Figure 9-17. Allowable solids loading for final clarifiers increase for better-settling sludge.

The yield is usually expressed as kilograms of SS produced per kilogram of BOD used.

The growth rate of microorganisms is

$$dx/dt = \mu X, \tag{9.13}$$

where μ is a proportionality constant (growth rate constant). Monod[6] demonstrated that for different substrate concentrations, the growth rate constant can be estimated as

$$\mu = \frac{\hat{\mu} S}{K_s + S} \tag{9.14}$$

and therefore

$$\frac{dS}{dT} = \frac{X}{Y}\mu = \left(\frac{X}{Y}\right)\left(\frac{\hat{\mu} S}{K_s + S}\right), \tag{9.15}$$

where

μ = growth rate constant (in s^{-1}),
$\hat{\mu}$ = maximum growth rate constant (s^{-1}), and
K_s = saturation constant (in mg/L).

The two constants $\hat{\mu}$ and K_s must be evaluated for each substrate and microorganism culture.

Figure 9-18 illustrates the application of biological process dynamics to the activated sludge process. The reactor has volume V and effluent flow 0. The reactor is

---

[6]Detailed discussion of the empirical Monod model of enzyme kinetics is beyond the scope of this text, but can be found in modern textbooks on biochemistry and on wastewater processing.

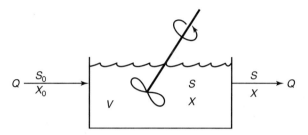

Figure 9-18. A biological reactor without microorganism recycle.

completely mixed: the influent is dispersed within the tank immediately upon introduction, so that there are no concentration gradients in the tank and the composition and water quality of the effluent are *exactly* those of the tank contents. In such a continuous reactor there are two retention times: for liquids and for solids. The liquid, or *hydraulic*, retention time, usually expressed in minutes, is

$$\bar{t} = \frac{V}{Q}. \tag{9.16}$$

The solids retention time, also known as *sludge age,* is the average time a solid (microorganism) particle stays in the system. Sludge age in wastewater treatment is also the *mean cell residence time* Θ_c and is defined as

$$\Theta_c = \frac{\text{Mass of SS contained by the aeration tank}}{\text{Mass rate of solids leaving the aeration tank}}.$$

For the system pictured in Fig. 9-18, the mean cell residence time is

$$\Theta_c = \frac{VX}{QX}, \tag{9.17}$$

which in this case is equal to $\bar{t}$. The mean cell residence time is usually expressed in days.

We now assume that, in the system under discussion, there are no organisms in the influent ($X_0 \sim= 0$), and that the reactor has attained a steady state, in which the microorganism growth rate is balanced by the rate of microorganism loss in the effluent. Death of microorganisms is ignored.

A mass balance for the system is

$$\begin{matrix}\text{Rate of change} \\ \text{in the reactor}\end{matrix} = \begin{matrix}\text{Rate of} \\ \text{inflow}\end{matrix} - \begin{matrix}\text{Rate of} \\ \text{outflow}\end{matrix} + \begin{matrix}\text{Net rate of growth} \\ \text{of microorganisms}\end{matrix}$$

$$\frac{dX}{dt}V = QX_0 - QX + \left(Y\frac{dS}{dt}\right)V. \tag{9.18}$$

In a steady-state system, $(dX/dt) = 0$, and $X_0 = 0$ is assumed. Substituting Eq. (9.15) into Eq. (9.18), and introducing the mean cell residence time Θ_c,

$$\frac{1}{\Theta_c} = \frac{\mu S}{K_s + S} \tag{9.19}$$

and

$$S = \frac{K_s}{\mu \Theta_c - 1}. \tag{9.20}$$

Equation (9.20) is important because it implies that the substrate concentration S is a function of both the kinetic constants (which are beyond our control for a given substrate) and the mean cell residence time. The mean cell residence time (or sludge age, as defined before) influences the substrate S and thus the treatment efficiency.

A system without microorganism recycle is not very efficient, since long hydraulic residence times are needed to prevent flushing out of the microorganisms. Successful activated sludge systems for wastewater treatment are based on microorganism recycling, as is shown in Fig. 9-19. Some simplifying assumptions are needed to model this system, however. We assume $X_0 = 0$, and further, that the microorganism separator is a perfect device and the effluent contains no microorganisms ($X_e = 0$). Steady-state conditions and perfect mixing are also assumed. Excess microorganisms, or waste activated sludge, are removed from the system at flow rate Q_w, and solids concentration $X_t \cdot X_t$ is the settler underflow concentration and the concentration of solids being recycled to the aeration tank. Finally, to simplify the model we make the obviously incorrect assumption that no substrate is removed in the settling tank and that the settling tank has zero volume, so that all of the microorganisms in the system are in the aeration tank. The only active volume is that of the aeration tank. The mean cell residence time in this case is

$$\Theta_c = \frac{\text{Mass of solids (microorganisms) in aeration tank}}{\text{Mass rate of solids leaving the system}} \tag{9.21}$$

$$\Theta_c = \frac{VX}{Q_w X_t + (Q - Q_w) X_t}, \tag{9.22a}$$

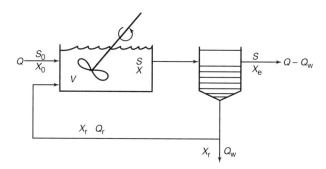

Figure 9-19. A biological reactor with microorganism recycle.

and since $X \cong 0$

$$\Theta_c = \frac{VX}{Q_w X_t} \tag{9.22b}$$

Note that in this case $\Theta_c \neq t$.

Substrate removal can be considered in terms of a *substrate removal velocity q*:

$$q = \frac{\text{Substrate mass removed}}{(\text{Microorganism mass under aeration})(\text{hydraulic residence time})}$$

$$q = \frac{S_0 - S}{X\bar{t}}. \tag{9.23}$$

The substrate removal velocity is a rational measure of the substrate removal activity, or the mass of BOD removed in a given time per mass of microorganisms doing the work, and is sometimes called the *process loading factor*. It is a useful operational and design tool.

Substrate removal velocity may also be derived from a mass balance of the system shown in Fig. 9-19:

$$\begin{array}{c}\text{Net rate of}\\\text{change of}\\\text{substrate}\end{array} = \begin{array}{c}\text{Rate of}\\\text{substrate}\\\text{inflow}\end{array} - \begin{array}{c}\text{Rate of}\\\text{substrate}\\\text{outflow}\end{array} - \begin{array}{c}\text{Rate of}\\\text{substrate}\\\text{utilization}\end{array}$$

$$\frac{dS}{dt}V = QS_0 - QS - qXV. \tag{9.24}$$

The rate of substrate utilization is

$$qXV = \frac{(\text{Substrate mass removed})(\text{Microorganism mass})(\text{reactor volume})}{(\text{Microorganism mass})(\text{time})(\text{reactor volume})}$$

$$qXV = \frac{\text{Substrate mass removed}}{\text{Time}}.$$

Solving for q gives

$$q = \frac{S_0 - S}{Xt}. \tag{9.25}$$

The Monod rate is defined in Eq. (9.13) as

$$\mu = \frac{dX}{dt}\frac{1}{X} = \frac{\text{Microorganism mass produced}}{(\text{Time})(\text{Microorganism mass})}.$$

The yield Y was previously defined as

$$Y = \frac{dX}{dS} = \frac{\text{Microorganism mass produced}}{\text{Substrate mass removed}}.$$

Using a mass balance in terms of microorganisms

Net rate of change in microorganism substrate	=	Rate of microorganism inflow	−	Rate of microorganism outflow	+	Rate of microorganism production

$$Y = \frac{dX}{dS} = QX_0 - Q_w X_t - (Q - Q_w)X_e + \mu XV. \tag{9.26}$$

Again assuming the steady state and $X_0 = X_e = 0$,

$$\mu = \frac{X_t Q_w}{XV}. \tag{9.27}$$

Note that μ is the reciprocal of the sludge age, or mean cell residence time so that

$$\Theta_c = \frac{XV}{X_r Q_w} = \frac{1}{\mu}. \tag{9.28}$$

Substituting the expression for μ from Eq. (9.14), the substrate removal velocity is

$$q = \frac{\mu}{Y} = \frac{\hat{\mu} S}{Y(K_s + S)} \tag{9.29}$$

Setting Eq. (9.29) equal to Eq. (9.23) and solving for $S_0 - S$, the reduction in BOD,

$$S_0 - S = \frac{\hat{\mu} S X \bar{t}}{Y(K_s + S)}. \tag{9.30}$$

Then from Eq. (9.29), the mean cell residence time (sludge age) is

$$\Theta_c = \frac{1}{(\mu/Y)Y} = \frac{1}{qY}, \tag{9.31}$$

and the concentration of microorganisms in the reactor (MLSS) is

$$X = \frac{S_0 - S}{\bar{t} q}. \tag{9.32}$$

EXAMPLE 9.6. An activated sludge system operates at a flow rate (0) of 4000 m³/day, with an incoming BOD(S_0) of 300 mg/L. A pilot plant showed the kinetic constants to be $Y = 0.5$ kg SS/kg BOD, $K_s = 200$ mg/L, $\mu = 2/\text{day}$. We need to design a treatment system that will produce an effluent BOD of 30 mg/L (90% removal). Determine (a) the volume of the aeration tank, (b) the MLSS, and (c) the sludge age. How much sludge will be wasted daily?

The MLSS concentration is usually limited by the ability to keep an aeration tank mixed and to transfer sufficient oxygen to the microorganisms. Assume in this case that $X = 4000$ mg/L. The hydraulic retention is then obtained by rearranging Eq. (9.30):

$$\bar{t} = \frac{0.5(300 - 30)(200 + 30)}{2(30)(4000)} = 0.129 \text{ day} = 3.1 \text{ h}.$$

The volume of the tank is then

$$V = \bar{t}Q = 4000(0.129) = 516 \text{ m}^3.$$

The sludge age is

$$\Theta_c = \frac{(4000 \text{ mg/L})(0.129 \text{ day})}{(0.5 \text{ kg SS/kg BOD})(300 - 30) \text{ mg/L}} = 3.8 \text{ days}.$$

Since

$$\frac{1}{\Theta_c} = \frac{\text{kg sludge wasted/day}}{\text{kg sludge in aeration tank}},$$

from Eq. (9.28)

$$X_r Q_w = \frac{XV}{\Theta_c} = \frac{(4000)(516)(10^3 \text{ L/m}^3)(1/10^6 \text{ kg/mg})}{3.8} = 543 \text{ kg/day}.$$

EXAMPLE 9.7. Using the same data as in the preceding example, what mixed liquor solids concentration would be necessary to attain 95% BOD removal, or $S = 15$ mg/L?

The substrate removal velocity is

$$q = \frac{(2 \text{ day}^{-1})15 \text{ mg/L}}{(0.5)(200 + 15) \text{ mg/L}} = 0.28 \text{ day}^{-1}$$

and

$$X = \frac{300 - 15}{(0.129)(0.28)} = 7890 \text{ mg/L}.$$

The sludge age would be

$$\Theta_c = \frac{1}{(0.28)(0.5)} = 7.1 \text{ days}.$$

Higher removal efficiency requires more microorganisms in the aeration tank.

A simple batch settling test can be used to estimate the return sludge pumping rates. After 30 min of settling, the solids in the cylinder are at a 55% concentration that would be equal to the expected return sludge solids, or

$$X_r = \frac{HX}{h} \tag{9.33}$$

where

X_r = the expected return suspended solids concentration (in mg/L),
X = MLSS (mg/L),
H = height of cylinder (m), and
h = height of settled sludge (m).

The mixed liquor solids concentration is a combination of the return solids diluted by the influent, or

$$X = \frac{Q_t X_t + Q X_0}{Q_t + Q}. \qquad (9.34)$$

If we again assume no solid is in the influent, or $X_0 = 0$, then

$$X = \frac{Q_t X_t}{Q_t + Q}. \qquad (9.35)$$

The performance of an activated sludge system often depends on the performance of the final clarifier. If this settling tank cannot achieve the required return sludge solids the MLSS will drop and the treatment efficiency will be reduced. Final clarifiers act as settling tanks for flocculent settling and as thickeners. Their design requires consideration of both solids loading and overflow rate. Solids loading, more fully explained in the next chapter, is expressed in terms of kilograms of solids per day applied to a surface area of square meters. Figure 9-17 shows some commonly used solids loadings for final clarifiers as a function of the SVI.

Secondary treatment of wastewater usually includes a biological step like activated sludge, which removes a substantial part of the BOD and the remaining solids. The typical wastewater that we began with now has the following approximate water quality:

	Raw wastewater	After primary treatment	After secondary treatment
BOD (mg/L)	250	175	15
SS (mg/L)	220	60	15
P (mg/L)	8	7	6

The effluent from secondary treatment meets the previously established effluent standards for BOD and SS. Only phosphorus content remains high. The removal of inorganic compounds, including inorganic phosphorus and nitrogen compounds, requires advanced or tertiary wastewater treatment.

TERTIARY TREATMENT

Both primary and secondary (biological) treatments are incorporated in conventional wastewater treatment plants. However, secondary treatment plant effluents often are

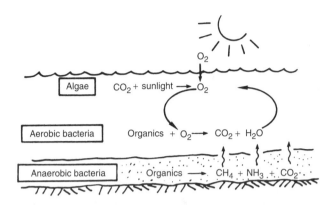

Figure 9-20. Schematic diagram of an oxidation pond and photograph of an aeration lagoon. (Photo courtesy of N. S. Nokkentved.)

not sufficiently clean. Some BOD and suspended solids remain, and neither primary nor secondary treatment is effective in removing phosphorus and other nutrients or toxic substances. A popular advanced treatment for BOD removal is the *polishing pond,* or *oxidation pond.* An oxidation pond and the reactions that take place in the pond are shown in Fig. 9-20. Oxidation ponds, as their name implies, are designed to be aerobic; hence, light penetration for algal growth is important and a large surface area is needed.

A sufficiently large oxidation pond may be the only treatment step for a small waste flow. When the rate of oxidation in a pond is too great and oxygen availability becomes limiting, the pond may be forcibly aerated by either diffusive or mechanical aerators. Such ponds are called *aerated lagoons* and are widely used in treating industrial effluent. Figure 9-20 shows an aerated lagoon serving a pulp and paper mill.

BOD may also be removed by activated carbon adsorption, which has the added advantage of removing some inorganic as well as organic compounds. Activated carbon adsorbs both chemically and physically. An activated carbon column is a completely enclosed tube, where dirty water is pumped in at the bottom and clear water exits at the top. Microscopic crevices in the carbon catch and hold colloidal and smaller particles. As the carbon column becomes saturated, the pollutants must be removed from the carbon in the tube and the carbon reactivated, usually by heating in the absence of oxygen. Reactivated or regenerated carbon is somewhat less efficient than virgin carbon, some of which must always be added to ensure effective performance.

Reverse osmosis, so designated because it uses semipermeable, or osmotic, membranes, is also finding acceptance as a treatment for various types of both organic and inorganic trace pollutants. The wastewater is forced through a semipermeable membrane that acts as a superfilter, rejecting dissolved as well as suspended solids.

Nitrogen in raw wastewater is mostly organic and ammonia nitrogen (called Kjeldahl nitrogen). Because nitrogen can exacerbate eutrophication in lakes and estuaries, it must often be reduced in wastewater treatment plants. The most widely used method of nitrogen removal is called biological nitrification/denitrification. First, the nitrogen is converted into inorganic nitrate nitrogen using microbes. This occurs in aeration tanks if the residence time is sufficiently high. Recall that carbonaceous BOD is exerted first, and only when the easily oxidized carbon compounds are used up will the nitrifiers begin the oxidation of nitrogen-containing compounds. If this occurs, the end product is nitrate nitrogen, NO_3^-. The process is thought to be in two stages, with various groups of microorganisms responsible for different stages. Simplified chemical equations describing this process are

$$2NH_4^+ + 3O_2 \xrightarrow{\text{Nitrosomonas}} 2NO_2^- + 2H_2O + 4H^+$$

$$2NO_2^- + O_2 \xrightarrow{\text{Nitrobacter}} 2NO_3^-.$$

These reactions are slow and require long retention times in the aeration tank as well as sufficient DO. The kinetics constants for the reactions are low, with very low yields, so that net sludge production is limited and washout is a constant danger.

In cases where the effluent does not enter lakes or rivers, the production of nitrate nitrogen is sufficient. In many cases, however, the nitrate nitrogen must be removed, and this is also accomplished biologically.

Once the ammonia has been oxidized to nitrate, it may be reduced by a broad range of facultative and anaerobic bacteria like *Pseudomonas.* This *denitrification* requires a source of carbon, and methanol (CH_3OH) is often used for that purpose. The reactions are

$$6NO_3^- + 2CH_3OH \rightarrow 6NO_2^- + 2CO_2I \pm 4H_2O$$

$$6NO_2^- + 3CH_3OH \rightarrow 3N_2 \uparrow + 3CO_2 \uparrow + 3H_2O + 6OH^-.$$

Another nutrient that must often be removed is phosphorus, possibly the most important chemical responsible for accelerated eutrophication.

Phosphate may be removed chemically or biologically. The most popular chemical methods use lime, $Ca(OH)_2$, and alum, $Al_2(SO_4)_3$. Under alkaline conditions, the calcium ion will combine with phosphate to form calcium hydroxyapatite, a white, insoluble precipitate that is settled out and removed from the wastewater. Insoluble $CaCO_3$ is also formed and removed and may be recycled by burning in a furnace:

$$CaCO_3 \rightarrow CO_2 + CaO.$$

Quicklime, CaO, is then slaked by adding water and forming lime, that may be reused:

$$CaO + H_2O \rightarrow Ca(OH)_2$$

The aluminum ion from alum precipitates as very slightly soluble aluminum phosphate,

$$Al^{3+} + PO_4^{3-} \rightarrow AlPO_4 \downarrow,$$

and also forms aluminum hydroxide

$$Al^{3+} + 3OH^- \rightarrow Al(OH)_3 \downarrow,$$

which forms sticky flocs that help to settle out *phosphates*. Alum is usually added in the final clarifier. The amount of alum needed to achieve a given level of phosphorus removal depends on the amount of phosphorus in the wastewater. The amount of sludge produced may be calculated by using stoichiometric relationships.

EXAMPLE 9.8. A wastewater contains 6.3 mg/L P, and it is found that an alum dosage of 13 mg/L as Al^{3+} achieves an effluent P concentration of 0.9 mg/L:

$$\begin{array}{cccc} & Al^{3+} & + \quad PO_4^{3-} & \rightarrow \quad AlPO_4 \\ \text{Molecular weights:} & 27 & 95 & 122 \quad \text{(or 31 as P)} \end{array}$$

or, from stoichiometry, 31 mg of P removal results in the formation of 122 mg aluminum phosphate sludge:

$$\left(5.4 \, \frac{mg}{L} \, \text{P removed}\right)\left(\frac{122}{31}\right) = 21.3 \, \frac{mg}{L} \, AlPO_4 \text{ sludge produced.}$$

The amount of Al^{3+} used to produce the $AlPO_4$ is

$$\left(5.4 \, \frac{mg}{L} \, \text{P removed}\right)\left(\frac{27 \, mg \, Al^{3+}}{(31 \, mg \, P)}\right) = 4.6 \, \frac{mg}{L} \, Al^{3+}$$

The excess aluminum, $13 - 4.7 = 8.3$ mg/L, probably formed aluminum hydroxide:

$$\begin{array}{cccc} & Al^{3+} & + \quad 3(OH^-) & \rightarrow \quad Al(OH)_3 \downarrow. \\ \text{Molecular weights:} & 27 & (3)(17) & 78 \end{array}$$

The production of Al(OH)$_3$ is

$$\left(8.3 \frac{mg}{L} Al^{3+}\right)\left(\frac{78 \text{ mg Al(OH)}_3}{(27 \text{ mgAl}^{3+})}\right) = 24 \frac{mg}{L} Al(OH)_3 \text{ sludge.}$$

The total sludge production is then

$$AlPO_4 + Al(OH)_3 = 21.3 + 24 = 45 \frac{mg}{L}.$$

Phosphorus also may be removed biologically. When the microorganisms in return activated sludge are starved by aeration after their removal in the final clarifier, they show a strong tendency to adsorb phosphorus almost instantly on introduction of wastewater. Such a "luxury uptake" is at a much higher rate than would be required eventually by metabolic activity. If the phosphorus-rich organisms are removed quickly and wasted, the excess phosphorus leaves with the waste activated sludge. Thus, both nitrogen and phosphorus can be removed by biological secondary treatment.

In a typical phosphorus removal system in a wastewater treatment plant, the mixed liquor in the first stage of the aeration tank is mixed but not aerated. The microorganisms react to this deprivation of oxygen by utilizing the phosphorus in the cells, thus gaining energy for sustaining life. When the microorganisms are then moved into the aerobic stage, the soluble phosphorus is rapidly absorbed within the cells in order to allow the microorganism to assimilate the dissolved organic material rapidly. This uptake of phosphorus is the "luxury uptake," and is caused by first solubilizing phosphorus, and then creating a condition where the microorganism try to rapidly assimilate the energy-containing organic matter. At this point, the microorganisms are settled out in the final clarifier and removed, taking the excess phosphorus with them. The waste activated sludge thus becomes enriched in phosphorus, which is beneficial if the sludge is applied to farmland.

The combined removal of nitrogen and phosphorus is often termed biological nutrient removal, or BNR. A number of treatment options are available for such a plant, but they all consist of a series of anoxic and aerobic basins.

With nitrogen and phosphorus removal, the effluent goal of the wastewater is attained:

	Raw wastewater	After primary treatment	After secondary treatment	After tertiary treatment
BOD (mg/L)	250	175	15	10
SS (mg/L)	220	60	15	10
P (mg/L)	8	7	6	0.5

Spraying secondary effluent on land and allowing soil microorganisms to degrade the remaining organic compounds provides an alternative to high-technology advanced

200 ENVIRONMENTAL ENGINEERING

wastewater treatment systems. Such *land treatment* systems have been used in Europe for many years and have recently been introduced in North America. They appear to represent a reasonable alternative to complex and expensive systems, especially for smaller communities. Irrigation is probably the most promising land treatment. Depending on the crop and the soil, about 1000 to 2000 ha of land are required for every cubic meter per second of wastewater flow. Nutrients like N and P that remain in secondarily treated wastewater are usually beneficial to crops.

CONCLUSION

A typical wastewater treatment plant, shown schematically in Fig. 9-21, includes primary, secondary, and tertiary treatment. The treatment and disposal of solids removed from the wastewater stream deserve special attention and is discussed in the next chapter.

Figure 9-22 is an aerial view of a typical wastewater treatment plant. Well-operated plants produce effluents that are often much less polluted than the receiving waters into which they are discharged. However, not all plants perform that well. Many wastewater treatment plants are only marginally effective in controlling water pollution, and plant operation is often to blame. Operation of a modern wastewater treatment plant

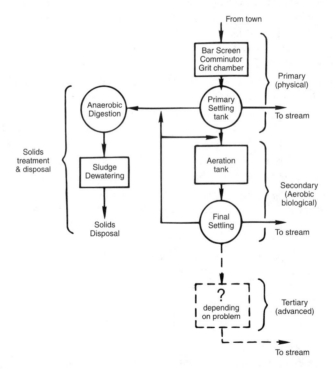

Figure 9-21. Block diagram of a complete wastewater treatment plant.

Wastewater Treatment 201

Figure 9-22. An aerial view of a secondary wastewater treatment plant. (Courtesy of Envirex.)

Figure 9-23. An unusual operating problem: a pickup truck in the primary clarifier. (Courtesy of Phillip Karr.)

is complex and demanding (though not often as demanding as the situation shown in Fig. 9-23). Unfortunately, operators have historically been poorly compensated, so that recruitment of qualified operators is difficult. All states now require licensing of operators, and operators' pay is improving. This is a welcome change, for it makes little sense to entrust unqualified operators with multimillion dollar facilities. Wastewater treatment requires proper plant design and proper operation. One without the other is a waste of money.

PROBLEMS

9.1 The following data were reported on the operation of a wastewater treatment plant:

	Influent(mg/L)	Effluent(mg/L)
BOD_5	200	20
SS	220	15
p	10	0.5

a. What percent removal was experienced for each of these?
b. What type of treatment plant will produce such an effluent? Draw a block diagram showing the treatment steps.

9.2 Describe the condition of a primary clarifier one day after the raw sludge pumps broke down.

9.3 Ponding, the excessive growth of slime on the rocks, is an operational problem with trickling filters. The excessive slime clogs spaces between the rocks so that water no longer flows through the filter. Suggest two cures for ponding.

9.4 Illegal connections are sometimes made to sanitary sewers. Suppose a family of four, living in a home with a roof area of $70 \times 40 \text{ ft}^2$, connects the roof drain to the sewer. If rain falls at the rate of 1 in./h, what percent increase will there be in the flow from their house over the dry weather flow? The dry weather flow is 50 gal per person per day.

9.5 Suppose you are an engineer hired to build a wastewater treatment plant. What would you choose as the five most important wastewater parameters to be tested? Why would you want to know these values? What tests would you run to determine influent and effluent characteristics, and why?

9.6 The influent and effluent data for a secondary treatment plant are:

	Influent(mg/L)	Effluent(mg/L)
BOD_5	200	20
SS	220	15
P	10	8

Calculate the removal efficiency. What is wrong with the plant?

9.7 Draw block diagrams of the unit operations necessary to treat the following wastes to effluent levels of $BOD_5 = 20$ mg/L, SS = 20 mg/L, P = 1 mg/L. Assume all BOD is due to *dissolved* chemicals, and that all SS are inert.

Waste	BOD (mg/L)	Suspended solids (SS, mg/L)	Phosphorus (mg/L)
Domestic	200	200	10
Chemical industry	40,000	0	0
Pickle cannery	0	300	1
Fertilizer mfg	300	300	200

9.8 The success of an activated sludge system depends on settling the solids in the final settling tank. If the sludge in a system started to bulk and not settle very well, and the SS concentration of the return activated sludge dropped from 10,000 to 4,000 mg/L,

  a. What will this do to the MLSS?
  b. What will this in turn do to the BOD removal, and why?

9.9 Assume that the volume of settled sludge in a 1-L cylinder after 30 mm is 300 mL both before and during a bulking problem, and the SVI is 100 and 250, respectively. What is the MLSS before and after the bulking problem? Does this agree with your answer to Problem 9.8?

9.10 A percolation test shows that the measured water level drops 5 in. in 30 min. What size percolation field would you need for a two-bedroom house? What size would you need if the water level dropped only 0.5 in.?

9.11 A 5-mgd conventional activated sludge plant has an influent BOD_5 of 200 mg/L. The primary clarifier removes 30% of the BOD. The plant's three aeration tanks are $20 \times 20 \times 10$ ft^3. What MLSS are necessary to attain 90% BOD removal?

9.12 An aeration system with a hydraulic retention time of 2.5 h receives a flow of 0.2 mgd at a BOD of 150 mg/L. The SS in the aeration tank are 4000 mg/L. The effluent BOD is 20 mg/L and the effluent SS are 30 mg/L. Calculate the F/M ratio for this system.

9.13 The MLSS in an aeration tank are 4000 mg/L. The flow from the primary settling tank is 0.2 m^3/s and the return sludge flow is 0.1 m^3/s. What must the return sludge SS concentration be to maintain the 4000 mg/L MLSS, assuming no net growth of biomass?

9.14 An industrial wastewater is composed of SS that are small beads of 0.1 mm diameter with a specific gravity of 2.65. Assume $\mu = 1.31$ cP.

  a. For how long would the wastewater have to remain in a settling tank 4 m deep in order to become clear?
  b. If a false floor is put in the settling tank so that it was now only 2 m deep, what fraction of the SS would be removed during the same time as calculated in (a)?
  c. What is the overflow rate for this tank?

9.15 A primary clarifier (settling tank) is 80 ft long, 30 ft wide, and 12 ft deep and receives a flow of 200,000 gallons per day (gpd).

  a. Calculate the overflow rate in gpd/ft^2.

b. What is the settling velocity in feet per day of the critical particle, assuming that the entire tank is used for clarification.

9.16 A rectangular primary clarifier (settling tank) is 10 m wide, 25 m long, and 5 m deep. The flow is $0.5 \, m^3/s$.

a. What is the design overflow rate?
b. Would all particles having a settling velocity of 0.01 cm/mm be removed?

9.17 A treatment plant receives wastewater at a flow rate of $0.5 \, m^3/s$. The total surface area of the primary clarifiers is $2700 \, m^2$ and the retention time is 3 h. Find the overflow rate and the depth of the clarifiers.

9.18 A transoceanic flight on a Boeing 747 with 430 persons aboard takes 7 h. Estimate the weight of water necessary to flush the toilets if each flush uses 2 gal. Make any assumptions necessary, and state them. What fraction of the total payload (people) would the flush water represent? How could this weight be reduced, since discharging, for obvious reasons, is illegal?

9.19 A family of 4 wants to build a house on a lot for which the percolation rate is 1.00 mm/mm. The county requires a septic tank hydraulic retention time of 24 h. Find the volume of the tank required and the area of the drain field. Sketch the system, including all dimensions.

9.20 Are the following design and expected performance for a primary clarifier reasonable? $Q\, 0.150 \, m^3/s$, influent SS 310 mg/L, raw primary sludge concentration 4% solids, SS removal efficiency = 60%, length = 10 m, width = 15 m, depth = 2.2 m. If not, suggest modifications.

9.21 A wastewater contains soluble phosphate at a concentration of 4 mg/L. How much ferric chloride would theoretically be necessary to precipitate this nutrient?

9.22 A community with a wastewater flow of 10 mgd is required to meet effluent standards of 30 mg/L for both BOD_5 and SS. Pilot plant results with the influent $BOD_5 = 250$ mg/L estimate the kinetic constants at $K_s = 100$ mg/L, $1 = 0.25$ day, and $Y = 0.5$. MLSS are maintained at 2000 mg/L. What is the hydraulic retention time, the sludge age, and the required tank volume?

Chapter 10

Sludge Treatment and Disposal

The field of wastewater treatment and engineering is littered with unique and imaginative processes for achieving high degrees of waste stabilization at attractive costs. In practice, few of these "wonder plants" have met expectations, often because they fail to pay sufficient attention to sludge treatment and disposal problems. Currently sludge treatment and disposal accounts for over 50% of the treatment costs in a typical secondary plant, prompting renewed interest in this none-too-glamorous, but essential aspect, of wastewater treatment.

This chapter is devoted to the problem of sludge treatment and disposal. The sources and quantities of sludge from various types of wastewater treatment systems are examined, followed by a definition of sludge characteristics. Such solids concentration techniques as thickening and dewatering are discussed next, concluding with considerations for ultimate disposal.

SOURCES OF SLUDGE

Sludges are generated from nearly all phases of wastewater treatment. The first source of sludge in a wastewater treatment facility is the suspended solids from the primary settling tank or clarifier. Ordinarily about 60% of the suspended solids entering the treatment facility become *raw primary sludge*, which is highly putrescible and very wet (about 96% water).

The removal of BOD is basically a method of wasting energy, and secondary wastewater treatment plants are designed to reduce the high-energy organic material that enters the treatment plant to low-energy chemicals. This process is typically accomplished by biological means, using microorganisms (the "decomposers" in ecological terms) that use the energy for their own life and procreation. Secondary treatment processes such as the popular activated sludge system are almost perfect systems. Their major fault is that the microorganisms convert too little of the high-energy organics to CO_2 and H_2O and too much of it to new organisms. Thus the system operates with an excess of these microorganisms, or *waste activated sludge*. As defined in the previous chapter, the mass of waste activated sludge per mass of BOD removed in secondary treatment is known as the *yield*, expressed as kilograms of suspended solids produced per kilogram of BOD removed.

Phosphorus removal processes also invariably end up with excess solids. If lime is used, calcium carbonates and calcium hydroxyapatites solids are formed. Similarly, aluminum sulfate produces solids in the form of aluminum hydroxides and aluminum phosphates. Even so-called "totally biological processes" for phosphorus removal end up with solids. The use of an oxidation pond or marsh for phosphorus removal is possible only if some organics (algae, water hyacinths, fish, etc.) are harvested periodically.

The quantities of sludge obtained from various wastewater treatment processes may be calculated as shown in Fig. 10-1. The symbols are defined as follows:

S_0 = influent BOD (kg/h, 5 days, 20°C),
X_0 = influent suspended solids (kg/h),
h = fraction of BOD not removed in primary clarifier,
i = fraction of BOD not removed in aeration or trickling filter,
X_e = plant effluent suspended solids (kg/h),
k = fraction of X_0 removed in primary clarifier,
j = fraction of solids not destroyed in digestion,
ΔX = net solids produced by biological action (kg/h), and
Y = yield = $\Delta X/\Delta S$, where $\Delta S = hS_0 - ihS_0$.

For domestic wastewater, typical values for these constants would be as follows:

$S_0 = (250 \times 10^{-3} \times Q)$ kg/h if Q is m³/h,
$S_0 = (250 \times 8.34 \times Q)$ kg/h if Q is mgd,
$X_0 = (220 \times 10^{-3} \times Q)$ kg/h if Q is m³/h,
$X_0 = (220 \times 8.34 \times Q)$ kg/h if Q is mgd,
$k = 0.6$,
$h = 0.7$,
$X_e = (220 \times 10^{-3} \times Q)$ m³g/h if Q is m³/h,
$X_e = (220 \times 8.34 \times Q)$ mgd if Q is mgd,
$j = 0.5$ for anaerobic,

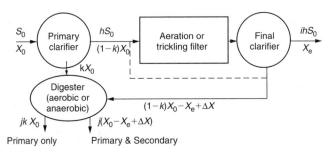

Figure 10-1. Schematic diagram of a generalized secondary treatment plant.

$j = 0.8$ for aerobic,
$i = 0.1$ for well-operated activated sludge,
$i = 0.2$ for trickling filters,
$Y = 0.5$ for activated sludge, and
$Y = 0.2$ for trickling filters.

If chemical sludges are also produced during the treatment process (e.g., from the precipitation of phosphorus), these quantities must be added to the total sludge estimated above.

EXAMPLE 10.1. A wastewater treatment plant has an influent BOD$_5$ of 250 mg/L, at a flow rate of 1570 m³/h (10 mgd), and influent suspended solids of 225 mg/L. The raw sludge produced would be equal to kX_0, where k is the fraction of suspended solids removed in the primary clarifier and X_0 is the influent suspended solids in kilograms/hour. If k is 0.6 (typical value), X_0 can be calculated as

$$(225 \text{ mg/L})(1570 \text{ m}^3/\text{h})(1000 \text{ L/m}^3)(10^{-6}) \text{ mg/kg} = 353 \text{ kg/h}$$

The raw sludge produced is then

$$0.6 \times 353 = 212 \text{ kg/h}.$$

CHARACTERISTICS OF SLUDGES

The important or relevant characteristics of sludges depend on what is to be done to the sludge. For example, if the sludge is to be thickened by gravity, its settling and compaction rates are important. On the other hand, if the sludge is to be digested anaerobically, the concentrations of volatile compounds, other organic solids, and heavy metals are important. Variability of the sludges is immensely important in the design of sludge handling and disposal operations. In fact, this variability may be stated in terms of three "laws":

1. No two wastewater sludges are alike in all respects.
2. Sludge characteristics change with time.
3. There is no "average sludge."

The first "law" of sludges reflects the fact that no two wastewaters are alike and that if the variable of treatment is added, the sludges produced will have significantly different characteristics.

The second "law" is often overlooked. For example, the settling characteristics of chemical sludges from the treatment of plating wastes (e.g., Pb(OH)$_2$, Zn(OH)$_2$, or Cr(OH)$_3$) vary with time simply because of uncontrolled pH changes. Biological sludges are, of course, continually changing, with the greatest change occurring when

Table 10-1. Sludge Characteristics

Sludge type	Physical				Chemical		
	Solids conc. (mg/L)[a]	Volatile solids (%)	Yield strength (dyne/cm^2)	Plastic viscosity (g/cm–s)	N as N (%)	P as P_2O_5 (%)	K as K_2O (%)
Water	—	—	0	0.01	—	—	—
Raw primary	60,000	60	40	0.3	2.5	1.5	0.4
Mixed digested	80,000	40	15	0.9	4.0	1.4	0.2
Waste activated	15,000	70	0.1	0.06	4.0	3.0	0.5
Alum, ppt	20,000	40	—	—	2.0	2.0	—
Lime, ppt	200,000	18	—	—	2.0	3.0	—

[a] Note that sludge with 10,000 mg/L solids is approximately 1% solids.

the sludge changes from aerobic to anaerobic (or vice versa). Not surprisingly, it is quite difficult to design sludge-handling equipment because the sludge may change in some significant characteristic in only a few hours.

The third "law" is constantly violated. Tables showing "average values" for "average sludges" are useful for illustrative and comparative purposes, and as such, are included in this chapter for general information; however, they should not be used for treatment design. Instead, you need to determine the specific and unique characteristics of the sludge that needs treatment. For illustrative purposes only, Table 10-1 shows characteristics of hypothetical, "average sludges." The first characteristic, solids concentration, is perhaps the most important variable, defining the volume of sludge to be handled and determining whether the sludge behaves as a liquid or a solid.

The second characteristic, volatile solids, is also important to sludge disposal. Disposal is difficult if the sludge contains high concentrations of volatile solids because gases and odors are produced as the sludge outgasses and the volatile substances are degraded. The volatile solids parameter is often interpreted as a biological rather than a physical characteristic, the assumption being that volatile suspended solids are a gross measure of the viable biomass. Another important parameter, especially in regard to ultimate disposal, is the concentration of pathogens, both bacteriological and viral. The primary clarifier seems to act as a viral and bacteriological concentrator, with a substantial fraction of these microorganisms existing in the sludge instead of the liquid effluent.

Rheological characteristics (degree of plasticity) are one of only a few truly fundamental physical parameters of sludges. Two-phase mixtures like sludges, however, are almost without exception non-Newtonian and *thixotropic*. Sludges tend to act as pseudoplastics, with an apparent yield stress and a plastic viscosity. The rheological behavior of a pseudoplastic fluid is defined by a rheogram shown as Fig. 10-2. The term *thixotropic* relates to the time dependence of the rheological properties. Sludges tend to act more like plastic fluids as the solids concentration increases. True plastic

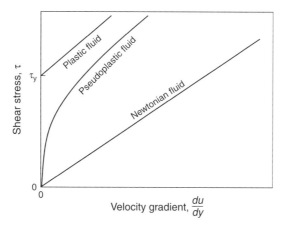

Figure 10-2. Rheograms for three different fluids.

fluids may be described by the equation

$$\tau = \tau_y + \eta \frac{du}{dy}, \tag{10.1}$$

where

τ = shear stress,
τ_y = yield stress,
η = plastic viscosity, and
du/dy = rate of shear, or the slope of the velocity (u)–depth (y) profile.

Although sludges are seldom true plastics, Eq. (10.1) may be used to approximate the rheogram.

The yield stress may vary from more than 40 dyne/cm² for 6% raw sludge to only 0.07 dyne/cm² for a thickened activated sludge. The large differences suggest that rheological parameters could be used for scale-up purposes. Unfortunately, rheological characteristics are difficult to measure and such analyses are not even included as standard methods (U.S. Environmental Protection Agency 1991).

The chemical composition of sludge is important for several reasons. The value of sludge as fertilizer is dependent on the availability of nitrogen, phosphorus, and potassium, as well as trace elements. A more important measurement, however, is the concentration of heavy metals and other toxic substances, which should be kept out of the food chain and general environment. The range of heavy metal concentrations in sludges is very large. For example, cadmium concentrations can vary from almost 0 to more than 1000 mg/kg. Industrial discharges are a major source of heavy metals and toxins in sludges; a single poorly operated industrial firm may contribute enough toxins to make the sludge worthless as a fertilizer. Although most engineers agree that it would be desirable to treat sludges at the plant to remove metals and toxins, it may not be cost-effective, or even possible, to remove high concentrations of heavy metals, pesticides, and other toxins from the sludge. Because of this, the best management approach would be to prevent or reduce the concentration of toxins in the influent.

SLUDGE TREATMENT

A great deal of money could be saved, and troubles averted, if sludge disposal could be done as the sludge is drawn off the main process train. Unfortunately, sludges have three characteristics that make such a simple solution unlikely: they are aesthetically displeasing, they are potentially harmful, and they have too much water.

The first two problems are often solved by stabilization, which may involve anaerobic or aerobic digestion. The third problem requires the removal of water, either by thickening or dewatering. The next three sections cover the topics of stabilization, thickening, and dewatering, followed by the considerations of ultimate sludge disposal.

Sludge Stabilization

The objective of sludge stabilization is to reduce the problems associated with sludge odor and putrescence, as well as reducing the hazard presented by pathogenic organisms. Sludge may be stabilized using lime, aerobic digestion, or anaerobic digestion.

Lime stabilization is achieved by adding lime, either as hydrated lime ($Ca(OH)_2$) or as quicklime (CaO) to the sludge, which raises the pH to 11 or above. This significantly reduces the odor and helps in the destruction of pathogens. The major disadvantage of lime stabilization is that its odor reduction is temporary. Within days the pH drops and the sludge once again becomes putrescible.

Aerobic digestion is a logical extension of the activated sludge system. Waste activated sludge is placed in dedicated tanks for a very long time, and the concentrated solids are allowed to progress well into the endogenous respiration phase, in which food is obtained only by the destruction of other viable organisms. Both total and volatile solids are thereby reduced. One drawback to this process is that aerobically digested sludges are more difficult to dewater than anaerobic sludges.

Sludge can also be stabilized by *anaerobic digestion*, as illustrated in Fig. 10-3. The biochemistry of anaerobic digestion is a staged process: solution of organic compounds by extracellular enzymes is followed by the production of organic acids by a large and hearty group of anaerobic microorganisms known, appropriately enough, as the *acid formers*. The organic acids are in turn degraded further by a group of strict anaerobes called *methane formers*. These microorganisms are the prima donnas of wastewater treatment, getting upset at the least change in their environment. The success of anaerobic treatment depends on maintenance of suitable conditions for the methane formers. Since they are strict anaerobes, they are unable to function in the presence of oxygen and are very sensitive to environmental conditions like pH, temperature, and the presence of toxins. A digester goes "sour" when the methane

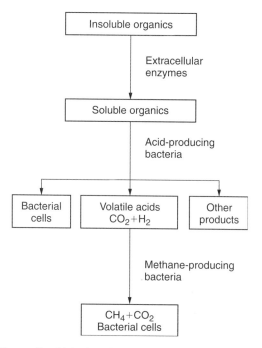

Figure 10-3. Generalized biochemical reactions in anaerobic sludge digestion.

formers have been inhibited in some way. The acid formers keep chugging away, making more organic acids, thus further lowering the pH and making conditions even worse for the methane formers. Curing a sick digester requires suspension of feeding and, often, massive doses of lime or other antacids.

Most treatment plants have both a primary and a secondary digester (Fig. 10-4). The primary digester is covered, heated, and mixed to increase the reaction rate. The temperature of the sludge is usually about 35°C (95°F). Secondary digesters are not mixed or heated, and are used for storage of gas produced during the digestion process (e.g., methane) and for concentrating the sludge by settling. As the solids settle, the liquid supernatant is pumped back to the main plant for further treatment. The cover of the secondary digester often floats up and down, depending on the amount of gas that has accumulated. The gas produced during anaerobic digestion contains enough in methane to be used as a fuel, and is frequently used to heat the primary digester. Large wastewater treatment facilities may produce enough methane to sell to local utility companies as natural gas. The city of Portland, OR is using an innovative approach to convert methane from anaerobic digestion into hydrogen for fuel cells. Unfortunately, many wastewater treatment facilities still "waste" (burn off) the gas from anaerobic digestion. As the availability of fossil fuel natural gas decreases, and prices increase, biologically produced methane may gain in economic value.

Anaerobic digesters are commonly designed on the basis of solids loading. Experience has shown that domestic wastewaters contain about 120 g (0.27 lb) of

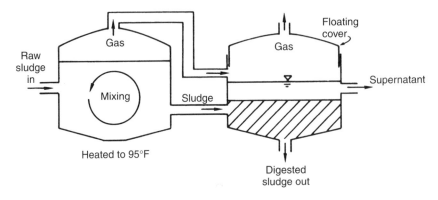

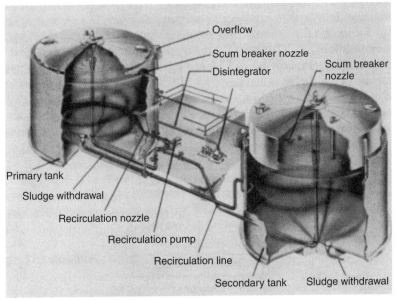

Figure 10-4. Anaerobic sludge digesters. (Photo courtesy of Dorr Oliver Inc.)

suspended solids per day per capita. The total suspended solids load entering the wastewater treatment facility can be estimated if you know the population served by the facility. Solids produced in secondary treatment must be added to the total load, along with any "special treatment" solids (e.g., phosphorus removal sludges). Once the solids production is calculated, the digester volume is estimated by assuming a reasonable loading factor such as 4 kg of dry solids/m^3 × day (0.27 lb/ft^3 × day). This loading factor is decreased if a higher reduction of volatile solids is desired.

EXAMPLE 10.2. Raw primary and waste activated sludge containing 4% solids is to be anaerobically digested at a loading of 3 kg/m^3 × day. The total sludge produced in the plant is 1500 kg of dry solids per day. Calculate the required volume of the primary digester and the hydraulic retention time.

The production of sludge requires

$$\frac{1500 \text{ kg/day}}{3 \text{ kg/m}^3\text{–day}} = 500 \text{ m}^3 \text{ digester volume}.$$

The total mass of wet sludge pumped to the digester is

$$\frac{1500 \text{ kg/day}}{0.04} = 37{,}500 \text{ kg/day}.$$

Since 1 L of sludge weighs about 1 kg, the volume of sludge is 37,500 L/day or 37.5 m^3/day, and the hydraulic residence time is

$$t = (500 \text{ m}^3)/(37.5 \text{ m}^3/\text{day}) = 13.3 \text{ days}.$$

The production of gas from digestion varies with the temperature, solids loading, solids volatility, and other factors. Typically, about 0.6 m^3 of gas/kg of volatile solids added (10 ft^3/lb) has been observed. This gas is about 60% methane and burns readily, usually being used to heat the digester and answer additional energy needs within the plant. An active group of methane formers operates at 35°C (95°F), and anaerobic digestion at this temperature has become known as *mesophilic digestion*. As the temperature is increased to about 45°C (115°F), another group of methane formers predominates, and this process is tagged *thermophilic digestion*. Although the latter process is faster and produces more gas, the elevated temperatures are more difficult and expensive to maintain.

Finally, a word of caution should be expressed about mixing in a primary digester. The assumption is invariably made that the tank is totally mixed, either mechanically or by bubbling gas through the tank. Unfortunately, digester mixing is quite difficult, and some studies have shown that on the average, only about 20% of the tank volume is well mixed!

All three stabilization processes reduce the concentration of pathogenic organisms to varying degrees. Lime stabilization achieves a high degree of sterilization, owing to the pH increase. Further, if quicklime (CaO) is used, the reaction is exothermic and the elevated temperatures assist in the destruction of pathogens. Aerobic digestion at ambient temperatures is not very effective in the destruction of pathogens. Anaerobic digesters have been well studied from the standpoint of pathogen viability because the elevated temperatures should result in substantial sterilization. Unfortunately, many pathogens survive the digestion process with little reduction in virulence. An anaerobic digester cannot, therefore, be considered a method of sterilization.

Sludge Thickening

Sludge thickening is a process in which the concentrations of solids is increased and the total sludge volume is decreased, but the sludge still behaves like a liquid instead of a solid. (In sludge dewatering, described later, the same objectives produce sludges that behave like a solid.) Thickening commonly produces sludge solids concentrations in the 3 to 5% range, whereas the point at which sludge begins to have the properties of a solid is between 15 and 20% solids. The sludge thickening process is gravitational, using the difference between particle and fluid densities to achieve greater compacting of solids.

The advantages of sludge thickening are substantial. When sludge with 1% solids is thickened to 5%, the result is an 80% reduction in volume (Fig. 10-5). Reducing from 1% solids to 20% solids, which might be achieved by sludge dewatering, would result in a 95% reduction in volume. The volume reduction translates into considerable savings in treatment, handling, and disposal costs.

Two types of nonmechanical thickening processes are presently in use: the *gravity thickener* and the *flotation thickener*. A typical gravity thickener is shown in Fig. 10-6. The influent, or feed, enters into the center of the tank and slowly moves toward the outer edge of the tank. The sludge solids settle and are removed through a drain in the bottom of the tank. The effluent flows over the weirs that encircle the tank and is usually returned to the primary or secondary treatment facility prior to discharging.

The thickening characteristics of sludges have for many years been described by the sludge volume index (SVI), described as

$$\text{SVI} = \frac{(1000)(\text{volume of sludge after 30 min, mL})}{\text{mg/L of SS}}. \qquad (9.11)$$

Treatment plant operators usually consider sludges with an SVI of less than 100 as well-settling sludges and those with an SVI of greater than 200 as potential problems.

The SVI is useful for estimating the settling ability of activated sludge, and is without question a valuable tool in running a secondary treatment plant. However, the SVI also has some drawbacks and potential problems. The SVI is not independent of solids concentration. Substantial increases in suspended solids can change the SVI

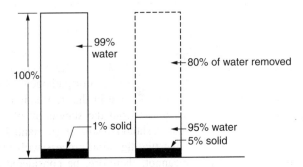

Figure 10-5. Volume reduction owing to sludge thickening.

Sludge Treatment and Disposal 215

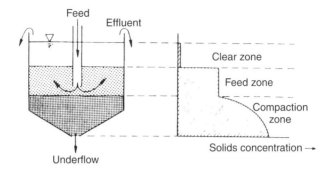

Figure 10-6. Gravity thickener. (Photo courtesy of Dorr Oliver Inc.)

severalfold. Further, for high levels of suspended solids, the SVI reaches a maximum value, above which no additional settling will occur. Take, for example, a lime sludge with a suspended solids concentration of 40,000 mg/L and suppose that it does not settle at all (the sludge volume after 30 min is 1000 mL). Calculating the SVI gives us

$$\text{SVI} = \frac{(1000)(1000)}{40,000} = 25.$$

How can a sludge that does not settle at all have an SVI of 25, when we originally noted that an SVI of less than 100 indicated a good settling sludge?

216   ENVIRONMENTAL ENGINEERING

The answer is that the maximum SVI for sludge with 40,000 mg/L of suspended solids concentration is 25. At 20,000 mg/L the maximum SVI is 50; at 10,000 mg/L the maximum SVI is 100; and so on. These maximum values can be plotted as in Fig. 10-7, which also shows how the SVIs for two different types of sludges might vary as their solids concentrations increase toward the maximum SVIs.

A better way to describe the settling characteristics of sludge is by using a flux plot. Flux plots are developed by running a series of settling tests at various solids concentrations and recording the sludge–water interface height with time. The velocity of the settling is calculated as the slope of the initial straight portion of the curve and plotted against the solids concentration. Multiplying a velocity by its corresponding solids concentration yields the solids flux, which is plotted against the concentration (see Fig. 10-8). Note that for concentration C_i, the interface settling velocity is v_i, and the solids flux at the concentration is $C_i v_i$ (kg/m^3 × m/h = kg/m^2 × h). The flux curve is a "signature" of the sludge thickening characteristics and may be used to design thickeners.

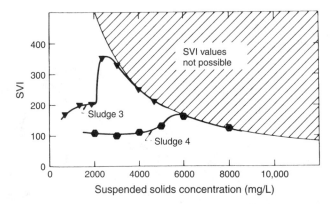

Figure 10-7. Maximum values of SVI.

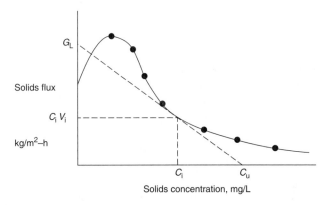

Figure 10-8. Calculation of limiting solids flux in gravitational thickening.

Designing a sludge thickener involves choosing a desired sludge solids concentration, C_u (estimated in laboratory tests), and drawing a line from this value tangent to the underside of the flux curve. The intercept at the ordinate is the limiting flux, G_L, or the solids flux that controls the thickening operation. In a continuous thickener it is impossible to pass more solids than G_L (at the stated concentration) through a unit area.

The units of solids flux are kg/m²–h (or other comparable units), and are defined as

$$G = \frac{Q_0 C_0}{A}, \qquad (10.2)$$

where

G = solids flux
Q_0 = interflow rate (m³/h),
C_0 = interflow solids concentration (mg/L), and
A = surface area of thickener (m²).

If G is the limiting flux, the necessary thickener area can be calculated as

$$A_L = \frac{Q_0 C_0}{G_L}, \qquad (10.3)$$

where the subscript L specifies the limiting area and flux.

This graphical procedure may be used to develop an optimal process. If the calculated thickener area is too large, a new, less ambitious underflow solids concentration may be selected and a new required area calculated.

EXAMPLE 10.3. What would be the required area for a thickener if in Fig. 10-8 C_u was 25,000 mg/L, G_L was 3 kg/m²–h, and the feed was 60 m³/h of sludge with 1% solids?

The limiting area is calculated as

$$A_L = \frac{Q_0 C_0}{G_L} = \frac{(60 \text{ m}^3/\text{h})(0.01)(1000 \text{ kg/m}^3)}{3 \text{ kg/m}^2\text{–h}} = 200 \text{ m}^2.$$

Suppose the C_u was 40,000 mg/L and the G_L was 1.8 kg/m²–day. Now what is the required area?

$$A_L = \frac{(60 \text{ m}^3/\text{h})(0.01)(1000 \text{ kg/m}^3)}{1.8 \text{ kg/m}^2\text{–h}} = 333 \text{ m}^2.$$

Note that the area requirement increases as the desired underflow concentration increases. In the absence of laboratory data, thickeners are designed on the basis of solids loading, which is another way of expressing the limiting flux. Gravity thickener design loadings for some specific sludges are shown in Table 10-2.

A flotation thickener, shown in Fig. 10-9, operates by forcing air under pressure to dissolve in the return flow, releasing the pressure as the return is mixed with the feed.

Table 10-2. Design Loadings for Gravity Thickeners[a]

Sludge	Design loading (kg solids/m^2–h)
Raw primary	5.2
Waste activated	1.2
Raw primary + waste activated	2.4
Trickling filter sludge	1.8

[a] 0.204 kg/m^2–h = lb/ft^2–h.

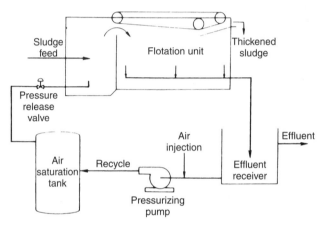

Figure 10-9. Flotation thickener.

As the air comes out of the solution, tiny bubbles attach themselves to solids and carry them upward to be scraped off as thickened sludge.

Sludge Dewatering

Unlike sludge thickening, where the treated sludge continues to behave as a liquid, dewatered sludge will behave like a solid after treatment. Dewatering is seldom used as an intermediate process unless the sludge is to be incinerated. Most wastewater plants use dewatering as a final method of volume reduction before ultimate disposal. In the United States, the usual dewatering techniques are sand beds, vacuum filters, pressure filters, belt filters, and centrifuges.

Sand beds have been used for a great many years and are still the most cost-effective means of dewatering when land is available. The beds consist of tile drains in gravel, covered by about 26 cm (10 in.) of sand. The sludge is poured on the sand to a depth of 8–12 in.; liquid is removed by seepage through the sand into the tile drains and evaporation. Although seepage into the sand results in a substantial loss of water, it only lasts for a few days. The sand pores are quickly clogged, and drainage into the sand ceases. The mechanism of evaporation takes over, and this process is actually responsible for the conversion of liquid sludge to solid. As the surface of the sludge dries, deep cracks develop that facilitate evaporation from lower layers in the sludge mat. In wet areas, sand beds may be enclosed under well-ventilated greenhouses to promote evaporation and prevent rain from falling into the beds.

For mixed digested sludge the usual design is to allow at least 3 weeks of drying time, depending on weather conditions and sludge depth. The addition of chemicals such as aluminum sulfate often increases the amount of water lost through drainage during the first few days of treatment, thereby shortening the overall drying time. Some engineers suggest allowing the sand bed to rest for a month after the sludge has been removed. This seems to be an effective means of increasing the drainage efficiency once the sand beds are again flooded.

Raw sludge will not drain well on sand beds and will usually have an obnoxious odor. Hence raw sludges are seldom dried on beds. Raw secondary sludges tend to seep through the sand or clog the sand pores so quickly that no effective drainage takes place. Aerobically digested sludges may be dried on sand, but are more difficult to dewater than anaerobically digested sludges.

If dewatering by sand beds is impractical, mechanical dewatering techniques must be used. Two general processes are used for mechanical dewatering: *filtration* and *centrifugation*.

Sludge filtration is usually done using a *pressure filter* or a *belt filter*. The pressure filter, shown in Fig. 10-10, uses positive pressure to force the water through a filter cloth. Typically, pressure filters are built as plate-and-frame filters, in which the sludge solids are captured and compressed between the plates and frames. The plates are then pulled apart to allow sludge cleanout. The belt filter, shown in Fig. 10-11, uses gravity drainage and a pressure filter. When the sludge is first introduced onto the moving belt, water drips through the belt and solids are retained. The belt then moves into the dewatering zone where the sludge is squeezed between two belts. Belt filters are quite effective in dewatering many different types of sludges and are widely used by small wastewater treatment plants.

The effectiveness of a filter in dewatering specific sludge is often measured using the "specific resistance to filtration" test. The resistance of sludge to filtration can be stated as

$$r = \frac{2PA^2 b}{\mu w}, \qquad (10.4)$$

where

r = specific resistance (m/kg),
P = vacuum pressure (N/m^2),
A = area of the filter (m^2),
μ = filtrate viscosity (N s/m^2),
w = cake deposit per volume of filtrate (for dry cakes this may be approximated as the feed solids concentration, kg/m^3), and
b = slope of the filtrate volume vs time/filtrate volume curve.

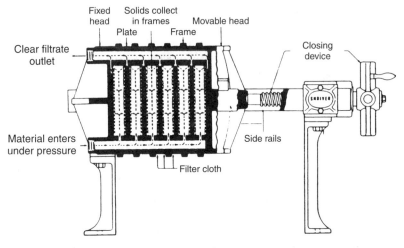

Figure 10-10. Pressure filters. (Photo courtesy of Envirex.)

The factor b may be determined from a simple test with a Büchner funnel (Fig. 10-12). The sludge is poured onto the filter, a vacuum is applied, and the volume of the filtrate is recorded against time. The plot of these data yields a straight line with slope b. Table 10-3 lists approximate ranges for specific resistance for different sludges.

EXAMPLE 10.4. Results of a Büchner funnel sludge filterability experiment are shown below. Calculate the specific resistance to filtration. Note that the first 2 min are ignored, partly because of storage of water in the filter and partly because the resistance of the filter becomes negligible after this time allowance.

Sludge Treatment and Disposal 221

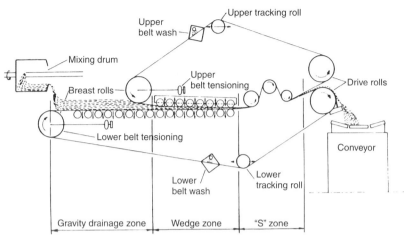

Figure 10-11. Belt filter.

Time (min)	Θ (s)	V (mL)	Corrected V (mL)	Θ/V (s/mL)
−2	—	0	—	—
0	0	1.5	0	—
1	60	2.8	1.3	46.3
2	120	3.8	2.3	52.3
3	180	4.6	3.1	58.0
4	240	5.5	4.0	60.0
5	300	6.1	4.6	65.2

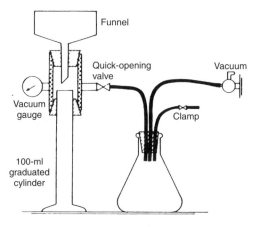

Figure 10-12. Büchner funnel test for determining specific resistance to filtration.

Table 10-3. Specific Resistance of Typical Sludges

Sludge type	Specific resistance (m/kg)
Raw primary	10–30 × 10^{14}
Mixed digested	3–30 × 10^{14}
Waste activated	5–20 × 10^{14}
Lime and biological sludge	1–5 × 10^{14}
Lime slurry	5–10 × 10^{13}
Alum	2–10 × 10^{13}

Other test variables are:

P = pressure = 10 psi = 703 g/cm^2,
μ = viscosity = 0.0011 P = 0.0011 N × s/m^2,
w = 0.075 g/mL = 75 kg/m^3, and
A = 44.2 cm^2 = 0.00442 m^2.

The data are plotted in Fig. 10-13. The slope of the line, b, is 5.73 s/cm^6 = 5.73 × (100 cm/m)^6 = 5.73 × 10^{12} s/m^6

$$r = \frac{(2)(6.9)(10^4)(0.00442)^2(5.73)(10^{12})}{(0.0011)(75)} = 1.86 \times 10^{13} \text{ mg/kg}.$$

Specific resistance to filtration can be used to estimate the required loadings of a filter. The *filter yield* is described as the mass of dewatered (dry) solids produced by

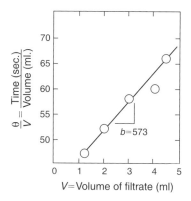

Figure 10-13. Laboratory filtration data from Example 10.4.

the filter per unit filter surface area per unit time,

$$Y_F = \left[\frac{2Pw}{\mu rt}\right]^{1/2}, \tag{10.5}$$

where

Y_F = filter yield (kg/m²–s),
P = filter pressure (N/m²),
w = feed solids concentration (kg/m³),
μ = filtrate viscosity (N–s/m²),
r = specific resistance to filtration (m/kg), and
t = time on the filter (s).

EXAMPLE 10.5. A sludge has a solids concentration of 4% and a specific resistance to filtration of 1.86×10^{13} m/kg. The pressure in a belt filter is expected to be 800 N/m² and the filtration time is 30 s. Estimate the belt area required for a sludge flow rate of 0.3 m³/s:

$$Y_F = \left[\frac{2(800)(40)}{(0.01)(1.86 \times 10^{13})(30)}\right]^{1/2} = 1.07 \times 10^{-4} \text{kg/m}^2 \cdot \text{s}.$$

This filter yield is approximately 8 lb/ft²–h, which is excellent production for a dewatering operation.

The Büchner funnel test for specific resistance is rather clumsy and time consuming. An indirect method for estimating how well sludge will filter uses the capillary suction time (CST) apparatus, developed in Great Britain. The CST device, illustrated in Fig. 10-14, allows the water to seep out of sludge (in effect, a filtration process)

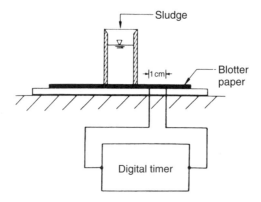

Figure 10-14. Capillary suction time apparatus.

and onto a blotter. The speed at which the water is taken up by the blotter is measured by timing the spread of water over a set distance. This time, in seconds, is correlated to specific resistance. A short time indicates a highly filterable sludge; a long time suggests slow, problematic filtration.

A major drawback of the CST test is that, unlike the specific resistance to filtration, the CST is dependent on the solids concentration. Using the Darcy equation it is possible to derive an expression that appears to be an excellent measure of sludge filterability,

$$\chi = \phi \left(\frac{\mu X}{t} \right), \quad (10.6)$$

where

χ = sludge filterability (kg² m⁴/s²),
ϕ = dimensionless instrument constant, specific for each CST apparatus,
μ = fluid (not sludge) viscosity (P),
X = suspended solids concentration (mg/L), and
t = capillary suction time (s).

The sludge filterability term (χ) can be used for design as specific resistance to filtration.

Centrifugation became popular in wastewater treatment only after organic polymers were available for sludge conditioning. Although the centrifuge will work on any sludge, most unconditioned sludges cannot be centrifuged with greater than 60 or 70% solids recovery. The centrifuge most widely used is the solid bowl decanter, which consists of a bullet-shaped body rotating on its axis. The sludge is placed into the bowl, and the solids settle out under about 500 to 1000 gravities (centrifugally applied) and are scraped out of the bowl by a screw conveyor (Fig. 10-15). Although laboratory tests are of some value in estimating centrifuge applicability, tests with continuous models are considerably better and highly recommended whenever possible.

Sludge Treatment and Disposal 225

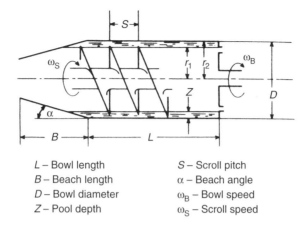

L – Bowl length
B – Beach length
D – Bowl diameter
Z – Pool depth

S – Scroll pitch
α – Beach angle
ω_B – Bowl speed
ω_S – Scroll speed

Figure 10-15. Solid bowl centrifuges. (Photo courtesy of Ingersoll Rand and I. Krüger.)

A centrifuge must be able to settle the solids first and then move them out of the bowl. Accordingly, two parameters have been suggested for scale-up between two geometrically similar machines. Settling characteristics are measured by the *sigma equation* (Ambler 1952). Without going into the derivation of this parameter, it is simply assumed that if two machines (1 and 2) are to have equal effects on the settling within the bowl, the following relationship must be true:

$$\frac{Q_1}{\Sigma_1} = \frac{Q_2}{\Sigma_2}. \tag{10.7}$$

The symbol Q represents the liquid flow rate into the machine, and Σ is a parameter composed of machine (not sludge!) characteristics. For solid bowl centrifuges, Σ is calculated as

$$\Sigma = \frac{V\omega}{g \ln(r_2/r_1)}, \qquad (10.8)$$

where

- ω = rotational speed of bowl (rad/s),
- g = gravitational constant (m/s^2),
- r_2 = radius from centerline to inside of bowl wall (m),
- r_1 = radius from centerline to surface of sludge (m), and
- V = liquid volume in the pool (m^3).

Thus if machine 1, at Σ_1 (calculated from machine parameters) and flow rate Q_1, produces a satisfactory result, it is expected that a second, geometrically similar machine with a larger Σ_2 will achieve equal dewatering performance at a flow rate of Q_2.

This analysis does not consider the movement of solids out of the bowl, which is a very important component of centrifugation. The solids movement may be calculated by the *beta equation* (Vesilind 1979) for two machines,

$$\frac{Q_1}{\beta_1} = \frac{Q_2}{\beta_2}. \qquad (10.9)$$

and

$$\beta = (\Delta\omega) S N \pi D z, \qquad (10.10)$$

where

- Q = solids per unit time (e.g., lbs/h),
- $\Delta\omega$ = difference in rotational speed between the bowl and conveyor, $\omega_B - \omega_s$ (rad/s),
- S = scroll pitch (distance between blades, m),
- N = number of leads,
- D = bowl diameter (m), and
- z = depth of sludge in bowl (m).

The scale-up procedure involves the calculation of Q_2 for liquid as well as solids throughout, and the lowest value governs the centrifuge capacity.

EXAMPLE 10.6. A solid bowl centrifuge (machine 1) is found to perform well if fed sludge containing 1% solids at 0.5 m^3/h. In order to scale-up to a larger machine (machine 2) it is necessary to determine the flow rate at which this geometrically similar machine would perform equally well. The machine variables are:

	Machine 1	Machine 2
Bowl diameter (cm)	20	40
Pool depth (cm)	2	4
Bowl speed (rpm)	4000	3200
Bowl length (cm)	30	72
Conveyor speed (rpm)	3950	3150
Pitch (cm)	4	8
Number of leads	1	1

Looking first at the settling (sigma) scale-up, the volume of the sludge in the pool may be approximated as

$$V = 2\pi \left(\frac{r_1 + r_2}{2}\right)(r_2 - r_1)L, \qquad (10.11)$$

where the variables are defined in Fig. 10-15. The volumes of the two machines are

$$V_1 = 2\pi \left(\frac{8 + 10}{2}\right)(10 - 8)(30) = 3400 \, \text{cm}^3$$

$$V_2 = 2\pi \left(\frac{16 + 20}{2}\right)(20 - 16)(72) = 32{,}600 \, \text{cm}^3.$$

Since

$$\omega = (\text{rpm})\left(\frac{1}{60}\right)\left(2\pi \, \frac{\text{rad}}{\text{rev}}\right)$$

$$\omega_1 = (4000)\left(\frac{1}{60}\right)(2\pi) = 420 \, \text{rad/s}$$

$$\omega_2 = (3200)\left(\frac{1}{60}\right)(2\pi) = 335 \, \text{rad/s}$$

$$\Sigma_1 = \frac{(420)^2}{980}\left(\frac{3400}{\ln(10/8)}\right) = 2.76 \times 10^6$$

$$\Sigma_2 = \frac{(335)^2}{980}\left(\frac{32{,}600}{\ln(20/16)}\right) = 16.8 \times 10^6.$$

228 ENVIRONMENTAL ENGINEERING

Thus

$$Q_2 = \frac{\Sigma_2}{\Sigma_1} Q_1 = \frac{16.8}{2.78}(0.5) = 3.2 \, \text{m}^3/\text{h}.$$

Next, looking at the solids loading (beta), the solids flow rate is

$$Q_1 = 0.5 \frac{\text{m}^3}{\text{h}}(0.01)(1000) \frac{\text{kg}}{\text{m}^3} = 5 \frac{\text{kg}}{\text{h}},$$

and from Eq. (10.9)

$$\beta_1 = (4000 - 3950)(4)(1)\pi(20)(2) = 25{,}120$$

$$\beta_2 = (3200 - 3150)(8)(1)\pi(40)(4) = 200{,}960$$

$$Q_2 = \frac{200{,}960}{25{,}120}(5) = 40 \frac{\text{kg}}{\text{h}},$$

corresponding to a flow rate of

$$\frac{40}{0.01}(10^{-3}) = 4 \frac{\text{m}^3}{\text{h}}.$$

The liquid loading thus governs, and the larger machine (machine 2) cannot be fed at a flow rate greater than 3.2 m²/h.

ULTIMATE DISPOSAL

Even after treatment, we are left with a large volume of sludge that needs a final resting place. The choices for ultimate disposal of sludge are limited to air, water, and land. Until quite recently, incineration ("air disposal") was viewed as an effective sludge reduction method, if not exactly an ultimate sludge disposal method (the residual ash still required disposal). However, strict controls on air pollution and increasing concern over global warming are making incineration an increasingly unlikely option. Disposal of sludges in deep water (such as oceans) is decreasing owing to adverse or unknown detrimental effects on aquatic ecology. Land disposal, particularly the use of sludge as fertilizer or soil conditioner, has historically been a favored disposal method, and is currently growing in popularity as other options become more problematic.

Incineration is actually not a method of disposal at all, but rather a sludge treatment step in which the organics are converted to H_2O and CO_2 and the inorganics drop out as a nonputrescible residue. Two types of incinerators have found use in sludge treatment: multiple hearth and fluid bed. The multiple-hearth incinerator, as the name implies, has several hearths stacked vertically, with rabble arms pushing the sludge progressively downward through the hottest layers and finally into the ash pit (Fig. 10-16).

Sludge Treatment and Disposal 229

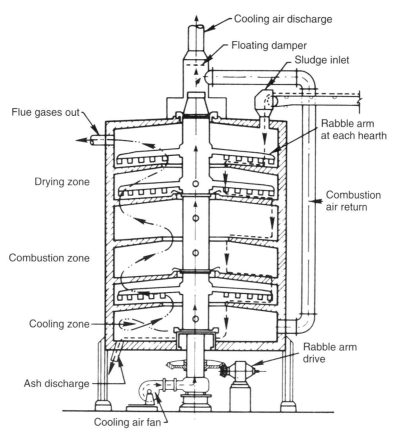

Figure 10-16. Multiple-hearth incinerator. (Courtesy of Nichols Engineering and Research Corp.)

The fluidized bed incinerator is full of hot sand suspended by air injection; the sludge is incinerated within the moving sand. Owing to the violent motion within the fluid bed, scraper arms are unnecessary. The sand acts as a "thermal flywheel," allowing intermittent operation. Despite a flurry of interest during the previous decade, sludge incineration is no longer considered the best available technology by many state regulatory agencies because of environmental concerns about atmospheric emissions and ash disposal.

The second method of disposal, land disposal, is becoming more popular, particularly in areas where there are restrictions on industrial contaminants entering the wastewater treatment. (Sludges contaminated with industrial chemicals may not be suitable for land application.) The ability of land to absorb sludge and to assimilate it depends on such variables as soil type, vegetation, rainfall, and slope. In addition, the important variable of the sludge itself will influence the capacity of a soil to assimilate sludge. Generally, sandy soils with lush vegetation, low rainfall, and gentle slopes have proven most successful. Mixed digested sludges have been spread from tank trucks, and activated sludges have been sprayed from both fixed and moving nozzles. The application rate has been variable, but 100 dry tons/acre-yr is not an unreasonable estimate. Most unsuccessful land application systems may be traced to overloading the soil. Given enough time, and the absence of toxic materials, soils will assimilate sprayed liquid sludge.

There has been some successful use of land application for sludge for fertilization, particularly in silviculture operations. Forests and tree nurseries are far enough from population centers to minimize aesthetic objections, and the variable nature of sludge is not so problematical in silviculture as in other agricultural applications. Sludge may also be treated as packaged fertilizer and plant food. The city of Milwaukee has pioneered the drying, disinfection, and deodorizing of sludge, which is packaged and marketed as the fertilizer Milorganite.

Transporting liquid sludge is often expensive, and volume reduction by dewatering is necessary. The solid sludge may then be deposited on land and disked into the soil. A higher application rate (tons/acre-yr) may be achieved by trenching, where 1-m^2 (3-ft^2) trenches are dug with a backhoe, and the sludge is deposited in the trench, then covered with soil.

In the past few years *chemical fixation*, which involves chemically bonding sludge solids so that the mixture "sets" in a few days, has found use in industries that have especially critical sludge problems. Although chemical fixation is expensive, it is often the only alternative for besieged industrial plants. The leaching from the solid seems to be minimal.

Sludge often contains compounds that are potentially harmful to vegetation and animals (including people), or that can cause degradation of surface water and groundwater supplies. Although most domestic sludges do not contain sufficient concentrations of toxins such as heavy metals to cause immediate harm to vegetation, the overall concentration of toxins or metals may bioaccumulate in plants and animals if the sludge is applied to the same land for an extended period of time. Because of this, sludges to be applied as fertilizer or soil conditioners must be tested to certify that they meet state and federal guidelines. It is possible to remove some toxins during sludge treatment, but the most effective means of controlling toxicity is to prevent the toxins from entering the sewerage system. Strongly enforced sewerage ordinances are necessary, particularly given the increasingly difficult problem of providing for the ultimate disposal of sludge.

CONCLUSION

Sludge disposal represents a major headache for many municipalities. Sludge represents the true residues of our civilization, and its composition reflects our style of living, our technological development, and our ethical concerns. "Pouring things down the drain" is our way of getting rid of all manner of unwanted materials, not recognizing that these materials often become part of the sludge that must, ultimately, be disposed of in the environment. All of us need to become more sensitive to these problems and keep potentially harmful materials out of our sewage system and out of sludge.

PROBLEMS

10.1 Using reasonable values, estimate the sludge production in a 10-mgd wastewater treatment plant. Estimate both sludge volume per time and dry solids per time.

10.2 A 1-L cylinder is used to measure the settleability of 0.5% suspended solids sludge. After 30 min, the settled sludge solids occupy 600 mL. Calculate the SVI.

10.3 What measures of "stability" would you need if a sludge from a wastewater treatment plant were to be

a. placed on the White House lawn?
b. dumped into a trout stream?
c. sprayed on a playground?
d. sprayed on a vegetable garden?

10.4 A sludge is thickened from 2000 to 17,000 mg/L. What is the reduction in volume, in percent?

10.5 A laboratory batch thickening test produces the following results:

Solids concentration (%)	Settling velocity (cm/min)
0.6	0.83
1.0	0.25
1.4	0.11
1.8	0.067
2.2	0.041

Calculate the required thickener area if the desired underflow concentration (C_u) is to be 3% solids and the feed to the thickener is 0.5 m^3/min at a solids concentration of 2000 mg/L.

10.6 A Büchner funnel test for specific resistance to filtration, conducted at a vacuum pressure of 6×10^4 N/m^2, with an area of 40 cm^2, and a solids concentration of 60,000 mg/L, yields the following data:

Time (min)	Filtrate volume (mL)
1	11
2	20
3	29
4	37
5	43

Calculate the specific resistance to filtration. Use $\mu = 0.011$ N–s/m^2. Is this sludge amenable to dewatering by vacuum filtration?

10.7 Two geometrically similar centrifuges have the following machine characteristics:

	Machine 1	Machine 2
Bowl diameter, cm	25	35
Pool depth $(r_2 - r_1)$, cm	3	4
Bowl length, cm	30	80

If the flow rates are 0.6 and 4.0 m³/h and the first machine operates at 1000 rpm, at what speed should machine 2 be operated to achieve similar performance, if only the liquid loading is important?

Chapter 11

Nonpoint Source Water Pollution

As rain falls and strikes the ground a complex runoff process begins, carrying with it dissolved and suspended material from the watershed into adjacent streams, lakes, and estuaries. Even before people entered the picture, sediments transported from the watershed would accumulate behind natural dams, along the inside curves of river bends, and at the mouths of streams.

Now view the world as it has been since the dawn of humankind — a busy place where human activities continue to influence our environment. For hundreds, and even thousands of years, these activities have included farming, harvesting trees, constructing buildings and roadways, mining and industrial production, and disposal of liquid and solid wastes. These activities have led to disruptions in the watershed vegetation and soils, increases in the amount of impervious surfaces (e.g., pavement and roads), introduction of agricultural chemicals, fertilizers, and animal wastes into the watershed, and the deposition of many types of atmospheric pollutants (e.g., hydrocarbons from automobile exhaust) in the watershed. The combination of these types of pollutants from diffuse, widespread sources is generally called *nonpoint source pollution.*

The kinds of nonpoint source pollutants entering a stream vary depending on the type of human activity in the watershed (Table 11-1). Runoff from agricultural regions typically contains elevated concentrations of suspended solids, dissolved salts and nutrients from fertilizers, biodegradable organic matter, pesticides, and pathogens from animal wastes. Activities that disrupt the vegetation cover or soil surface, such as construction and silvaculture, will contribute suspended sediments and sediment-bound nutrients such as phosphorus to surface runoff. Runoff from silvaculture sites may also contain herbicides that were applied to control the growth of undesirable plants. Urban runoff, one of the worst sources of nonpoint source pollution, often contains high concentrations of suspended and dissolved solids; nutrients and pesticides from landscaped areas; toxic metals, oil and grease, and hydrocarbons from roads; pathogens from pet wastes and leaking septic tanks; and synthetic organics such as detergents, degreasers, chemical solvents, and other compounds that accumulate on impervious surfaces or are carelessly poured down storm drains.

Water movement is the prime mode of transport for nonpoint source pollutants, whether they are dissolved in water or suspended in surface runoff. The concentrations of soluble pollutants such as road deicing salts, acids from abandoned mine sites,

Table 11-1. Major Nonpoint Source Pollution Categories[a]

Category	Suspended solids	Dissolved solids	High BOD	Nutrients	Toxic metals	Pesticides	Pathogens	Synthetic organics/ hydrocarbons[b]
Agriculture	***	***	***	***	*	***	***	n
Construction	***	n	*	**	n	n	n	n
Urban runoff[c]	***	**	**	***	***	***	***	***
Mining[d]	**	**	n	n	***	n	n	n
Silvaculture	***	n	*	**	n	**	n	n

[a] *** = potentially high pollutant source; ** = moderate source; * = low source; n = negligible source.
[b] Includes industrial solvents and reagents, detergents, oil and grease, petroleum hydrocarbons, and other organics not normally found in surface runoff.
[c] Includes residential and urban runoff from buildings, roads, impervious surfaces, and landscaped areas.
[d] Primarily abandoned mine sites; active mine sites are usually regulated as point sources.

water-soluble pesticides, and nitrogen-based plant nutrients (ammonia, nitrite, and nitrate) are a function of the contact time between the pollutant and water. Longer contact times allow more pollutants to be dissolved in the water. Soluble pollutants are often more concentrated in groundwater than surface runoff, particularly in agricultural regions where pesticide and nitrogen concentrations in local wells may exceed safe levels for drinking water. Insoluble pollutants include suspended sediments, along with most metals, microbial pathogens, most forms of phosphorus, and many pesticides and organics that are insoluble in water or are physically or chemically bound to sediment particles.

SEDIMENT EROSION AND THE POLLUTANT TRANSPORT PROCESS

Sediment erosion is the result of a complex interaction between the timing, intensity, and duration of precipitation and the structure of the landscape. Because many nonpoint source pollutants are transported with suspended sediments, the factors that contribute to soil erosion also contribute to nonpoint source pollution.

Soil erosion can be separated into four categories: *rainsplash*, *sheetwash*, *gullying*, and *stream/channel* erosion (Table 11-2). Rainsplash erosion results when rain strikes the ground with enough force to splatter and dislodge soil particles. The energy in a raindrop is determined by droplet size, velocity of fall, and the intensity characteristics of the particular storm. Soil stability (size, shape, composition, and strength of soil aggregates and soil clods) and slope are also important in determining how readily the raindrop dislodges soil particles. Vegetation cover greatly reduces or eliminates rainsplash erosion by preventing raindrops from landing on exposed soil surfaces and by reducing evaporation, which keeps the soil moist and makes the soil particles less susceptible to detachment.

Sheetwash erosion can occur when the rate of precipitation exceeds the rate at which water can infiltrate into the soil. Under these conditions water accumulates at the soil surface and forms irregular sheets that flow downhill, carrying sediment along with the surface runoff. The intensity and duration of precipitation are important factors

Table 11-2. Types of Erosion That May Contribute to Nonpoint Source Pollution

Type of erosion	Description
Rainsplash	From direct impact of falling raindrops; detached soil and chemicals can be transported downhill.
Sheetwash	From raindrop splash and runoff in wide, thin layers of surface water; wall of water transports pollutants downhill.
Gullying	From concentrated rivulets of water cutting 5- to 10-cm rills in soil; rills grow into gullys.
Stream/channel	From confluence of rills and gullys; increased volumes and velocities cause erosion along streams and river beds.

in determining when sheetwash erosion will occur. A gentle summer shower is less likely to result in sheetwash erosion than a sudden, torrential downpour. Slope and soil characteristics are also important, along with antecedent precipitation and vegetation cover. Surface runoff occurs more readily in saturated soils or very dry soils that have developed a hard, impervious crust. Soils with a healthy vegetation cover are less likely to experience sheetwash erosion, even when the soils are saturated or very dry, because the cover helps slow surface runoff and hold soil particles in place.

As surface runoff flows downhill across an unconsolidated surface such as exposed soil, the water collects into small channels (rills), and eventually forms wider, deeper erosion gullys. When water is confined to a distinct channel, whether a rill, gully, or stream channel, the runoff moves more rapidly and erosion becomes more intense within the confines of the channel. The channel will develop sinuous bends and curves, with areas of active erosion along the outer edge of the bend and depositional areas along the inside of the bend. Only a portion of the sediment detached and transported from upland regions in a watershed is actually carried all the way into a lake or estuary. In many cases, significant portions of the sediments are deposited at the base of slopes, in the floodplain, or in depositional regions along stream banks.

Estimating Sediment Erosion

Estimating the amount of sediment that will be transported out of a watershed is an important part of measuring and abating nonpoint source pollution. Elevated sediment concentrations are common in streams passing through agricultural regions, construction sites, freshly logged forests, and residential areas. Along with the sediments, each nonpoint pollution source has its own characteristic array of pollutants: pathogens, phosphorus, and pesticides from agricultural regions; phosphorus and herbicides from cleared forests; and phosphorus, pathogens, metals, and petroleum hydrocarbons from residential areas.

Because many urban pollutants are transported with sediment, and because sediment itself is an important pollutant, mathematical models are often used to predict sediment erosion and transport to a body of water. Most sediment transport models include adjustments for variation in soil characteristics, slope, vegetative cover, percent impervious surfaces, precipitation patterns, and other relevant watershed characteristics.

One method for estimating the sediment transport from sheet and rill erosion is a loading function based on the universal soil loss equation (USLE)

$$A = R \times K \times \text{LS} \times C \times P, \tag{11.1}$$

where

A = soil loss (ton/acre–year),
R = rainfall erosivity index,
K = soil erodibility factor,

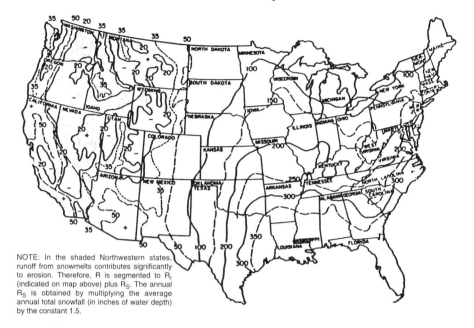

Figure 11-1. Average annual rainfall erosivity indices (R) for the USA.

LS = topographic factor,
C = vegetation cover factor, and
P = erosion control practice factor.

The rainfall erosivity index (R) is calculated by summing the product of rainfall energy during an individual storm event (E), measured in foot–tons per acre per inch of rainfall, and the maximum 30-min rainfall intensity (I), measured in inches per hour, for a specific period of time such as one year or a season:

$$R = \frac{\sum_{i=1}^{n} E_i I_{30i}}{100}. \tag{11.2}$$

Figure 11-1 shows average rainfall erosivity indices for the USA.

The soil erodibility factor, K, is the average soil loss, in tons per acre, per 100 ft–tons per acre of rainfall erosivity standardized for a plot of land with 9% slope, 72.6 ft long, under continuous cultivation (Fig. 11-2). The topographic factor, LS, combines both the slope-length and slope-steepness factors and reflects the effect of both length of flow and steepness of slope on soil loss (Fig. 11-3). These two factors together account for the capability of runoff to detach and transport soil material. The vegetation cover factor (C) and the erosion-control practice factor (P) adjust for the influences of different vegetation types and erosion-control techniques (Tables 11-3 and 11-4). Both are calculated as ratios compared to the soil quantity eroded from clean-tiled soil under identical slope and rainfall conditions. Local Soil Conservation Service offices keep records of regional values for C and P.

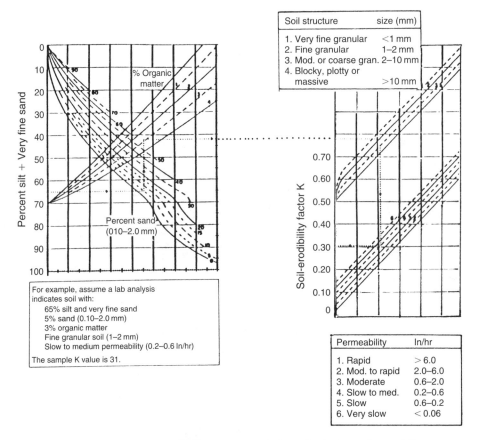

Figure 11-2. Soil erodibility nomograph for calculating values of K.

EXAMPLE 11.1. An 830-acre watershed is located 5 miles south of Indianapolis in central Indiana. Given the following information, compute the projected sediment from sheet and rill erosion in terms of average daily loading: 180 acres of cropland; corn in conventional tillage; average annual yield of 40–45 bu/acre; cornstalks remain in field after harvest; contour strip-cropped planting method; soil is Fayette silt loam; slope is 6%; slope length is 250 ft.

The factors for cropland in Eq. (11.1) are:

$R = 200$ from Fig. 11-1,
$K = 0.37$ from Fig. 11-2,
$LS = 1.08$ from Fig. 11-3,
$C = 0.49$ from Table 11-3 and USDA Soil Conservation Field Office,
$P = 0.25$ from Table 11-4, and
$S_d = 0.60$ from USDA Soil Conservation Field Office.

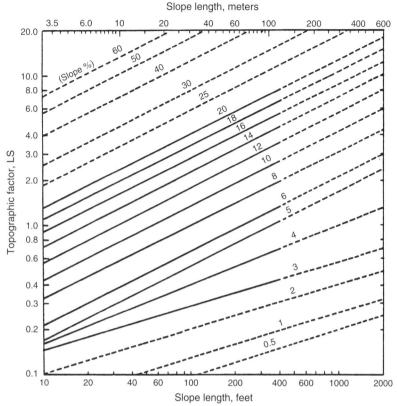

Figure 11-3. Topographic factor (LS) for areas east of the Mississippi River.

Then from Eq. (11.1), the annual soil for the cropland would be

$$A_{annual} = 200 \times 0.37 \times 1.08 \times 0.49 \times 0.25 \times 0.60$$

$$A_{annual} = 5.87 \text{ tons/acre–year,}$$

and the average daily soil loss would be

$$A_{daily} = \frac{5.86 \text{ tons/acre–year}}{365 \text{days/year}}$$

$$A_{daily} = 0.016 \text{ tons/acre–day.}$$

Calculating for watershed loading,

$$A_{\text{watershed daily average}} = 180 \text{ acres} \times 0.016 \text{ tons/acre–day}$$

$$A_{\text{watershed daily average}} = 2.9 \text{ tons/day.}$$

Table 11-3. Vegetation Cover Factors (*C*) for Various Types of Ground Cover[a]

Land use groups	Examples	Range of *C* values
Permanent vegetation	Protected woodland Prairie Permanent pasture Sodded orchard Permanent meadow	0.0001–0.45
Established meadows	Alfalfa Clover Fescue	0.0004–0.3
Small grains	Rye Wheat Barley Oats	0.07–0.5
Large-seeded legumes	Soybeans Cowpeas Peanuts Field peas	0.1–0.65
Row crops	Cotton Potatoes Tobacco Vegetables Corn Sorghum	0.1–0.70
Fallow	Summer fallow Plowing-to-crop growth period	1.0

[a] *Source*: USEPA (1976), p. 59.

The USLE was developed by Agricultural Research Service (U.S. Department of Agriculture) scientists W. Wischmeir and D. Smith in 1965 to predict soil loss in agriculture regions. Because the original USLE was developed for agricultural land, the equation factors are best defined for pastures and cropland. A newer version, the Revised Universal Soil Loss Equation (RUSLE), is available as a computer model from the U.S. Department of Agriculture (USDA) National Sedimentation Laboratory web site (http://www.sedlab.olemiss.edu/rusle). The RUSLE has been updated to estimate soil loss wherever ground is broken, such as during construction or strip mining, and is used to estimate soil erosion for a variety of soil management scenarios including agricultural best management practices, mining and construction activities, land reclamation projects, silvaculture activities, and residential storm water treatment options. Used alone, the RUSLE projects relative changes in sediment yields if different nonpoint source control technologies are designed and constructed. When integrated

Table 11-4. P Values for Erosion Control Practices on Cropland[a]

Slope	Type of control practice				
	Up- and downhill	Cross-slope farming without strips	Contour farming	Cross-slope farming with strips	Contour strip-cropping
2.0-7	1.0	0.78	0.50	0.37	0.25
7.1-12	1.0	0.80	0.60	0.45	0.30
12.1-18	1.0	0.90	0.80	0.60	0.40
18.1–24	1.0	0.95	0.90	0.67	0.45

[a] *Source*: USEPA (1976), p. 64.

into a watershed-level hydrologic model or landscape-level GIS model (e.g., HSPF), the RUSLE can be used to estimate sediment erosion and loading on a much larger scale.

PREVENTION AND MITIGATION OF NONPOINT SOURCE POLLUTION

There has been a growing emphasis on prevention and mitigation of nonpoint source pollution during the past decade. Many counties now require agricultural "farm plans" that provide site-specific guidance for minimizing agricultural nonpoint source pollution. Similarly, many cities are now required to have discharge permits for storm drains that empty into natural water bodies. The following section will describe current practices for preventing and mitigating nonpoint source pollution from three types of human activities: agricultural practices, construction, and urban storm water runoff. As indicated in Table 11-1, these three sources are associated with a wide range of pollutants, including suspended solids, nutrients, toxic metals, pesticides, human pathogens, and organics.

Agricultural Nonpoint Source Pollution

Runoff from agricultural areas may contain high concentrations of suspended sediments, dissolved and suspended nutrients (phosphorus and nitrogen), biodegradable organic matter, pesticides (herbicides, insecticides, fungicides), and pathogens from animal waste. If biosolids from municipal wastewater treatment facilities have been used as a soil conditioner or fertilizer, the runoff may also contain traces of toxic metals and other residues associated with municipal sludges (see Chap. 10).

Pollution control in agricultural areas is aimed at preventing or reducing sediment erosion; controlling pesticide runoff; improving the efficiency of fertilizer and irrigation water use, and reducing the amount lost to groundwater seepage and surface water runoff; improving riparian and conservation buffers; restricting animal access to streams; and improving manure management techniques. The U.S. Environmental

Protection Agency (USEPA) and the USDA have developed a training module for learning agricultural management practices that protect water quality (http://www.epa.gov/watertrain/agmodule). The USEPA/USDA training module describes the following eight basic types of agricultural management practices: conservation tillage, crop nutrient management, pest management, conservation buffers, irrigation water management, grazing management, animal feeding operations management, and erosion and sediment control.

Conservation tillage involves leaving crop residues from previous plantings in place rather than plowing the field before replanting. This practice not only reduces soil erosion, it helps hold nutrients and previously applied pesticides in the field and conserves soil moisture.

Crop nutrient management is designed to increase the efficiency of applying crop fertilizers by measuring soil nutrient levels (particularly nitrogen), plant chlorophyll concentrations (this helps determine the nitrogen needs for a particular crop), soil organic matter concentrations, and irrigation water nutrient concentrations. In addition to reducing nonpoint pollution, careful crop nutrient management can save money and improve the overall crop yield.

Pest management incorporates the concept of *integrated pest management* (IPM), which uses a combination of approaches to control plant pests. Chemical pesticides are still used in IPM, but only sparingly, and only in conjunction with natural controls such as selecting resistant crops, timing harvests and rotating crops to upset the pest's life cycle, and using biological controls such as natural predators.

Conservation buffers can be as simple as leaving strips of untilled vegetation to reduce soil erosion in a plowed field, or as advanced as replanting a riparian corridor with native vegetation to provide shade for the stream and appropriate habitat for local wildlife. The main purpose of a conservation buffer is to use permanent vegetation to improve the environment, which may include stabilizing soils, reducing nonpoint pollution to streams, protecting crops and livestock, improving the aesthetic environment, and providing wildlife habitat.

Irrigation water management is designed to reduce nonpoint pollution associated with irrigation practices and reduce the energy costs associated with transport and application of irrigation water. Most water used for irrigation must be transported from distant sources or pumped from deep aquifers. Careful timing and efficient use of irrigation water can reduce costs, as well as reduce nonpoint source pollution. Irrigation water picks up large amounts of dissolved solids, pesticides, and other pollutants as it moves through the soil. Subsurface drains are often installed under irrigated fields to collect the excessively salty drainage water before it contaminates local aquifers.

Grazing management involves adjusting the number and kinds of animals in a pasture to limit soil erosion and water pollution problems associated with overgrazing. This practice is closely associated with *animal feeding operations management*, which includes a grazing management plan along with exclusion of animals from sensitive habitats such as stream banks and appropriate animal waste management practices. Waste management is a particular problem for large feedlots, dairies, poultry yards, and other areas with high animal densities. The amount of manure generated at the site may exceed the local need for crop fertilizers. Converting the animal waste into a commercial product (e.g., methane, dry fertilizer, or soil conditioner) is an option, but may not be economically self-sustaining for small- to medium-sized facilities.

Erosion and sediment control includes a wide variety of agricultural practices designed to protect the soil from being suspended and transported by water or wind. Conservation tillage, conservation buffers, and grazing management (discussed earlier) are all part of the attempt to reduce or eliminate soil erosion in agricultural areas. Other approaches include terracing and contour farming, diverting runoff into retention/detention basins, maintaining natural or constructed wetlands (which act as a sediment trap), conditioning soil to increase infiltration rates, maintaining wind buffers, and creating a site-specific erosion control plan that accounts for the specific needs of a particular agricultural site.

Construction Nonpoint Source Pollution

Erosion of soil at construction sites will not only cause water quality problems offsite, but may be regarded as the loss of a valuable natural resource. Homebuyers expect a landscaped yard, and lost topsoil is often costly for the contractor to replace. Builders of houses, highways, and other construction view soil erosion as a process that must be controlled in order to maximize economic return. Under the USEPA's 1999 storm water management regulations, construction activities that disturb one acre or more are required to have NPDES storm water permits (see discussion of Urban Nonpoint Source Pollution).

Controlling nonpoint pollution at construction sites involves careful planning prior to, during, and after the construction process. Prior to construction the site should be evaluated to identify natural features that will affect drainage and soil erosion. An erosion control plan should be developed to identify specific mitigation techniques that will be used during construction, and long-term storm water and erosion control features that will become part of the final constructed site. Environmentally sensitive areas, including steep slopes, critical wildlife habitat, and natural waterways such as wetlands, ponds, and intermittent streams, must be identified, and clearing in those areas must follow local construction and erosion control ordinances. Construction traffic, particularly heavy equipment, should be kept away from the root zone of trees and other vegetation that you want to preserve. If the site will have a septic system, construction traffic should be directed away from the absorption field to avoid soil compaction.

The next step in reducing sediment pollution from construction sites is to install erosion control devices prior to clearing the site. There are many possible techniques for controlling erosion and reducing sediment transport during construction. One approach is to cover the soil using biodegradable mulch, plastic mesh, reseeded vegetation (or existing vegetation), or other materials that prevent precipitation from landing on bare soil. Sediments can also be trapped on-site using sediment traps, catch basins, vegetation filter strips, silt fences, straw bales, or gravel roadways. Storm water can be deflected away from the site or channeled through the site in such a way that sediment erosion is minimized. Even with erosion control devices in place, runoff from construction sites usually contains higher concentrations of suspended sediments than runoff from established sites, so it is important to protect existing storm drains by placing filter bags, straw bales, silt fences, or other filtering devices around nearby storm drain inlets.

The final step in controlling nonpoint pollution from construction sites is to revegetate the site. This should be done as quickly as possible, and may involve several steps if the site is to be landscaped. Professional landscaping often requires bringing in additional topsoil, grading the landscaped area, as well as seeding and planting, fertilizing, and mulching. If the landscaping process is delayed, a temporary soil cover should be used to protect the valuable topsoil and prevent erosion and sediment pollution.

Urban Nonpoint Source Pollution

In 1983, the Environmental Protection Agency published the results from the Nationwide Urban Runoff Program (NURP), a comprehensive study of the water quality in storm water runoff at 81 sites in the United States (USEPA 1983). The NURP study revealed that storm water runoff from urban areas contained higher concentrations of many different kinds of pollutants, especially metals, nutrients (nitrogen and phosphorus), oxygen-demanding substances, and suspended solids. Subsequent studies have added pathogenic microorganisms, oil and petroleum hydrocarbons, pesticides, and a variety of synthetic organics to the list of urban storm water pollutants. Table 11-5 lists the major categories of urban pollutants and typical watershed sources for each type of pollutant.

Although urban runoff has long been recognized as a major source of pollution, regulatory control of urban nonpoint sources of pollution has lagged behind point source regulations. During the past decade, however, many countries have begun major efforts to reduce or eliminate pollution in storm water runoff from urban areas. In 1990 the USEPA introduced new rules that direct cities with populations of at least 100,000 to begin managing storm water runoff as point source discharges that require NPDES discharge permits. In December 1999, this requirement was extended to include cities with populations of at least 10,000.

The current storm water NPDES permits have six required control measures: *public education and outreach*, and *public participation/involvement* to inform citizens about the sources of storm water pollution and encourage citizen participation in developing and implementing pollution control measures; *illicit discharge detection and elimination* to identify and correct illegal discharges; *construction site runoff control* and *post-construction runoff control* to reduce soil erosion and other types of pollution associated with construction activities; and *pollution prevention/good housekeeping* to reduce pollution in runoff from municipal operations.

Although we often think of pollution abatement as requiring a technological approach, pollutants in storm runoff can also be reduced through planning and

Table 11-5. Sources of Contaminants in Urban Storm water Runoff[a]

Contaminant	Typical sources
Sediment/floatables	Streets, lawns, driveways, roads, construction activities, atmospheric deposition, drainage channel erosion
Pesticides/herbicides	Residential lawns and gardens, roadsides, utility right-of-ways, commercial and industrial landscaped areas, soil wash-off
Organic materials	Residential lawns and gardens, commercial landscaping, animal wastes
Metals	Automobiles, bridges, atmospheric deposition, industrial areas, soil erosion, corroding metal surfaces, combustion processes
Oil/grease/hydrocarbons	Roads, driveways, parking lots, vehicle maintenance areas, gas stations, illicit dumping to storm drains
Microbial pathogens	Lawns, roads, leaky sanitary sewer lines, sanitary sewer cross-connections, animal waste, septic systems
Nutrients (N and P)	Lawn fertilizer, atmospheric deposition, automobile exhaust, soil erosion, animal waste, detergents

[a] From "Preliminary Data Summary of Urban Storm Water Best Management Practices," EPA-821-R-99-012, USEPA, Washington, DC, August 1999.

regulatory efforts, by enforcement of existing construction regulations, and through public education. Street litter, for example, may be reduced by passage and enforcement of anti-littering regulation, public education about the polluting effects of litter, installation of litter collection devices, or street sweeping or vacuuming. Transportation residues like oil, gas, and grease from cars, and particulates from deteriorating road surfaces can be reduced by selecting road surfaces less susceptible to deterioration, instituting automobile exhaust inspection programs, educating the public about the pollution control benefits of keeping automobiles tuned up and operating properly, or using clean-up technologies such as oil and grease separators and sedimentation basins.

There are three major steps that can be taken to mitigate urban nonpoint source pollution: first, *reduce surface runoff* from urban sites; second, use *source control* to reduce the amount of pollution picked up by runoff; and finally, use appropriate technology (*best management practices* or BMPs) to remove or treat pollutants in the runoff. Urban areas have a higher percentage of impervious surfaces such as roads, sidewalks, driveways, and parking lots, and often have more disturbed soil compared to nonurban watersheds. Because of this, a greater percentage of precipitation falling in urban areas becomes surface runoff. Pollution control techniques that reduce the amount of surface runoff or increase infiltration rates (e.g., infiltration trenches, rain barrels, dry wells, porous driveways, vegetative covers) can be effective at reducing erosion and pollution transport.

246   ENVIRONMENTAL ENGINEERING

In addition to reducing surface runoff, we can emphasize source control to reduce the amount of pollutants deposited on impervious surfaces. Source control can be attained by increasing public education, providing waste disposal sites (e.g., pet waste disposal stations, hazardous waste disposal sites), introducing appropriate planning and regulatory policies, and enforcing pollution control regulations. Street sweeping, the oldest and least expensive source control technique, is still used in most cities. Street sweeping reduces sediment loading in runoff but fails to pick up the finer particles that are often the most significant source of pollution (fine silts and clays carry a disproportionately large amount of biodegradable and toxic substances, metals, and nutrients compared to coarser sediments). Street vacuuming is more efficient in collecting the small particles, but is more expensive and often ineffective during wet periods. Street flushing is an effective way to clean street surfaces, but only if the runoff is retained in catch basins, and the basins are cleaned periodically to remove refuse and other solids.

The final step in mitigating urban nonpoint source pollution is to use appropriate structural and nonstructural BMPs to reduce the concentration of pollutants in storm runoff (Table 11-6).

Table 11-6. Structural and Nonstructural Storm Water BMPs[a]

Structural BMPs	Examples
Infiltration systems	Infiltration basins, porous pavement, infiltration trenches, infiltration wells
Detention systems	Detention basins, underground vaults, pipes, and tanks
Retention systems	Wet ponds, retention tanks, tunnels, vaults, and pipes
Constructed wetlands	Artificial wetlands, wetland basins, wetland channels
Filtration systems	Sand filters, misc. media filters
Vegetated systems	Grass filter strips, vegetated swales
Nonstructural BMPs	
Automotive product and household hazardous material disposal	
Commercial and retail space good housekeeping	
Industrial good housekeeping	
Modified use of fertilizers, pesticides, and herbicides	
Lawn debris management	
Animal waste disposal	
Maintenance practices (e.g., catch basin cleaning)	
Illicit discharge detection and elimination	
Education and outreach programs	
Storm drain inlet stenciling	
Minimizing directly connected impervious surfaces	
Low-impact development and land use planning	

[a]From "Preliminary Data Summary of Urban Storm Water Best Management Practices," EPA-821-R-99-012, USEPA, Washington, DC, August 1999.

Infiltration systems reduce the total amount of surface runoff and the amount of sediment transport by increasing the amount of precipitation that infiltrates into the soil. *Detention systems* reduce pollutant transport by retarding the rate of runoff and by encouraging the settling of suspended solids. Most *retention systems* are wet ponds that are sized to hold a typical (e.g., 6-month) storm event. The ponds may be lined or unlined. The unlined ponds increase infiltration and can help with groundwater recharge; however, unlined ponds are not appropriate if the surface runoff contains pollutants that could contaminate a drinking water aquifer or in areas where a high water table would result in groundwater flowing into the pond. *Constructed wetlands* are similar to retention systems and wet ponds in that they have permanent standing water, but have the added advantage that they incorporate biotic functions such as nutrient uptake and microbial degradation of pollutants. *Filtration systems* use sand, soil, organic material, carbon, or other materials to filter out pollutants. Filter systems can be incorporated into existing storm drainage structures by adding underground vaults. *Biofilters* and other vegetated systems such as grassy swales and filter strips can be used to treat shallow flow or sheet flow by increasing infiltration and reducing sediment transport. Large *bioretention* systems add biological treatment such as nutrient uptake and microbial degradation of pollutants. There are also a large number of relatively new, commercially developed storm water treatment systems that incorporate one or more of the above BMPs. The USEPA's Environmental Technology Verification Program is currently evaluating the performance of a number of these commercial systems; their results are posted at http://www.epa.gov/etv/prgm.

Most of the structural BMPs are designed to remove sediments and pollutants that are transported in surface runoff along with sediments (e.g., phosphorus, pathogens, metals). Many of the nonstructural BMPs focus on reducing pesticides, hydrocarbons, commercial and industrial chemicals such as detergents and solvents, pet wastes, and fertilizers. Public education programs, along with appropriate enforcement of pollution control regulations, can be very effective at keeping these pollutants out of urban storm runoff.

Pollutant removal efficiencies vary widely (Table 11-7) depending on the treatment technology, site-specific considerations such as local weather patterns, maintenance of the system, and design constraints. In some instances, expensive treatment systems that offer no measurable reduction in storm water pollutant transport have been installed. The reasons for such failures are occasionally predictable. An undersized retention basin or wet pond may not have a long enough retention time to allow settling of suspended sediments. Poorly maintained systems may accumulate so much sediment that filters become clogged, sedimentation basins fill, and plants in artificial wetlands smother as their roots are covered with silt. Sometimes, however, a BMP may fail to perform for no apparent reason. Urban storm water treatment technology has grown rapidly in the past 20 years. Many of the systems that have been installed have not had follow-up testing to determine the level of effectiveness at that site or to develop local criteria for improved pollutant removal. This will undoubtedly be an active research area for environmental engineers and water quality professionals for the next decade.

Table 11-7. Typical Pollutant Removal Efficiencies for BMPs

	TSS	TP	TN	Metals	Pathogens
Catch basin	60–97	na	na	na	na
Catch basin + sand filter	70–90	na	30–40	50–90	na
Constructed wetlands	50–90	0–80	0–40	30–95	na
Grassed swale	20–40	20–40	10–30	10–60	na
Infiltration basin/trench	50–99	50–100	50–100	50–100	75–98
Porous pavement	60–90	60–90	60–90	60–90	na
Sand filters	60–90	0–80	20–40	40–80	40
Vegetated filter strip	40–90	30–80	20–60	20–80	na
Wet pond	50–90	20–90	10–90	10–95	na

From EPA 1993.

CONCLUSION

Controlling nonpoint source pollution in our waterways is currently one of the biggest challenges facing engineers, regulators, and scientists. Many nonpoint pollutants originate from common, everyday human activities, like driving a car, putting in a new rose bed, walking the dog, or building a house. Because of its widespread nature, nonpoint source pollution is difficult to contain, even harder to eliminate, and costly to mitigate. Source control through public education, community planning, and regulatory guidelines can be very effective, but often requires substantial changes in human behavior. Technological approaches, such as storm water BMPs, can help reduce nonpoint source pollution, but rarely if ever eliminate it entirely. Add to this, it is difficult to predict whether a pollution reduction effort will be successful because of the large variability in performances seen in the existing storm water treatment systems. Given the rapid changes occurring in nonpoint source pollution treatment technology, it is important to continue gathering information on what works, what does not work, and what factors contribute to successful nonpoint source pollution reduction.

PROBLEMS

11.1 Use Table 11-1 compare the relative concentrations of nonpoint source pollutants coming from construction sites vs those from urban runoff. How would you rank these same pollutants in terms of harm to the environment? Compare and discuss both rankings.

11.2 Contact the Soil Conservation Service agent in your area and calculate the soil loss from 500 acres of cropland using the USLE. Assume:

—conventional tillage
—cornstalks remain in field after harvest

—farmer uses contour stripped planting
—soil is Fayette silt loam
—slope = 7%
—slope length = 200 ft.

11.3 On-site wastewater treatment and disposal systems are criticized for creating nonpoint source water pollution. Describe the pollutants that would be generated from on-site wastewater treatment and identify BMPs (or other approaches) that could be used to control the pollutants.

11.4 Discuss the wastewater treatment technologies in Chap. 9 that are particularly applicable to help control pollution from urban storm water runoff.

11.5 Calculate the cost of controlling runoff from the construction of a 5-mile highway link across rolling countryside near the city or town where you live. Assume that hay bales are sufficient along the construction site if they are coupled with burlap barriers at key locations. No major collection or treatment works are required. Document your assumptions, including labor and materials charges, as well as time commitments.

11.6 Design a water quality monitoring plan to assess the effectiveness of a wet pond constructed to treat urban runoff. Assume that the pond was sized to contain a 6-month storm event, the pond discharges into a lake that is the drinking water source for a city of 100,000, and the watershed draining into the wet pond is predominantly residential. Describe the pollutants you would measure, the frequency of monitoring, and the hydrologic conditions that you would select for sampling.

Optional: contact a local water testing laboratory and determine the approximate cost for your monitoring plan, including labor and transportation costs, sample analysis fees, and report production costs.

Chapter 12

Solid Waste

Solid wastes other than hazardous and radioactive materials are considered in this chapter. Such solid wastes are often called *municipal solid waste* (MSW) and consist of all the solid and semisolid materials discarded by a community. The fraction of MSW produced in domestic households is called *refuse*. Until fairly recently, refuse was mostly food wastes, but new materials such as plastics and aluminum cans have been added to refuse, and the use of kitchen garbage grinders has decreased the food waste component. Most of the 2000 new products created each year by American industry eventually find their way into MSW and contribute to individual disposal problems.

The components of refuse are *garbage* or food wastes; *rubbish*, including glass, tin cans, and paper; and *trash*, including larger items like tree limbs, old appliances, and pallets that are not usually deposited in garbage cans.

The relationship between solid waste and human disease is intuitively obvious but difficult to prove. If a rat is sustained by an open dump, and that rat sustains a flea that transmits murine typhus to a human, the absolute proof of the pathway would require finding the particular rat and flea, an obviously impossible task. Nonetheless, we have observed more than 20 human diseases associated with solid waste disposal sites, and there is little doubt that improper solid waste disposal is a health hazard.

Disease vectors are the means by which disease organisms are transmitted, and water, air, and food may all be vectors. The two most important disease vectors related to solid wastes are rats and flies. Flies are such prolific breeders that 70,000 flies can be produced in 1 ft^3 of garbage, and carry many diseases like bacillary dysentery. Rats not only destroy property and infect by direct bite, but carry insects like fleas and ticks that may also act as vectors. The plagues of the Middle Ages were directly associated with the rat populations.

Public health is also threatened by infiltration of leachate from MSW disposal into groundwater, and particularly into drinking water supplies. Leachate is formed when rainwater collects in landfills, pits, waste ponds, or waste lagoons, and stays in contact with waste material long enough to leach out and dissolve some of its chemical and biochemical constituents. Leachate may be a major groundwater and surface water contaminant, particularly where there is heavy rainfall and rapid percolation through the soil.

In this chapter, the quantities and composition of MSW are discussed, and a brief introduction is given to disposal options and the specific problems of litter. Disposal is

252   ENVIRONMENTAL ENGINEERING

discussed further in Chap. 13, and Chap. 14 is devoted to the problems and promises of recovery of energy and materials from refuse.

QUANTITIES AND CHARACTERISTICS OF MUNICIPAL SOLID WASTE

The quantities of MSW generated in a community may be estimated by one of three techniques: input analysis, secondary data analysis, or output analysis. Input analysis estimates MSW based on use of a number of products. For example, if 100,000 cans of beer are sold each week in a particular community, the MSW, including litter, can be expected to include 100,000 aluminum cans per week. The estimation technique is highly inaccurate except in small and isolated communities.

Secondary data may be used to estimate solid waste production by some empirical relationship. For example, one study (Shell and Shure 1972) concluded that solid waste generation could be predicted as

$$W = 0.01795S - 0.00376F - 0.00322D + 0.0071P - 0.0002I + 44.7, \quad (12.1)$$

where

W = waste generated (tons),
S = number of stops made by the MSW pickup truck,
F = number of families served,
D = number of single family dwellings,
P = population, and
I = adjusted income per dwelling unit (dollars).

Models like this one are inherently inaccurate and may have no general application.

When possible, solid waste generation should be measured by output analysis, that is, by weighing the refuse dumped at the disposal site, either with truck scales or with portable wheeled scales. Refuse must generally be weighed in any case, because fees for use of the dump (called *tipping fees*) depend on weight dumped. Daily weight of refuse varies with the day of the week and the week of the year. Weather conditions also affect refuse weight, since moisture content can vary between 15 and 30%. If every truckload cannot be weighed, statistical methods must be used to estimate the total quantity from sample truckload weights.

Characteristics of Municipal Solid Waste

Refuse management depends on both the characteristics of the site and the characteristics of the MSW itself: gross composition, moisture content, particle size, chemical composition, and density.

Gross composition may be the most important characteristic affecting MSW disposal, or the recovery of materials and energy from refuse. Composition varies from one community to another, as well as with time in any one community. Refuse composition

Table 12-1. Average Annual Composition of MSW in the United States

	As generated		As disposed	
	Millions of tons	%	Millions of tons	%
Paper	37.2	36.7	44.9	41.5
Glass	13.3	13.1	13.5	12.5
Metal				
Ferrous	8.8	8.7	8.8	8.1
Aluminum	0.9	0.9	0.9	0.8
Other nonferrous	0.4	0.4	0.4	0.4
Plastics	6.4	6.3	6.4	5.9
Rubber and leather	2.6	2.6	3.4	3.1
Textiles	2.1	2.1	2.2	2.0
Wood	4.9	4.8	4.9	4.5
Food waste	22.8	22.5	20.0	18.5
Miscellaneous	1.9	1.9	2.8	2.6
Total	101.3		108.2	

is expressed either "as generated" or "as disposed," since moisture transfer takes place during the disposal process and thereby changes the weights of the various fractions of refuse. Table 12-1 shows typical components of average U.S. refuse. The numbers in Table 12-1 are useful only as guidelines; each community has characteristics that influence its solid waste production and composition.

The moisture content of MSW may vary between 15 and 30%, and is usually about 20%. Moisture is measured by drying a sample at 77°C (170°F) for 24 h, weighing, and calculating as

$$M = \frac{w - d}{w} \times 100, \qquad (12.2)$$

where

M = moisture content, percent,
w = initial, wet weight of sample, and
d = final, dry weight of sample.

Particle size distribution is particularly important in refuse processing for resource recovery, and is discussed further in Chap. 14.

The chemical composition of typical refuse is shown in Table 12-2. The use of both proximate and ultimate analysis in the combustion of MSW and its various fractions is discussed further in Chaps. 13 and 14. The density of MSW varies depending upon location, season, humidity, and so on. Table 12-3 shows some typical MSW densities.

Table 12-2. Proximate and Ultimate Chemical Analysis of MSW

	Proximate analysis (%)	Ultimate analysis (%)
Moisture	15–35	15–35
Volatile matter	50–60	
Fixed carbon	3–9	
Noncombustibles	15–25	
Higher heat value	3000–6000 Btu/lb	
Carbon		15–30
Hydrogen		2–5
Oxygen		12–24
Nitrogen		0.2–1.0
Sulfur		0.02–0.1

Source. U.S. Department of Health, Education, and Welfare, *Incinerator Guidelines*, 1969.

Table 12-3. Refuse Densities

	kg/m^3	lb/yd^3
Loose refuse	60–120	100–200
Dumped refuse from a collection vehicle	200–240	350–400
Refuse in a collection vehicle	300–400	500–700
Refuse in a landfill	300–540	500–900
Baled refuse	470–700	800–1200

COLLECTION

In the United States, and in most other industrialized countries, solid waste is collected by trucks. These are usually packers, trucks that carry hydraulic rams to compact the refuse to reduce its volume and can thus carry larger loads (Fig. 12-1). Collections are facilitated by the use of containers that are emptied into the truck with a mechanical or hydraulic mechanism. Commercial and industrial containers, "dumpsters," either are emptied into the truck or are carried by truck to the disposal site (Fig. 12-2). Collection is an expensive part of waste management, and many new devices and methods have been proposed in order to cut costs.

Garbage grinders reduce the amount of garbage in refuse. If all homes had garbage grinders, the frequency of collection could be decreased. Garbage grinders are so ubiquitous that in most communities garbage collection is needed only once a week. Garbage grinders put an extra load on the wastewater treatment plant, but sewage is relatively dilute and ground garbage can be accommodated easily both in sewers and in treatment plants.

Pneumatic pipes have been installed in some small communities, mostly in Sweden and Japan. The refuse is ground at the residence and sucked through underground lines. Walt Disney World in Florida also has a pneumatic pipe system in which the collection

Solid Waste 255

Figure 12-1. Packer truck used for residential refuse collection.

stations scattered throughout the park receive the refuse and pneumatic pipes deliver the waste to a central processing plant (Fig. 12-3). There are no garbage trucks in the Magic Kingdom.

Kitchen garbage compactors can reduce collection and MSW disposal costs and thus reduce local taxes, but only if every household has one. A compactor costs about as much as other large kitchen appliances, but uses special high-strength bags, so that the operating cost is also a consideration. At present they are beyond the means of many households. Stationary compactors for commercial establishments and apartment houses, however, have already had significant influence on collection practices.

Transfer stations are part of many urban refuse collection systems. A typical system, as shown in Fig. 12-4, includes several stations, located at various points in a city, to which collection trucks bring the refuse. The drive to each transfer station is relatively short, so that workers spend more time collecting and less time traveling. At the transfer station, bulldozers pack the refuse into large containers that are trucked to the landfill or other disposal facility. Alternatively, the refuse may also be baled before disposal.

Cans on wheels, often provided by the community, are widely used for transfer of refuse from the household to the collection truck. As shown in Fig. 12-5, the cans are pushed to the curb by the householder and emptied into the truck by a hydraulic lift. This system saves money and has reduced occupational injuries dramatically. Garbage collection workers suffer higher lost-time accident rates than other municipal or industrial workers.

Route optimization may result in significant cost saving as well as increased effectiveness. Software is available for selecting least-cost routes and collection frequencies. Route optimization is not new. It was first addressed by the mathematician Leonard Euler in 1736. He was asked to design a parade route for the city of Königsberg in East Prussia (now Kaliningrad in Russia) in such a way that the parade would not

Figure 12-2. Containerized collection system. (Courtesy of Dempster Systems.)

cross any bridge over the River Pregel more than once (Fig. 12-6). Euler showed that such a route was not possible, and, in a further generalization, that in order to arrive back at the starting point by such an *Euler's tour*, an even number of *nodes* had to be connected by an even number of *links*. The objective of garbage collection truck routing is to create a Euler's tour and thereby eliminate *deadheading*, or retracing a link without additional collection.

Although sophisticated routing programs are available, it is often just as easy to develop a route by common sense or *heuristic* means. Some heuristic rules for routing

Solid Waste 257

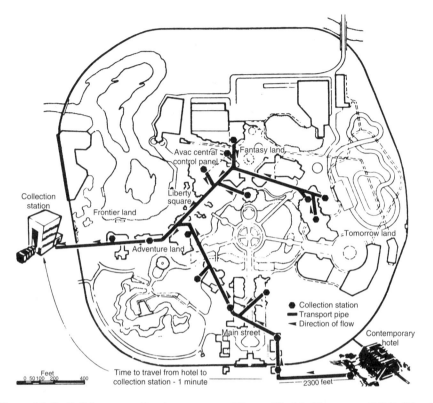

Figure 12-3. Solid waste collection system at Disney World. (Courtesy of AVAC Inc.)

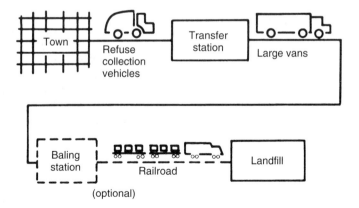

Figure 12-4. Transfer station method of solid waste collection.

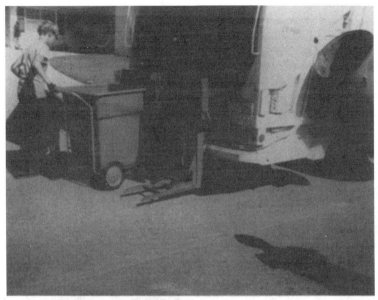

Figure 12-5. The "green can" system of solid waste collection.

trucks are (Liebman *et al.* 1975, Shuster and Schur 1974):

- Routes should not overlap.
- Routes should be compact and not fragmented.
- The starting point of the route should be as close to the truck garage as possible.

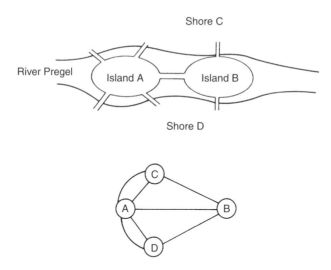

Figure 12-6. The seven bridges of Königsberg; the Euler routing problem.

- Heavily traveled streets should be avoided during rush hours.
- One-way streets that cannot be traversed in one line should be looped from the upper end of the street.
- Dead-end streets should be collected when the truck is on the right side of the street.
- Collection should be downhill on hills, so the truck can coast.
- Long straight paths should be routed before looping clockwise.
- For certain block patterns, standard paths, as shown in Fig. 12-7, should be used.
- U-turns should be avoided.

Figure 12-7 shows three examples of heuristic routing. In the first two, each side of the street is to be collected separately; in the third example, both sides of the street are collected at once.

DISPOSAL OPTIONS

Ever since the Romans invented city dumps, municipal refuse has been disposed of outside the city walls. As cities and suburbs grew, and metropolitan areas grew contiguous, and as the use of "throwaway" packages and containers increased, finding a place for MSW disposal became a critical problem. Many cities in the United States encouraged "backyard burning" of trash, in order to reduce MSW volume and disposal cost. Building codes in many cities mandate the installation of garbage grinders in new homes. Cities like Miami, FL, which has no landfill sites at all, built MSW incinerators.

Increasing urban air pollution has resulted in prohibition of backyard burning, even of leaves and grass clippings, and de-emphasis of municipal incineration. Increased

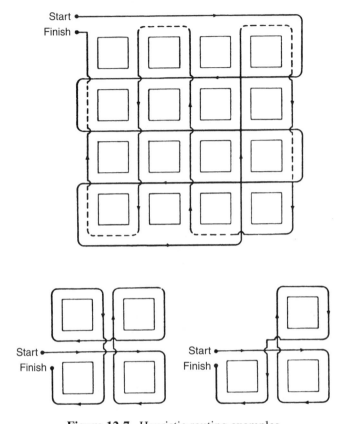

Figure 12-7. Heuristic routing examples.

residential development of land that was once forested or agricultural and changes in forest management practices have resulted in increases in forest and grass fires, and ultimately have led to a complete prohibition of backyard burning in almost all communities. Spontaneous dump fires and the spread of disease from dumps led to the prohibition of open dumps after 1980, in conformance with the Resource Conservation and Recovery Act (RCRA) of 1976. The sanitary landfill has become the most common method of disposal, because it is reasonably inexpensive and is considered relatively environmentally sound.

Unfortunately, landfilling is not the ultimate solution to the solid waste disposal problem. Although modern landfills are constructed so as to minimize adverse effects on the environment, experience has shown that they are not fail-safe. Moreover, the cost of landfilling is increasing rapidly, as land becomes scarce and refuse must be transported further and further from where it is generated. Rising public "environmental consciousness" is making waste processing and reclamation of waste material and energy appear increasingly attractive. Options for resource recovery are discussed further in Chap. 14.

LITTER

Litter is unsightly, a breeding ground for rats and other rodents, and hazardous to wildlife. Deer and fish, attracted to aluminum can pop-tops, ingest them and die in agony. Plastic sandwich bags are mistaken for jellyfish by tortoises, and birds strangle themselves in the plastic rings from six-packs.

Anti-litter campaigns and attempts to increase public awareness have been ongoing for many years. Bottle manufacturers and bottlers encourage voluntary bottle return. The popularity of "Adopt-a-road" programs has also sharply increased littering awareness, and has the potential to reduce roadside litter.

Restrictive beverage container legislation is a more drastic assault on litter. The Oregon "Bottle Law" prohibits pop-top cans and discourages the use of nonreturnable glass beverage bottles. The law operates by placing an artificial deposit value on all carbonated beverage containers so that it is in the user's interest to bring them back to the retailer for a deposit. The retailer in turn must recover the money from the manufacturer and sends all of the bottles back to the bottling company. The bottling company must now either discard these bottles, send them back to the bottle manufacturer, or refill them. In any case, it becomes more efficient for the manufacturer to either refill or recover the bottles rather than to throw them away. The beverage industry is thus forced to rely more heavily on returnable containers, reducing the one-way containers such as steel cans or plastic bottles. Such a process saves money, materials, and energy, and has the added effect of reducing litter.

CONCLUSION

The solid waste problem has three facets: source, collection, and disposal. The first is perhaps the most difficult. A "new economy" of reduced waste, increased longevity instead of planned obsolescence, and thriftier use of natural resources is needed. Collection and disposal of refuse are discussed in the next chapter.

PROBLEMS

12.1 Walk along a stretch of road and collect the litter in two bags, one for beverage containers only and one of everything else. Calculate: (a) the number of items per mile, (b) the number of beverage containers per mile, (c) weight of litter per mile, (d) weight of beverage containers per mile, (e) percent of beverage containers by weight, and (f) percent of beverage containers by count. If you are working for the bottle manufacturers, would you report your data as (e) or (f)? Why?

12.2 How would a tax on natural resource withdrawal affect the economy of solid waste management?

12.3 What effect do the following have on the quantity and composition of MSW: (a) garbage grinders, (b) home compactors, (c) nonreturnable beverage containers, and (d) a newspaper strike? Make quantitative estimates of the effects.

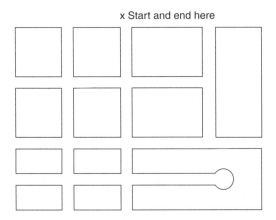

Figure 12-8. Route for Problem 12.7.

12.4 Drive along a measured stretch of road or highway and count the pieces of litter visible from the car. (Do this with one person driving and another counting!) Then walk along the same stretch and pick up the litter, counting the pieces and weighing the full bags. What percent of the litter by piece (and by weight if you have enough information) is visible from the car?

12.5 On a map of your campus or your neighborhood, develop an efficient route for refuse collection, assuming that every blockface must be collected.

12.6 Using a study hall, lecture hall, or student lounge as a laboratory, study the prevalence of litter by counting the items in the waste receptacles vs the items improperly disposed of. Vary the conditions of your laboratory in the following way (you may need cooperation from the maintenance crew):

- Day 1: normal conditions (baseline)
- Day 2: remove all waste receptacles except one
- Day 3: add additional receptacles (more than normal).

If possible, do several experiments with different numbers of receptacles. Plot the percent of material properly disposed of vs the number of receptacles, and discuss the implications.

12.7 Using heuristic routing, develop an efficient route for the map shown in Fig. 12-8 if (a) both sides of the street are to be collected together or (b) one side of the street is collected at a time.

Chapter 13

Solid Waste Disposal

Disposal of solid wastes is defined as placement of the waste so that it no longer impacts society or the environment. The wastes are either assimilated so that they can no longer be identified in the environment, as by incineration to ash, or they are hidden well enough so that they cannot be readily found. Solid waste may also be processed so that some of its components may be recovered, and used again for a beneficial purpose. Collection, disposal, and recovery are all part of the total solid waste management system, and this chapter is devoted to disposal.

DISPOSAL OF UNPROCESSED REFUSE IN SANITARY LANDFILLS

The only two realistic options for disposal are in the oceans and on land. Because the environmental damage done by ocean disposal is now understood, the United States prohibits such disposal by federal law, and many developed nations are following suit. This chapter is therefore devoted to a discussion of land disposal.

Until the mid-1970s, a solid waste disposal facilities was usually a *dump* in the United States and a *tip* (as in "tipping") in Great Britain. The operation of a dump was simple and inexpensive: trucks were simply directed to empty loads at the proper spot on the dump site. The piled-up volume was often reduced by setting the refuse on fire, thereby prolonging the life of the dump. Rodents, odor, insects, air pollution, and the dangers posed by open fires all became recognized as serious public health and aesthetic problems, and an alternative method of refuse disposal was sought. Larger communities frequently selected incineration as the alternative, but smaller towns could not afford the capital investment required and opted for land disposal.

The term *sanitary landfill* was first used for the method of disposal employed in the burial of waste ammunition and other material after World War II, and the concept of burying refuse was used by several Midwestern communities. The sanitary landfill differs markedly from open dumps: open dumps are simply places to deposit wastes, but sanitary landfills are engineered operations, designed and operated according to acceptable standards (Fig. 13-1).

Sanitary landfilling is the compaction of refuse in a lined pit and covering of the compacted refuse with an earthen cover. Typically, refuse is unloaded, compacted with bulldozers, and covered with compacted soil. The landfill is built up in units called *cells* (Fig. 13-2). The daily cover is between 6 and 12 in. thick depending on

264   ENVIRONMENTAL ENGINEERING

Figure 13-1. The sanitary landfill.

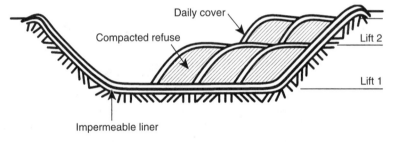

Figure 13-2. Arrangement of cells in an area-method landfill.

soil composition (Fig. 13-3), and a final cover at least 2 ft thick is used to close the landfill. A landfill continues to subside after closure, so that permanent structures cannot be built on-site without special foundations. Closed landfills have potential uses as golf courses, playgrounds, tennis courts, winter recreation, or parks and greenbelts. The sanitary landfilling operation involves numerous stages, including siting, design, operation, and closing.

Siting Landfills

Siting of landfills is rapidly becoming the most difficult stage of the process since few people wish to have landfills in their neighborhoods. In addition to public acceptability, considerations include:

- *Drainage*: Rapid runoff will lessen mosquito problems, but proximity to streams or well supplies may result in water pollution.

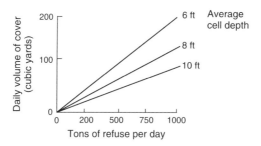

Figure 13-3. Daily volume of cover versus refuse disposal rate.

- *Wind*: It is preferable that the landfill be downwind from any nearby community.
- *Distance from collection*.
- *Size*: A small site with limited capacity is generally not acceptable since finding a new site entails considerable difficulty.
- *Rainfall patterns*: The production of leachate from the landfill is influenced by the weather.
- *Soil type*: Can the soil be excavated and used as cover?
- *Depth of the water table*: The bottom of the landfill must be substantially above the highest expected groundwater elevation.
- *Treatment of leachate*: The landfill must be proximate to wastewater treatment facilities.
- *Proximity to airports*: All landfills attract birds to some extent, and are therefore not compatible with airport siting.
- *Ultimate use*: Can the area be used for private or public use after the landfilling operation is complete?

Although daily cover helps to limit disease vectors, a working landfill still has a marked and widespread odor during the working day. The working face of the landfill must remain uncovered while refuse is added and compacted. Wind can pick material up from the working face, and the open refuse attracts feeding flocks of birds. These birds are both a nuisance and a hazard to low-flying aircraft using nearby airports. Odor from the working face and the truck traffic to and from the landfill make a sanitary landfill an undesirable neighbor to nearby communities.

Early sanitary landfills were often indistinguishable from dumps, thereby enhancing the "bad neighbor" image. In recent years, as more landfills have been operated properly, it has even been possible to enhance property values with a closed landfill site, since such a site must remain open space. Acceptable operation and eventual enhancement of the property are understandably difficult to explain to a community.

Design of Landfills

Modern landfills are designed facilities, much like water or wastewater treatment plants. The landfill design must include methods for the recovery and treatment of the

266   ENVIRONMENTAL ENGINEERING

leachate produced by the decomposing refuse, and the venting or use of the landfill gas. Full plans for landfill operation must be approved by the appropriate state governmental agencies before construction can begin.

Since landfills are generally in pits, the soil characteristics are of importance. Areas with high groundwater would not be acceptable, as would high bedrock formations. The management of rainwater during landfilling operations as well as when the landfill is closed must be part of the design.

Operation of Landfills

The landfill operation is actually a biological method of waste treatment. Municipal refuse deposited as a fill is anything but inert. In the absence of oxygen, anaerobic decomposition steadily degrades the organic material to more stable forms. This process is very slow and may still be going on as long as 25 years after the landfill closes.

The liquid produced during decomposition, as well as water that seeps through the groundcover and works its way out of the refuse, is known as *leachate*. This liquid, though relatively small in volume, contains pollutants in high concentration. Table 13-1 shows typical leachate composition. Should leachate escape the landfill, its effects on the environment may be severe. In a number of instances, leachate has polluted nearby wells to a degree that they ceased to be sources of potable water.

The amount of leachate produced by a landfill is difficult to predict. The only available method is water balance: the water entering a landfill must equal the water flowing out of the landfill, or leachate. The total water entering the top soil layer is

$$C = P(1 - R) - S - E, \tag{13.1}$$

where

C = total percolation into the top soil layer (mm),
P = precipitation (mm),

Table 13-1. Typical Sanitary Landfill Leachate Composition

Component	Typical value
BOD_5	20,000 mg/L
COD	30,000 mg/L
Ammonia nitrogen	500 mg/L
Chloride	2,000 mg/L
Total iron	500 mg/L
Zinc	50 mg/L
Lead	2 mg/L
Total polychlorinated biphenyl (PCB) residue	1.5 μg/L
pH	6.0

Solid Waste Disposal   267

Table 13-2. Percolation in Three Landfills[a]

Location	Precipitation, P (mm)	Runoff coefficient, R	Evapotranspiration, E (mm)	Percolation, C (mm)
Cincinnati	1025	0.15	568	213
Orlando	1342	0.07	1172	70
Los Angeles	378	0.12	334	0

[a]D. G. Tenn, K. J. Haney, and T. V. Degeare, *Use of the Water Balance Method for Predicting Leachate Generation from Solid Waste Disposal Sites* (U.S. Environmental Protection Agency, OSWMP, SW-168, Washington, DC 1975).

R = runoff coefficient,
S = storage (mm), and
E = evapotranspiration (mm).

The percolation for three typical landfills is shown in Table 13–2.

Using these figures it is possible to predict when landfills produce leachate. Clearly, Los Angeles landfills may virtually never produce leachate. Leaching through a 7.5-m (25-ft) deep landfill in Orlando, FL, might take 15 years, while a 20-m (65-ft) deep landfill in Cincinnati can produce leachate after only 11 years. Leachate production depends on rainfall patterns as well as on total amount of precipitation. The figures given for Cincinnati and Orlando are typical of the "summer thunderstorm" climate that exists in most of the United States. The Pacific Northwest (west of the Pacific Coast Range) has a maritime climate, in which rainfall is spread more evenly through the year. Seattle landfills produce leachate at approximately twice the rate of Cincinnati landfills, although the annual rainfall amount is approximately the same.

Gas is a second by-product of a landfill. Since landfills are anaerobic biological reactors, they produce CH_4 and CO_2. Gas production occurs in four distinct stages, as illustrated in Fig. 13-4. The first stage is aerobic and may last from a few days to several months, during which time aerobic organisms are active and affect the decomposition. As the organisms use up the available oxygen, the landfill enters the second stage, at which anaerobic decomposition begins, but at which methane-forming organisms have not yet become productive. During the second stage, the acid formers cause a buildup of CO_2. The length of this stage varies with environmental conditions. The third stage is the anaerobic methane production stage, during which the percentage of CH_4 progressively increases, as does the landfill interior temperature to about 55°C (130°F). The final, steady-state condition occurs when the fractions of CO_2 and CH_4 are about equal, and microbial activity has stabilized. The amount of methane produced from a landfill may be estimated using the semi-empirical relationship (Chian 1977)

$$CH_aO_bN_c + \left(\tfrac{1}{4}\right)(4 - a - 2b + 3c)H_2O$$
$$\rightarrow \tfrac{1}{8}[4 - a + 2b + 3c]CO_2 + (4 + a - 2b - 3c)CH_4. \quad (13.2)$$

Equation (13.2) is useful only if the chemical composition of the waste is known.

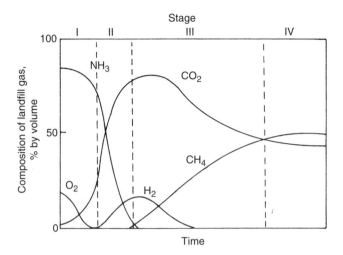

Figure 13-4. States in the decomposition of organic matter in landfills.

The rate of gas production from sanitary landfills may be controlled by varying the particle size of the refuse by shredding before placing the refuse in the landfill, and by changing the moisture content. Gas production may be minimized with the combination of low moisture, large particle size, and high density. Unwanted gas migration may be prevented by installing escape vents in the landfill. These vents, called "tiki torches," are kept lit and the gas is burned off as it is formed. Improper venting may lead to dangerous accumulation of methane. In 1986, a dozen homes near the Midway Landfill in Seattle were evacuated because potentially explosive quantities of methane had leaked through underground fissures into the basements. Venting of the accumulated gas, so that the occupants could return to their homes, took three weeks.

Since landfills produce considerable quantities of methane, landfill gas can be burned to produce electric power. Alternatively, the gas can be cleaned of CO_2 and other contaminants and used as pipeline gas. Such cleaning is both expensive and troublesome. The most reasonable use of landfill gas is to burn it as is in some industrial application like brickmaking.

Closure and Ultimate Use of Landfills

Municipal landfills must be closed according to state and federal regulations. Such closure includes the permanent control of leachate as well as gas, and the placement of an impermeable cap. The cost of closure is very high and must be incorporated in the tipping fee during the life of the landfill. This is one of the primary factors responsible for the dramatic increase in landfill tipping fees.

Solid Waste Disposal 269

Figure 13-5. A motel built on a landfill that experienced differential settling.

Biological aspects of landfills as well as the structural properties of compacted refuse limit the ultimate uses of landfills. Landfills settle unevenly, and it is generally suggested that nothing at all be constructed on a landfill for at least two years after closure, and that no large permanent structures ever be built. With poor initial compaction, about 50% settling can be expected in the first five years. The owners of the motel shown in Fig. 13-5 learned this the hard way.

Landfills should never be disturbed. Disturbance may cause structural problems, and trapped gases can present a hazard. Buildings constructed on landfills should have spread footings (large concrete slabs) as foundations, although some have been constructed on pilings that extend through the fill onto rock or some other strong material.

VOLUME REDUCTION BEFORE DISPOSAL

Refuse is bulky and does not compact easily, so that volume requirements of landfills are significant. Where land is expensive, the costs of landfilling may be high. Accordingly, various ways to reduce refuse volume have been found effective.

In the right circumstances, burning of refuse in waste-to-energy facilties (discussed in the next chapter) is an effective treatment of municipal solid waste. Burning reduces the volume of waste by a factor of 10 to 20, and the ash is both more stable and more compactable than the refuse itself.

Pyrolysis is combustion in the absence of oxygen. The residues of pyrolysis, combustible gas, tar, and charcoal, have economic value but have not yet found acceptance

270   ENVIRONMENTAL ENGINEERING

as a raw material. The tar contains water that must be removed; the charcoal is full of glass and metal that must be separated. These separations render the by-products too expensive to be competitive. Pyrolysis reduces the volume considerably, produces a stable end product, and has few air pollution problems. On a large scale, such as for some of our larger cities, pyrolysis as a method of volume reduction has significant advantages over incineration. Pyrolysis may also be used for sludge disposal, thus solving two major solid waste problems for a community. Such systems, however, remain to be proven in full-scale operation.

Another method of volume reduction is baling. Solid waste is compressed into desk-sized blocks that can then be handled with fork lifts and stacked in the landfill depression. Because of the high density of the refuse (on the order of 2000 lb/yd^3), the rate of decomposition is slow and odor is reduced. Baled refuse does not therefore require daily cover, further saving landfill space. Local and state regulations may, however, require baled refuse landfills to provide daily cover, which substantially reduces the cost advantages of baling.

CONCLUSION

This chapter begins by defining the objective of solid waste disposal as the placement of solid waste so that it no longer impacts society or the environment. At one time, this was fairly easy to achieve: dumping solid waste over city walls was quite adequate. In modern civilization, however, this is no longer possible, and adequate disposal is becoming increasingly difficult.

The disposal methods discussed in this chapter are only partial solutions to the solid waste problem. Another solution would be to redefine solid waste as a resource and use it to produce usable goods. This idea is explored in the next chapter.

PROBLEMS

13.1 Suppose that the municipal garbage collectors in a town of 10,000 go on strike, and as a gesture to the community, your college or university decides to accept all city refuse temporarily and pile it on the football field. If all the people dumped refuse into the stadium, how many days must the strike continue before the stadium is filled to 1 yard deep? Assume the density of the refuse as 300 lb/yd^3, and assume the dimensions of the stadium as 120 yards long and 100 yards wide.

13.2 If a town has a population of 100,000, what is the daily production of wastepaper?

13.3 What would be some environmental impacts and effects of depositing dewatered (but sloppy wet) sludge from a wastewater treatment plant into a sanitary landfill?

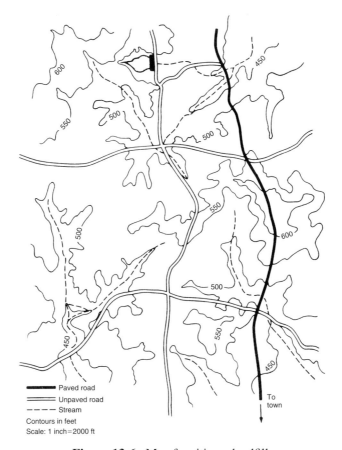

Figure 13-6. Map for siting a landfill.

13.4 If all of the readily decomposable organic refuse in a landfill were cellulose, $C_6H_{10}O_5$, how many cubic meters of gas (CO_2 and CH_4) would be produced per kilogram of refuse?

13.5 Choose a place for a 25-acre landfill on the map shown in Fig. 13-6. What other information would you need? Justify your selection of the site.

Chapter 14

Reuse, Recycling, and Resource Recovery

Finding new sources of energy and materials is becoming increasingly difficult. Concurrently, we are finding it more and more difficult to locate solid waste disposal sites, and the cost of disposal is escalating exponentially. As a result, society's interest in reuse, recycling, and recovery of materials from refuse has grown.

Reuse of materials involves either the voluntary continued use of a product for a purpose for which it may not have been originally intended, such as the reuse of coffee cans for holding nails, or the extended use of a product, such as retreading automobile tires. In materials reuse the product does not return to the industrial sector, but remains within the public or consumer sector.

Recycling is the collection of a product by the public and the return of this material to the industrial sector. This is very different from reuse, where the materials do not return for remanufacturing. Examples of recycling are the collection of newspapers and aluminum cans by individuals and their collection and eventual return to paper manufacturers or aluminum companies. The recycling process requires the participation of the public, since the public must perform the separation step.

Recovery differs from recycling in that the waste is collected as mixed refuse, and then the materials are removed by various processing steps. For example, refuse can be processed by running it under a magnet that is supposed to remove the steel cans and other ferrous materials. This material is then sold back to the ferrous metals industry for remanufacturing. Recovery of materials is commonly conducted in a *materials recovery facility* (MRF, pronounced "murph"). The difference between recycling and recovery is that in the latter the user of the product is not asked to do any separation, while in the former that crucial separation step is done voluntarily by a person who gains very little personal benefit from going to the trouble of separating out waste materials. Recycling and recovery, the two primary methods of returning waste materials to industry for remanufacturing and subsequent use, are discussed in more detail in the next section.

RECYCLING

Two incentives could be used to increase public participation in recycling. The first is regulatory, in that the government *dictates* that only separated material will be picked up. This type of approach has had only limited success in democracies like the United States, because dictation engenders public resentment.

A more democratic approach to achieve cooperation in recycling programs is to appeal to the sense of community and to growing concern about environmental quality. Householders usually respond very positively to surveys about prospective recycling programs, but the *active* response, or participation in source separation, has been less enthusiastic.

Participation can be increased by making source separation easy. The city of Seattle has virtually 100% participation in its household recycling program because the separate containers for paper, cans, and glass are provided, and the householder only needs to put the containers out on the curb. The city of Albuquerque sells, for ten cents each, large plastic bags to hold aluminum and plastic containers for recycling. The bags of recyclables, and bundled newspapers, are picked up at curbside along with garbage. Municipal initiatives like this are costly, however.

A major factor in the success or failure of recycling programs is the availability of a market for the pure materials. Recycling can be thought of as a chain, which can be pulled by the need for post-consumer materials, but which cannot be pushed by the collection of such materials by the public. A recycling program therefore includes, by necessity, a market for the materials collected; otherwise, the separated materials will end up in the landfill along with the mixed unseparated refuse.

In recent years there has been a strong indication that the public is willing to spend the time and effort to separate materials for subsequent recycling. What has been lacking has been the markets. How can these be created? Simply put, markets for recycled materials can be created by public demand. If the public insists, for example, on buying only newspapers that have been printed on recycled newsprint, then the newspapers will be forced in their own interest to use recycled newsprint, and this will drive up and stabilize the price of used newsprint.

Knowing this, and sensing the mood of the public, industry has been quick to produce products that are touted as being from "recycled this" and "recycled that." Most often, the term "recycled" is used incorrectly in such claims, since the material used has never been in the public sector. Paper, for example, has for years included fibers produced during the production of envelopes and other products. This wastepaper never enters the public sector, but is an industrial waste that gets immediately used by the same industry. This is not "recycling," and such products will not drive the markets for truly recycled materials. The public should become more knowledgeable about what are and are not legitimate recycled products, and the government may force industries to adopt standards for the use of such terms as "recycled."

RECOVERY

Most processes for separation of the various materials in refuse rely on a characteristic or property of the specific materials, and this characteristic is used to separate the material from the rest of the mixed refuse. Before such separation can be achieved,

Reuse, Recycling, and Resource Recovery 275

however, the material must be in separate and discrete pieces, a condition clearly not met by most components of mixed refuse. An ordinary "tin can" contains steel in its body, zinc on the seam, a paper wrapper on the outside, and perhaps an aluminum top. Other common items in refuse provide equally or more challenging problems in separation.

The separation process can be facilitated by decreasing the particle size of refuse, thus increasing the number of particles and achieving a greater number of "clean" particles. The size reduction step, although not strictly materials separation, is commonly a first step in a solid waste processing facility.

Size Reduction

Size reduction, or *shredding*, is brute force breaking of particles of refuse by swinging hammers in an enclosure. Two types of shredders are used in solid waste processing: the vertical and horizontal hammermills, as shown in Fig. 14-1. In vertical hammermills, the refuse enters the top and must work its way past the rapidly swinging hammers, clearing the space between the hammer tips and the enclosure. Particle size is controlled by adjusting this clearance. In the horizontal hammermill, the hammers swing over a grate that may be changed, depending on the size of product required.

The solid waste processing facility in Fig. 14-2 has a conveyor belt leading up to a vertical shredder, with a control room above and to the left. The hammers inside the shredder are shown in Fig. 14-3. As the hammers reduce the size of the refuse components, they are themselves worn down. Typically, a set of hammers such as those shown can process 20,000 to 30,000 tons of refuse before having to be replaced (Vesilind 1980).

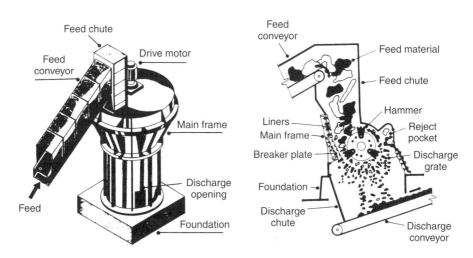

Figure 14-1. Vertical and horizontal hammermills.

Figure 14-2. A shredding facility showing the conveyor belt leading to a vertical shredder.

Shredder Performance and Design

The performance of a shredder is assessed by the degree of particle size reduction achieved and then compared with the cost in power and shredder wear. The particle size distribution of a nonhomogeneous material such as municipal solid waste (MSW) is expressed by the Rosin–Rammler distribution function (Rosin and Rammler 1933)

$$Y = 1 - e^{-(x/x_c)^n}, \qquad (14.1)$$

where

- Y = the cumulative fraction of particles (by weight) less than size x,
- n = a constant, and
- x_c = the *characteristic particle size*: the size at which 63.2% of the particles (by weight) are smaller ($1 - 1/e = 0.632$).

Reuse, Recycling, and Resource Recovery 277

Figure 14-3. Inside a vertical hammermill. The hammers have been worn down by the shredding process. (Photo courtesy of W. A. Worrell.)

Equation (14.1) may also be expressed as

$$\ln \frac{1}{1-Y} = \left(\frac{x}{x_c}\right)^n,$$

so that a plot like that of Fig. 14-4 may be made using log–log coordinates. Often, shredder performance is specified on the basis of "90% of the particles (by weight) passing a given size." The conversion from characteristic size to 90% passing is possible by recognizing that

$$x_c = \frac{x}{\left[\ln\left(\frac{1}{1-Y}\right)\right]^{\frac{1}{n}}}. \tag{14.2}$$

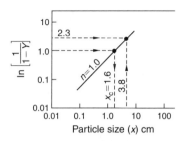

Figure 14-4. Rosin–Rammler particle size distribution equation. In this sample $n = 1.0$ and $x_c = 1.6$.

Table 14-1. Rosin–Rammler Exponents for Typical Shredder Installation[a]

Location	n	x_c (cm)
Washington, DC	0.689	2.77
Wilmington, DE	0.629	4.16
Charleston, SC	0.823	4.03
St. Louis, MO	0.995	1.61
Pompano Beach, FL	0.587	0.67

[a] From Stratton and Alter (1978).

If $Y = 0.90$, or 90%, Eq. (14.2) reduces to

$$x_c = \frac{x_{90}}{(2.3)^{1/n}}, \tag{14.3}$$

where x_{90} is the size at which 90% of the particles pass. Ordinarily, $n = 1.0$, in which case $x_{90} = 2.3 x_c$. Table 14-1 gives some typical values of x_c and n.

In addition to particle size reduction, shredder performance must be measured in terms of power use, expressed as specific energy (kilowatt hours/ton of refuse processed). Figure 14-5 illustrates how solids feed rate, shredder speed, moisture content, and size of particles in the feed are interrelated.

A semi-empirical equation often used to estimate the power requirements for shredders was developed by Bond (Bond 1952). The specific energy W required to reduce a unit weight of material 80% finer than some diameter L_F to a product 80% finer than some diameter L_P, where both L_F and L_P are in micrometers (μm), is expressed as

$$W = 10 W_i \left[\frac{1}{\sqrt{L_P}} - \frac{1}{\sqrt{L_F}} \right], \tag{14.4}$$

where W_i is the *Bond work index*, a factor that is a function of the material processed for a given shredder and a function of shredder efficiency for a given material. W has the dimension of kilowatt hours/ton (kwh/ton) if L_P and L_F are in micrometers. The factor 10 also corrects for the dimensions. Typical values of the Bond work index for MSW and other specific materials are given in Table 14-2.

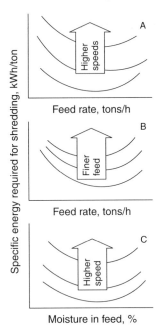

Figure 14-5. (A) Effect of feed rate and rotor speed on specific energy used in shredding. (B) Effect of feed rate and feed particle size on specific energy. (C) Correlation between moisture content in the refuse and motor speed.

Table 14-2. Typical Bond Work Indices[a]

Location	Material	Bond work index (kWh/ton)
Washington, DC	MSW	463
Wilmington, DE	MSW	451
Charleston, SC	MSW	400
St. Louis, MO	MSW	434
Pompano Beach, FL	MSW	405
Washington, DC	Paper	194
Washington, DC	Steel cans	262
Washington, DC	Aluminum cans	654
Washington, DC	Glass	8

[a] From Stratton and Alter (1978).

The Rosin–Rammler particle size distribution equation (Eq. (14.1)) may be combined with the Bond work index concept. Equation (14.1) may be rewritten

$$(1 - Y) = e^{-(x/x_c)^n}, \qquad (14.1)$$

where $(1 - Y)$ is the cumulative fraction *larger* than some size x. If

$$0.2 = \exp\left[\frac{L_P}{x_c}\right]^n,$$

then $x = L_P$, the screen size through which 80% of the product passes. Solving for L_P,

$$L_P = x_c(1.61)^{1/n}.$$

Substituting this expression back into Eq. (14.4),

$$W = \frac{10W_i}{\left[x_c(1.61)^{1/n}\right]^{1/2}} = \frac{10W_i}{[L_P]^{1/2}}. \quad (14.5)$$

EXAMPLE 14.1. Assuming $W_i = 400$ and x_c of the product is 1.62 cm, $n = 1$, and $L_F = 25$ cm (about 10 in.; a realistic estimate for raw refuse), find the expected power requirements for a shredder processing 10 tons/h.
From Eq. (14.5)

$$W_i = \frac{10(400)}{\left[(16,200)(1.61)^{1/1}\right]^{1/2}} = \frac{10(400)}{[(25,000)]^{1/2}} = 16.7 \text{ kWh/ton}.$$

(Note: x_c and L_F must be in μm.)
The power requirement is $(16.7 \text{ kWh/ton})(10 \text{ tons/h}) = 167 \text{ kW}$.

General Expressions for Material Recovery

In separating any one material from a mixture, the separation is termed *binary* because only two outputs are sought. When a device is to separate more than one material from a mixture, the process is called *polynary* separation.

Figure 14-6 shows a binary separator receiving a mixed feed of x_0 and y_0. The objective is separation of the x fraction: the first exit stream is to contain the x component, but the separation is not perfect and contains an amount of contamination y_1. This stream is called the *product* or *extract*, while the second stream, containing mostly y but also some x, is the *reject*. The percent of x recovered in the first output stream, $R_{(x_1)}$, may be expressed as

$$R_{(x_1)} = \left(\frac{x_1}{x_0}\right) 100. \quad (14.6)$$

$R_{(x_1)}$ alone does not describe the performance of the binary separator adequately. If the separator were turned *off* all of the feed would go to the first output, the extract would be $x_0 = x_1$, making $R_{(x_1)} = 100\%$. However, in this case there would have

Figure 14-6. Definition sketch of a binary separator.

been no separation. Accordingly, the *purity* of the extract stream as a percent may be defined as

$$P_{(x_1)} = \left(\frac{x_1}{x_1 + y_1}\right) 100. \tag{14.7}$$

A separator might extract only a small amount of pure x, so that the recovery $R_{(x_1)}$ would also be very small. The performance of a materials separator is assessed by both recovery and purity, and may thus be characterized by an additional parameter, the separator *efficiency* $E_{(x,y)}$, as (Worrell 1979)

$$E_{(x,y)} = \left(\frac{x_1}{x_0} \times \frac{y_2}{y_0}\right)^{1/2} \times 100. \tag{14.8}$$

EXAMPLE 14.2. A binary separator, a magnet, is to separate a product, ferrous materials, from a feed stream of shredded refuse. The feed rate to the magnet is 1000 kg/h, and contains 50 kg of ferrous materials. The product stream weighs 40 kg, of which 35 kg are ferrous materials. What is the percent recovery of ferrous materials, their purity, and the overall efficiency?

The variables in Eqs. (14.6), (14.7), and (14.8) are

$$x_0 = 50 \text{ kg} \qquad y_0 = 1000 - 50 = 950 \text{ kg}$$
$$x_1 = 35 \text{ kg} \qquad y_1 = 40 - 35 = 5 \text{ kg}$$
$$x_2 = 50 - 35 = 15 \text{ kg} \qquad y_2 = 950 - 5 = 945 \text{ kg}$$

$$R_{(x_1)} = \left(\frac{35}{50}\right) 100 = 70\%$$

$$P_{(x_1)} = \left(\frac{35}{35+5}\right) 100 = 88\%.$$

Then

$$E_{(x,y)} = \left(\frac{35}{50}\right)\left(\frac{945}{950}\right) 100 = 70\%.$$

Screens

Screens separate material solely by size and do not identify the material by any other property. Consequently, screens are most often used in materials recovery as a classification step before a materials separation process. For example, glass can be sorted (technically but perhaps not economically) into clear and colored fractions by optical coding. This process, however, requires that the glass be of a given size, and screens may be used for the necessary separation.

The *trommel*, shown in Fig. 14-7, is the most widely used screen in materials recovery. The charge inside the trommel behaves in three different ways, depending on the speed of rotation. At slow speeds, the trommel material is *cascading*: not being lifted but simply rolling back. At higher speed, *cataracting* occurs, in which centrifugal force carries the material up to the side and then it falls back. At even higher speeds, *centrifuging* occurs, in which material adheres to the inside

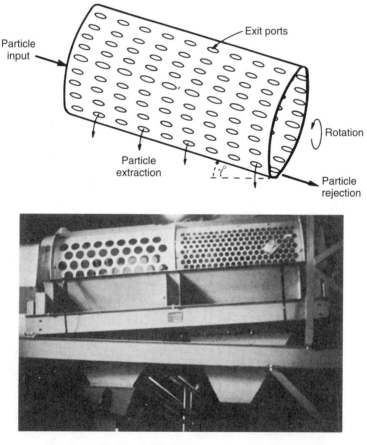

Figure 14-7. Trommel screen.

of the trommel. Obviously, the efficiency of a trommel is enhanced when the particles have the greatest opportunity to drop through the holes, and this occurs during cataracting. Trommel speed is often designed as some fraction of the critical speed, defined as that rotational speed at which the materials will just begin centrifuging, and calculated as

$$\eta_c = \sqrt{\frac{g}{4\pi^2 r}}, \quad (14.9)$$

where

η_c = critical speed (rotations/s),
g = gravitational acceleration (cm/s^2), and
r = radius of drum (cm).

EXAMPLE 14.3. Find the critical speed for a 3-m-diameter trommel:

$$\eta_c = \sqrt{\frac{980}{4\pi^2(150)}} = 0.407 \text{ rotations/s}.$$

Air Classifiers

Materials may be separated by their aerodynamic properties. In shredded MSW, most of the aerodynamically less dense materials are organic, and most of the denser materials are inorganic; thus air classification can produce a refuse-derived fuel (RDF) superior to unclassified shredded refuse.

Most air classifiers are similar to the unit pictured in Fig. 14-8. The fraction escaping with the air stream is the *product* or *overflow*, the fraction falling out the

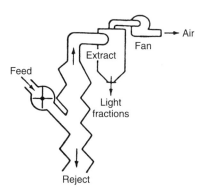

Figure 14-8. Air classifier.

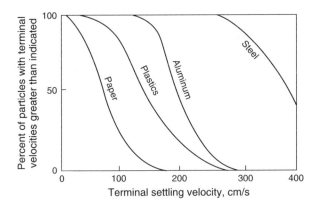

Figure 14-9. Terminal settling velocities of various components of MSW. No single air speed will result in complete separation of organic material (plastics and paper) from inorganic (steel and aluminum). Idealized curves.

bottom is the *reject* or *underflow*. The recovery of organic materials by air classification is adversely influenced by two factors:

- Not all organic materials are aerodynamically less dense, nor are all inorganic materials denser.
- Perfect classification of more and less dense materials is difficult because of the stochastic nature of material movement in the classifier.

The first factor is illustrated in Fig. 14-9, in which the terminal settling velocity (the air velocity at which particles will just begin to rise with the air stream) is plotted against the fraction of particles of various materials. Complete separation of organic from inorganic material can never occur, regardless of the chosen air velocity.

The second factor is illustrated in Fig. 14-10, which shows the efficiency, as defined by Eq. (14.9), vs the feed rate to the classifier. As solids loading increases, more particles that would have exited with the less dense fraction may be caught in the underflow stream, and vice versa. Recovery of organic material in the product stream may be plotted against the recovery of inorganic material in the reject stream. The ideal performance curve may be obtained from the terminal settling plots like Fig. 14-9 and by calculating $R_{(x_1)}$ and $R_{(y_2)}$ for a continuous classifier at different speeds. The closer the two curves, the more effective the air classifier. Moreover, as may be seen in Fig. 14-10, the performance deteriorates as the loading on the air classifier is increased. The throat of the classifier can become clogged at high feed rates, rendering a clean separation impossible.

EXAMPLE 14.4. An air classifier operates with an air velocity of 200 cm/s, and the feed contains equal amounts (by weight) of paper, plastics, aluminum, and steel, having

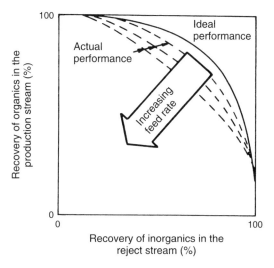

Figure 14-10. Actual and ideal performance of air classifiers. Idealized curves.

terminal settling curves as shown in Fig. 14-9. What would be the recovery of organic material, and what would be the purity of the recovered product?

Since each component is 25% by weight of the feed, $x_0 = 25\% + 25\% = 50\%$ and $y_0 = 25\% + 25\% = 50\%$. From Fig. 14-9, at 200 cm/s air velocity, the fractions of the components in the overflow (product) are:

Paper	100%
Plastics	80%
Aluminum	50%
Steel	0%

Thus the total percentage of organic materials in the product (x_1 in Eqs (14.6)–(14.8)) is

$$(100)\left(\tfrac{1}{4}\right) + (80)\left(\tfrac{1}{4}\right) = 45\%,$$

---

and the percentage of inorganic materials in the product (y_1 in Eq. (14.7)) is

$$(50)\left(\tfrac{1}{4}\right) + (0)\left(\tfrac{1}{4}\right) = 12.5\%.$$

From Eq. (14.6), the recovery of organic materials is

$$R_{\text{org}} = \frac{45}{25 + 25}(100) = 90\%,$$

and the purity is

$$P_{\text{orgproduct}} = \frac{45}{45 + 12.5}(100) = 78\%.$$

Magnets

Ferrous material may be removed from refuse using magnets, which continually extract the ferrous material and reject the remainder. Figure 14-11 shows two types of magnets. With the belt magnet, recovery of ferrous material is enhanced by placing the belt close to the refuse; however, such placement decreases the purity of the product. The depth of refuse on the belt can also pose difficulties, since the heavy ferrous particles tend to settle to the bottom of the refuse on the conveyor and are then further from the magnet than other refuse components.

Separation Equipment

Countless other unit operations for materials handling and storage have been tried. Jigs have been used for removing glass, froth flotation has been successfully employed to separate ceramics from glass, eddy current devices have recovered aluminum in commercial quantities, and so on. As recovery operations evolve, more and better materials separation and handling equipment will be introduced. Figure 14-12 is a diagram of a typical MRF.

Energy Recovery from the Organic Fraction of Municipal Solid Waste

The organic fraction of refuse is a useful secondary fuel. Shredded and separated by classification, it may be used in existing electric generating plants either as a fuel to

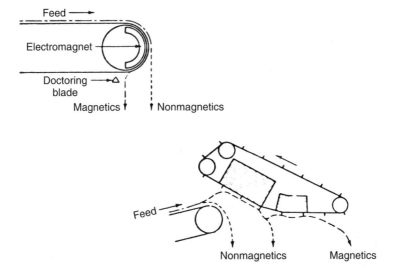

Figure 14-11. Two types of magnets used for resource recovery.

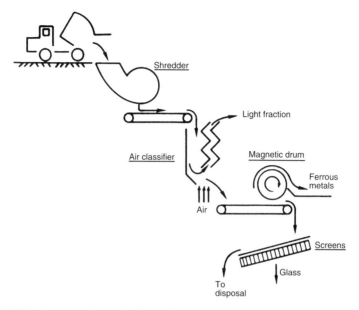

Figure 14-12. Diagram of a typical materials separation facility for refuse processing.

supplement coal or as the sole fuel in separate boilers. Fuel made from refuse is called refuse-derived fuel, or RDF. Combustion of organic material is assumed to proceed by the reaction

$$(HC)_x + O_2 \rightarrow xCO_2 + \frac{x}{2}H_2O + \text{heat}.$$

However, not all hydrocarbons are oxidized completely to carbon dioxide and water, and other components of the fuel, like nitrogen and sulfur, are also oxidized, by the reactions

$$N_2 + O_2 \rightarrow 2NO$$
$$2NO + O_2 \rightarrow 2NO_2$$
$$S + O_2 \rightarrow SO_2$$
$$2SO_2 + O_2 \rightarrow 2SO_3.$$

As discussed in Chap. 19, NO_2 is an important component in the formation of photochemical smog. SO_2 is damaging to health and vegetation, and the reaction product SO_3 forms "acid rain" by the reaction

$$SO_3 + H_2O \rightarrow H_2SO_4.$$

Stoichiometric oxygen is the theoretical amount of oxygen required for combustion (in terms of air, it is *stoichiometric air*) and is calculated from the chemical reaction, as in the following example.

EXAMPLE 14.5. If carbon is combusted as

$$C + O_2 \to CO_2 + \text{heat},$$

how much air is required per gram of carbon? One mole of oxygen is required for each mole of carbon used. The atomic weight of carbon is 12 g/g–atom and the molecular weight of O_2 is $2 \times 16 = 32$ g/mole. Hence 1 g of C requires

$$\frac{32}{14} = 2.28 \text{ g } O_2.$$

Air is 23.15% O_2 by weight; total amount of air required to combust 1 g of C is

$$\frac{2.28}{0.2315} = 9.87 \text{ g air.}$$

The yield of energy from combustion is measured as the calories of heat liberated per unit weight of material burned. This is the *heat of combustion* or, in engineering terms, the *heat value*. Heat value is measured using a calorimeter, in which a small sample of fuel is placed in a water-jacketed stainless steel bomb under high pressure of pure oxygen and then fired. The heat generated is transferred to the water in the water jacket, and the rise in water temperature is measured. Knowing the mass of the water, the energy liberated during combustion can be calculated. In SI units the heat value is expressed as kilojoules (kJ) per kilograms; in British units, as British thermal units (Btu) per pound. Table 14-3 lists some heats of combustion for common hydrocarbons, and gives some typical values for refuse and RDF. The rate at which heat goes into a boiler is sometimes called the *heat rate*.

In designing any combustion operation, the system must be analyzed for thermal as well as materials balance. Materials balance requires that the total mass of the inputs (air and fuel) must equal the mass of the outputs (stack emissions and bottom ash).

Table 14-3. Typical Values of Heats of Combustion

Fuel	Heat of combustion	
	kJ/kg	Btu/lb
Carbon (to CO_2)	32,800	14,100
Hydrogen	142,000	61,100
Sulfur (to SO_2)	9300	3980
Methane	55,500	23,875
Residual oil	41,850	18,000
Raw refuse	9300	4000
RDF (air classified)	18,600	8000

Reuse, Recycling, and Resource Recovery   289

For thermal balance the heat input must equal the heat output plus losses. Example 14.6 shows calculations for a typical thermal balance.

EXAMPLE 14.6. A processed refuse containing 20% moisture and 60% organic material is fed to a boiler at a rate of 1000 kg/h. From a calorimetric analysis of the refuse, a dry sample was determined to have a heat value of 19,000 kJ/kg. Calculate the thermal balance for this system.

The heat from combustion of the RDF is

$$H_{comb} = (19,000 \text{ kJ/kg})(1000 \text{ kg/h}) = 19 \times 10^6 \text{ kJ/h}.$$

The organic fraction of the RDF contains hydrogen, which is combusted to water. Therefore, the heat of combustion value includes the latent heat of vaporization of water, because the water formed is vaporized during the combustion. Since this heat is absorbed by the water formed, it is a heat loss. Assuming that the organic constituents of the RDF are 50% hydrogen (by weight), and given that the latent heat of vaporization of water is 2,420 kJ/kg, the heat loss from the vaporization is

$$H_{vsp} = (1000 \text{ kg/h})(0.6 \text{ organic})(0.5 \text{ H})(2420 \text{ kJ/kg}) = 0.726 \times 10^6 \text{ kJ/h}.$$

The RDF also contains moisture that is vaporized:

$$H_{mois} = (1000 \text{ kg/h})(0.2 \text{ moisture})(2420 \text{ kJ/kg}) = 0.484 \times 10^6 \text{ kJ/h}.$$

There is heat loss associated with radiation, usually assumed as 5% of the heat input. Not all of the organic material will combust. Assume that the ashes contain 10% of the organic material, so that the heat loss is

$$H_{rad} = (19 \times 10^6 \text{ kJ/h})(0.05) = 0.95 \times 10^6 \text{ kJ/h}$$

$$H_{noncom} = (10/60)(19 \times 10^6 \text{ kJ/h}) = 3.17 \times 10^7 \text{ kJ/h}.$$

The stack gases also contain heat, which is usually assumed to be the difference between the heat of combustion and the other losses calculated:

$$\text{Heat input} = \text{Heat output}$$

$$H_{comb} = H_{vap} + H_{mois} + H_{rad} + H_{noncom} + H_{stack}$$

$$19 \times 10^6 = (0.726 + 0.484 + 0.95 + 3.17)10^6 + H_{stack}$$

$$H_{stack} = 13.67 \times 10^6 \text{ kJ/h}.$$

Some of the 13.67×10^6 kJ/h may be recovered by running cold water into the boiler through the water wall tubes and producing steam. If 2000 kg/h of steam at a temperature of 300°C and a pressure of 4000 kPa is required and the temperature of the boiler water is 80°C, calculate the heat loss in the stack gases.

Heat in the boiler water is

$$H_{wat} = (2000 \text{ kg/h})(80 + 273)\text{K}(0.00418) \text{ kJ/kg-K} = 2951 \text{ kJ/h}.$$

where 0.00418 kJ/kg–K is the specific heat of water. The heat in the steam, at 300°C and 4000 kPa, is 2975 kJ/kg, so that

$$H_{steam} = (2000 \text{ kg/h})(2975) \text{ kJ/kg} = 5.95 \times 10^6 \text{ kJ/h}.$$

The heat balance then yields

$$H_{comb} + H_{wat} = H_{vap} + H_{mois} + H_{rad} + H_{noncom} + H_{steam} + H_{stack}$$

$$(19 + 0.0002)10^6 = (0.726 + 0.484 + 0.95 + 3.17 + 5.95)10^6 + H_{stack}$$

$$H_{stack} = 7.72 \times 10^6 \text{ kJ/h}.$$

Both aerobic and anaerobic decomposition can extract useful products biochemically from RDF. In the anaerobic system, refuse is mixed with sewage sludge and the mixture is digested. Operational problems have made this process impractical on a large scale, although single household units that combine human excreta with refuse have been used.

Aerobic decomposition of refuse is better known as *composting* and results in the production of a useful soil conditioner that has moderate fertilizer value. The process is exothermic and at a household level has been used as a means of producing hot water for home heating. On a community scale, composting may be a mechanized operation, using an aerobic digester (Fig. 14-13) or a low-technology operation using long rows of shredded refuse known as *windrows* (Fig. 14-14). Windrows are usually about 3 m (10 ft) wide at the base and 1.5 m (4 to 6 ft) high. Under these conditions, known as *static pile composting* (Fig. 14-15), sufficient moisture and oxygen are available to support aerobic life. The piles must be turned periodically to allow sufficient oxygen to penetrate all parts of the pile or, alternatively, air can be blown into the piles.

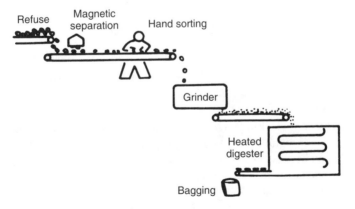

Figure 14-13. Mechanical composting operations.

Reuse, Recycling, and Resource Recovery   291

Figure 14-14. Windrow composting.

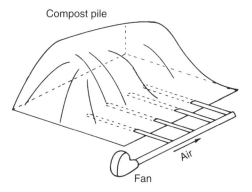

Figure 14-15. Static pile composting.

Temperatures within a windrow approach 140°F, entirely because of biological activity. The pH will approach neutrality after an initial drop. With most wastes, additional nutrients are not needed. The composting of bark and other materials, however, is successful only with the addition of nitrogen and phosphorus. Moisture must usually be controlled because excessive moisture makes maintenance of aerobic conditions difficult, while a dearth of moisture inhibits biological life. A moisture content of 40 to 60% is considered desirable.

There has been some controversy over the use of inoculants, freeze-dried cultures used to speed up the process. Once the composting pile is established, which requires about two weeks, the inoculants have not proved to be of any significant value.

Most MSW contains sufficient organisms for successful composting, and "mystery cultures" are not needed.

The endpoint of a composting operation is reached when the temperature drops. The compost should have an earthy smell, similar to peat moss, and a dark brown color. Compost is an excellent soil conditioner, but is not yet widely used by U.S. farmers. Inorganic fertilizers are cheap and easy to apply, and most farms are located where soil conditions are good. As yet, plentiful food supplies in developed countries does not dictate the use of marginal cropland where compost would be of real value.

CONCLUSION

The solid waste problem must be addressed from the point of view of source control as well as disposal. Many reuse and recycling methods are still in the exploratory stage, but need development as land for disposal grows scarcer and more expensive, and refuse continues to accumulate. We are still, unfortunately, years away from the development and use of fully recyclable and biodegradable materials. The only truly disposable package available today is the ice-cream cone.

PROBLEMS

14.1 Estimate the critical speed for a trommel screen, 2 m in diameter.

14.2 What air speed is required in an air classifier to suspend a spherical piece of glass 1 mm in diameter? (See Eq. 9.4; assume $\rho_s = 2.65 \text{ g/cm}^3$, $C_D = 2.5$, $\mu = 2 \times 10^{-4}$ P, and $\rho = 0.0012 \text{ g/cm}^3$.)

14.3 Refer to Fig. 14-16. By replotting the curves, estimate the Rosin–Rammler constants n and x_c. Compare these with the data in Table 14-1.

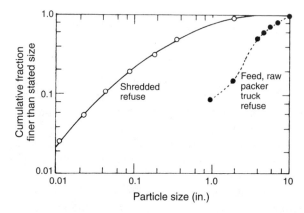

Figure 14-16. Feed and product curves in shredding, for Problems 14.3 and 14.4.

14.4 If the Bond work index is 400 kWh/ton, what is the power requirement to process 100 ton/h of the refuse described in Fig. 14-16?

14.5 Estimate the heating value of MSW based on the ultimate analysis shown in Table 12-1.

14.6 A power plant burns 100 ton/h of coal. How much air is needed if 50% excess air is used?

14.7 An air classifier performance is:

	Organics (kg/h)	Inorganics (kg/h)
Feed	80	20
Product	60	10
Reject	20	10

Calculate the recovery, purity, and efficiency.

14.8 Using the data in Example 14.6, what percent of the heating value can be saved by reducing the moisture content of refuse to 0?

14.9 In Example 14.6, what fraction of the coal is "wasted" owing to the loss of heat in the stack gases (if steam is produced)?

14.10 What is the "wasted" coal in Problem 14.9 heat equivalent in barrels of oil? The heating value of a barrel of oil is 6×10^6 kJ.

Chapter 15

Hazardous Waste

For centuries, chemical wastes have been the by-products of developing societies. Disposal sites were selected for convenience and placed with little or no attention to potential impacts on groundwater quality, runoff to streams and lakes, and skin contact as children played hide-and-seek in a forest of abandoned 55-gal drums. Engineering decisions here historically were made by default; lack of planning for handling or processing or disposal at the corporate or plant level necessitated "quick and dirty" decision by mid- and entry-level engineers at the end of production processes. These production engineers solved disposal problems by simply piling or dumping these waste products "out back."

Attitudes in the United States began to change in the 1960s, 1970s, and 1980s. Air, water, and land are now no longer viewed as commodities to be polluted with the problems of cleanup freely passed to neighboring towns or future generations. Governments have responded to public concerns with revised local zoning ordinances, updated public health laws, and new major Federal Clean Air and Clean Water Acts. In 1976, the Federal Resource Conservation and Recovery Act (RCRA) was enacted to give the U.S. Environmental Protection Agency (EPA) specific authority to regulate the generation, transportation, and disposal of dangerous and hazardous materials. The law was strengthened in 1984 with passage of the Hazardous and Solid Waste Amendments to RCRA. In the 1990s we found that engineering knowledge and expertise had not kept pace with this awakening to the necessity to manage hazardous wastes adequately. This chapter discusses the state of knowledge in the field of hazardous waste engineering, tracing the quantities of wastes generated in the nation from handling and processing options through transportation controls, to resource recovery, and ultimate disposal alternatives.

MAGNITUDE OF THE PROBLEM

Over the years, the term "hazardous" has evolved in a confusing setting as different groups advocate many criteria for classifying a waste as "hazardous." Within the federal government, different agencies use such descriptions as toxic, explosive, and radioactive to label a waste as hazardous. Different states have other classification systems, as did the National Academy of Sciences and the National Cancer Institute. These systems are displayed in Table 15-1. Selected classification criteria are described in more detail in Table 15-2.

Table 15-1. Historical Definitions of Hazardous Waste

	Definition Criteria											
	Toxico-logical	Flamma-bility	Explo-sives	Corro-sive	Reacti-vity	Oxidizing material	Radio-active	Irritant	Strong sanitizer	Biocon-centration	Carcinogenic, mutagenic	Quan-tity
15 U.S. Code, Section 1261 (commerce)	x	x		x				x	x			x
Code of Federal Regulations v. 16, Part 1500	x	x		x			x	x	x			x
Food, Drug and Cosmetic Act, 21 U.S. Code	x											x
Code of Federal Regulations v. 49, Parts 100–199 (transportation)	x	x	x	x		x	x	x		x	x	x
Code of Federal Regulations v. 40, Part 162 (pesticides)	x	x										
Code of Federal Regulations v. 40, Part 162 (ocean dumping)	x						x			x	x	x
NIOSH Toxic Substances List (occupational health)	x										x	
Drinking Water Standards, 42 U.S. Code	x						x			x	x	x
Federal Water Pollution Control Act, 42 U.S. Code	x									x	x	x
Clean Air Act Section 112, 42 U.S. Code	x						x				x	
California State List	x	x	x	x	x			x	x			
National Academy of Sciences	x	x			x							
Battelle Memorial Institute	x	x	x		x	x		x		x	x	
Department of Defense — Army	x	x	x	x	x	x	x					
Department of Defense — Navy	x						x					
National Cancer Institute	x										x	

Table 15-2. Classification Criteria for Hazardous Waste

Criterion	Description
Bioconcentration	The process by which living organisms concentrate an element or compound in levels in excess of the surrounding environment.
LD_{50} (lethal dose 50)	A calculated dose of a chemical substance that is expected to kill 50% of a population exposed through a route other than respiration (mg/kg of body weight).
LC_{50}	A calculated concentration of a chemical substance that, when following the respiratory route, will kill 50% of a population during a 4-h exposure period (ambient concentration in mg/L).
Phytotoxicity	The ability of a chemical substance to cause poisonous reactions in plants.

The federal government attempted to impose a nationwide classification system under the implementation of RCRA, in which a hazardous waste is defined by the degree of instability, corrosivity, reactivity, or toxicity. This definition includes acids, toxic chemicals, explosives, and other harmful or potentially harmful waste. In this chapter, this is the applicable definition of hazardous waste. Radioactive wastes are excluded (except in Department of Transportation regulations). Such wastes obviously are hazardous, but their generation, handling, processing, and disposal differ from chemically hazardous wastes. Moreover, all radioactive materials, as well as health protection from ionizing radiation, have been regulated by a separate and independent government agency: the Atomic Energy Commission from 1954 until 1974, and the Nuclear Regulatory Commission since 1974. The radioactive waste problem is addressed separately in Chap. 16.

Given this somewhat limited definition, more than 60 million metric tons, by wet weight, of hazardous waste are generated annually throughout the United States. More than 60% is generated by the chemical and allied products industry. The machinery, primary metals, paper, and glass products industries each generate between 3 and 10% of the nation's total. Approximately 60% of the hazardous waste is liquid or sludge. Major generating states, including New Jersey, Illinois, Ohio, California, Pennsylvania, Texas, New York, Michigan, Tennessee, and Indiana contribute more than 80% of the nation's total production of hazardous waste, and the waste's majority is disposed of on the generator's property.

A hasty reading of these hazardous waste facts points to several interesting, though shocking, conclusions. Most hazardous waste is generated and inadequately disposed of in the eastern portion of the country. In this region, the climate is wet with patterns of rainfall that permit infiltration or runoff to occur. Infiltration permits the transport of hazardous waste into groundwater supplies, and surface runoff leads to the contamination of streams and lakes. Moreover, most hazardous waste is generated and disposed of in areas where people rely on aquifers for drinking water. Major aquifers and well withdrawals underlie areas where the wastes are generated. Thus, the hazardous waste problem is compounded by two considerations: the wastes are generated and disposed of in areas where it rains and in areas where people rely on aquifers for supplies of drinking water.

WASTE PROCESSING AND HANDLING

Waste processing and handling are key concerns as a hazardous waste begins its journey from the generator site to a secure long-term storage facility. Ideally, the waste can be stabilized, detoxified, or somehow rendered harmless in a treatment process similar to the following:

Chemical Stabilization/Fixation. In this process, chemicals are mixed with waste sludge, the mixture is pumped onto land, and solidification occurs in several days or weeks. The result is a chemical nest that entraps the waste, and pollutants such as heavy metals may be chemically bound in insoluble complexes. Asphalt-like compounds form "cages" around the waste molecules, while grout and cement form actual chemical bonds with the trapped substances. Chemical stabilization offers an alternative to digging up and moving large quantities of hazardous waste, and is particularly suitable for treating large volumes of dilute waste. Proponents of these processes have argued for building roadways, dams, and bridges with a selected cement as the fixing agent. The adequacy of the containment offered by these processes has not been documented, however, as long-term leaching and defixation potentials are not well understood.

Volume Reduction. Volume reduction is usually achieved by incineration, which takes advantage of the large organic fraction of waste being generated by many industries, but may lead to secondary problems for hazardous waste engineers: air emissions in the stack of the incinerator and ash production in the base of the incinerator. Both by-products of incineration must be addressed in terms of risk, as well as legal and economic constraints (as must all hazardous waste treatment, for that matter). Because incineration is often considered a very good method for the ultimate disposal of hazardous waste, we discuss it in some detail later in this chapter.

Waste Segregation. Before shipment to a processing or long-term storage facility, wastes are segregated by type and chemical characteristics. Similar wastes are grouped in a 55-gal drum or group of drums, segregating liquids such as acids from solids such as contaminated laboratory clothing and equipment. Waste segregation is generally practiced to prevent undesirable reactions at disposal sites and may lead to economics of scale in the design of detoxification or resource recovery facilities.

Detoxification. Many thermal, chemical, and biological processes are available to detoxify chemical wastes. Options include:

- neutralization
- ion exchange
- incineration
- pyrolysis

- aerated lagoons
- waste stabilization ponds

These techniques are specific; ion exchange obviously does not work for every chemical, and some forms of heat treatment may be prohibitively expensive for sludge that has a high water content.

Degradation. Methods that chemically degrade some hazardous wastes and render them less hazardous exist. Chemical degradation is a form of chemical detoxification. Waste-specific degradation processes include hydrolysis, which destroys organophosphorus and carbonate pesticides, and chemical dechlorination, which destroys some polychlorinated pesticides. Biological degradation generally involves incorporating the waste into the soil. Landfarming, as it has been termed, relies on healthy soil microorganisms to metabolize the waste components. Landfarming sites must be strictly controlled for possible water and air pollution that results from overactive or underactive organism populations.

Encapsulation. A wide range of material to encapsulate hazardous waste is available. Options include the basic 55-gal steel drum (the primary container for liquids), clay, plastics, and asphalt; these materials may also be implemented to solidify the waste. Several layers of different materials are often recommended for the outside of the drum, such as an inch or more of polyurethane foam to prevent corrosion.

TRANSPORTATION OF HAZARDOUS WASTES

Hazardous wastes are transported across the nation on trucks, rail flatcars, and barges. Truck transportation and particularly small-truck transportation is a highly visible and constant threat to public safety and the environment. There are four basic elements in the control strategy for the movement of hazardous waste from a generator — a strategy that forms the basis of U.S. Department of Transportation (USDOT) regulation of hazardous materials transportation as set forth in Vol. 49, Parts 170–180 of the Code of Federal Regulations.

Haulers. Major concerns over hazardous waste haulers include operator training, insurance coverage, and special registration of transport vehicles. Handling precautions include workers wearing gloves, face masks, and coveralls, as well as registration of handling equipment to control future use of the equipment and avoid situations in which hazardous waste trucks today are used to carry produce to market tomorrow. Schedules for relicensing haulers and checking equipment are part of an overall program for ensuring proper transport of hazardous wastes. The Chemical Manufacturer's Association and the USDOT operate a training program for operators of long-distance vehicles hauling hazardous materials.

Hazardous Waste Manifest. The concept of a cradle-to-grave tracking system has long been considered key to proper management of hazardous waste. This "bill of lading" or "trip ticket" ideally accompanies each barrel of waste and describes the content of each barrel to its recipient. Copies of the manifest are submitted to generators and state officials so all parties know that each waste has reached its desired

300   ENVIRONMENTAL ENGINEERING

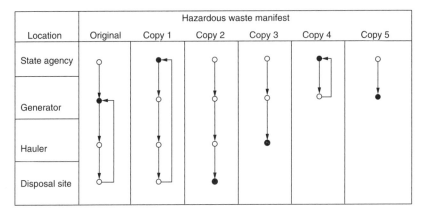

Figure 15-1. Possible routing of copies of a hazardous waste manifest.

destination in a timely manner. This system serves four major purposes: (1) it provides the government with a means of tracking waste within a given state and determining quantities, types, and locations where the waste originates and is ultimately disposed of; (2) it certifies that wastes being hauled are accurately described to the manager of the processing/disposing facility; (3) it provides information for recommended emergency response if a copy of the manifest is not returned to the generator; and (4) it provides a database for future planning within a state. Figure 15-1 illustrates one possible routing of copies of a selected manifest. In this example, the original manifest and five copies are passed from the state regulatory agency to the generator of the waste. Copies accompany each barrel of waste that leaves the generating site, and are signed and mailed to the respective locations to indicate the transfer of the waste from one location to another.

Packaging. Regulations of USDOT prescribe the design and construction of packages used to transport all hazardous materials, whether they are considered waste or usable materials. Corrosive material, flammable material, volatile material, material that, if released, would be toxic by inhalation, and any material that, if released during transportation, would pose a threat to human health and safety or to the environment must be packaged for transportation according to regulations.

Labeling and Placarding. Before a waste is transported from a generating site, each container is labeled and the transportation vehicle is placarded. Announcements that are appropriate include warnings for explosives, flammable liquids, corrosive material, strong oxidizers, compressed gases, and poisonous or toxic substances. Multiple labeling is desirable if, for example, a waste is both explosive and flammable. These labels and placards warn the general public of possible dangers and assist emergency response teams as they react in the event of a spill or accident along a transportation route.

Hazardous Waste   301

Accident and Incident Reporting. Accidents involving hazardous wastes must be reported immediately to state regulatory agencies and local health officials. Accident reports that are submitted immediately and indicate the amount of materials released, the hazards of these materials, and the nature of the failure that caused the accident may be instrumental in containing the spilled waste and cleaning the site. For example, if liquid waste can be contained, groundwater and surface water pollution may be avoided. USDOT maintains a database of hazardous materials accident and incident reports on the website of the Bureau of Transportation Statistics (www.bts.gov).

RECOVERY ALTERNATIVES

Recovery alternatives are based on the premise that one person's waste is another person's prize. What may be a worthless drum of electroplating sludge to the plating engineer may be a silver mine to an engineer skilled in metals recovery. In hazardous waste management, two types of systems exist for transferring this waste to a location where it is viewed as a resource: *hazardous waste materials transfers* and *hazardous waste information clearinghouses*. In practice, one organization may display characteristics of both of these pure systems.

The rationale behind both transfer mechanisms is illustrated in Fig. 15-2. An industrial process typically has three outputs: (1) a principal product, which is sold to a consumer; (2) a useful by-product available for sale to another industry; and (3) waste, historically destined for ultimate disposal. Waste transfers and clearinghouses act to minimize this flow of waste to a landfill or to ocean burial by directing it to a previously unidentified industry or firm that perceives the waste as a resource. As the regulatory and economic climate of the nation evolves, these perceptions may continue to change and more and more waste may be economically recovered.

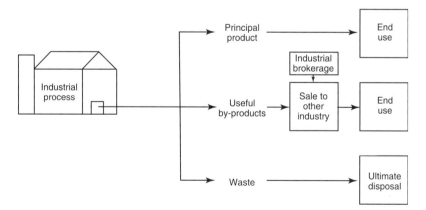

Figure 15-2. Rationale for hazardous waste clearinghouses and exchanges.

Information Clearinghouses

The pure clearinghouse has limited functions. These institutions offer a central point for collecting and displaying information about industrial wastes. Their goal is to introduce interested potential trading partners to each other through the use of anonymous advertisements and contacts. Clearinghouses generally do not seek customers, negotiate transfers, set prices, process materials, or provide legal advice to interested parties. One major function of a clearinghouse is to keep all data and transactions confidential so trade secrets are not compromised.

Clearinghouses are also generally subsidized by sponsors, either trade or governmental. Small clerical staffs are organized in a single office or offices spread throughout a region. Little capital is required to get these operations off the ground, and annual operation expenses are relatively low.

The value of clearinghouse operations should not be overemphasized. Often they are only able to operate in the short term; they evolve from an organization with many listings and active trading to a business with minimal activity as plant managers make their contacts directly with waste suppliers and short-circuit the system by eliminating the clearinghouse.

Materials Exchanges

In comparison with the clearinghouse concept, a pure materials exchange has many complex functions. A transfer agent within the exchange typically identifies generators of waste and potential users of the waste. The exchange will buy or accept waste, analyze its chemical and physical properties, identify buyers, reprocess the waste as needed, and sell it at a profit.

The success of an exchange depends on several factors. Initially, a highly competent technical staff is required to analyze waste flows, and design and prescribe methods for processing the waste into a marketable resource. The ability to diversify is critical to the success of an exchange. Its management must be able to identify local suppliers and buyers of their products. Additionally, an exchange may even enter the disposal business and incinerate or landfill waste.

Although exchanges have been attempted with some success in the United States, a longer track record exists in Europe. Belgium, Switzerland, Germany, most of the Scandinavian countries, and the United Kingdom all have experienced some success with exchanges. The general characteristics of European waste exchanges include:

- operation by the national industrial associations,
- services offered without charge,
- waste availability made known through published advertisements,
- advertisements discussing chemical and physical properties, as well as quantities, of waste, and
- advertisements coded to maintain confidentiality.

Five wastes are generally recognized as having transfer value: (1) wastes having a high concentration of metals, (2) solvents, (3) concentrated acids, (4) oils, and (5) combustibles for fuel. That is not to say these wastes are the only transferable items. Four-hundred tons per year of foundry slag containing 50 to 60% metallic Al, 150 m^3/yr of 90% methanol with trace mineral acids, and 4 tons of deep frozen cherries were transformed from waste to resource in one European exchange. One person's waste may truly be another person's valued resource.

HAZARDOUS WASTE MANAGEMENT FACILITIES

Siting Considerations

A wide range of factors must be considered in siting hazardous waste management facilities. Some of these are determined by law: for example, RCRA prohibits landfilling of flammable liquids. Socioeconomic factors are often the key to siting. Joseph Koppel (Koppel 1985) has coined the acronym LULU — locally undesirable land use — for a facility that no one wants nearby but that is going to be put somewhere. Certainly, hazardous waste facilities are LULUs.

In selecting a site, all of the relevant "-ologies" must be considered: hydrology, climatology, geology, and ecology, as well as current land use, environmental health, and transportation. EPA also requires risk analysis under regulations promulgated under RCRA (Chap. 3).

Hydrology. Hazardous waste landfills should be located well above historically high groundwater tables. Care should be taken to ensure that a location has no surface or subsurface connection, such as a crack in confining strata, between the site and a water course. Hydrologic considerations limit direct discharge of wastes into groundwater or surface water supplies.

Climatology. Hazardous waste management facilities should be located outside the paths of recurring severe storms. Hurricanes and tornadoes disrupt the integrity of landfills and incinerators, and cause immediate catastrophic effects on the surrounding environment and public health in the region of the facility. In addition, areas of high air pollution potential should be avoided in site selection processes. These areas include valleys where winds or inversions act to hold pollutants close to the surface of the earth, as well as areas on the windward side of mountain ranges, i.e., areas similar to the Los Angeles area where long-term inversions are prevalent.

Geology. A disposal or processing facility should be located only on stable geologic formations. Impervious rock, which is not littered with cracks and fissures, is an ideal final liner for hazardous waste landfills.

Ecology. The ecological balance must be considered as hazardous waste management facilities are located in a region. Ideal sites in this respect include areas of low fauna and flora density, and efforts should be made to avoid wilderness areas, wildlife refuges, and animal migration routes. Areas with unique plants and animals, especially endangered species and their habitat, should also be avoided.

Alternative Land Use. Areas with low ultimate land use should receive prime consideration. Areas with high recreational use potential should be avoided because of the increased possibility of direct human contact with the wastes.

Transportation. Transportation routes to facilities are a major consideration in siting hazardous waste management facilities. USDOT guidelines suggest the use of interstate and limited-access highways whenever possible. Other roads to the facilities should be accessible by all-weather highways to minimize spills and accidents during periods of rain and snowfall. Ideally, the facility should be close to the generation of the waste in order to reduce the probability of spills and accidents as wastes are transported.

Socioeconomic Factors. Factors that could make or break an effort to site a hazardous waste management facility fall under this major heading. Such factors, which range from public acceptance to long-term care and monitoring of the facility, are:

1. Public control over the opening, operation, and closure of the facility. Who will make policy for the facility (Slovic 1987)?

2. Public acceptance and public education programs. Will local townspeople permit it?

3. Land use changes and industrial development trends. Does the region wish to experience the industrial growth induced by such facilities?

4. User fee structures and recovery of project costs. Who will pay for the facility? Can user charges be used to induce industry to reuse, reduce, or recover the resources materials in the waste?

5. Long-term care and monitoring. How will postclosure maintenance be guaranteed and who will pay?

All are critical concerns in a hazardous waste management scheme.

The term *mixed waste* refers to mixtures of hazardous and radioactive wastes; for example, organic solvents used in liquid scintillation counting are an excellent example. Siting a mixed waste facility is difficult because the laws and regulations governing handling of chemically hazardous waste overlap and sometimes conflict with those governing handling of radioactive waste.

Incinerators

Incineration is a controlled process that uses combustion to convert a waste to a less bulky, less toxic, or less noxious material. The principal products of incineration from a volume standpoint are carbon dioxide, water, and ash, but the products of primary concern because of their environmental effects are compounds containing sulfur, nitrogen, and halogens. When the gaseous combustion products from an incineration process contain undesirable compounds, a secondary treatment such as afterburning, scrubbing, or filtration is required to lower concentrations to acceptable levels before atmospheric release. The solid ash products from the incineration process are also a major concern and must reach adequate ultimate disposal.

Hazardous Waste 305

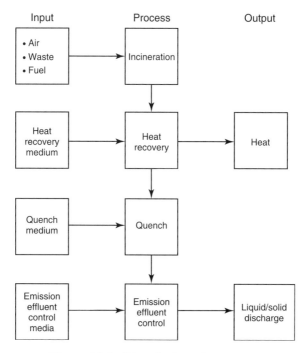

Figure 15-3. Waste incineration system.

Figure 15-3 shows the schematic of a generalized incineration system that includes the components of a waste incineration system. The actual system, which may contain one or more of the components, is usually dependent upon the individual waste's application requirement. The advantages of incineration as a means of disposal for hazardous waste are:

1. Burning wastes and fuels in a controlled manner has been carried on for many years and the basic process technology is available and reasonably well developed. This is not the case for some of the more exotic chemical degradation processes.

2. Incineration is broadly applicable to most organic wastes and can be scaled to handle large volumes of liquid waste.

3. Incineration is the best known method for disposal of "mixed waste" (see previous description).

4. Large expensive land areas are not required.

The disadvantages of incineration include:

1. The equipment tends to be more costly to operate than many other alternatives, and the process must meet the stringent regulatory requirements of air pollution control.

2. It is not always a means of ultimate disposal in that normally an ash remains that may or may not be toxic but that in any case must be disposed of properly and with minimal environmental contamination.

3. Unless controlled by applications of air pollution control technology, the gaseous and particulate products of combustion may be hazardous to health or damaging to property.

The decision to incinerate a specific waste will therefore depend first on the environmental adequacy of incineration as compared with other alternatives, and second on the relative costs of incineration and the environmentally sound alternatives.

The variables that have the greatest effect on the completion of the oxidation of wastes are waste combustibility, residence time in the combustor, flame temperature, and the turbulence present in the reaction zone of the incinerator. The combustibility is a measure of the ease with which a material may be oxidized in a combustion environment. Materials with a low flammability limit, low flash point, and low ignition and autoignition temperatures may be combusted in a less severe oxidation environment, i.e., at a lower temperature and with less excess oxygen.

Of the three "T's" of good combustion, time, temperature, and turbulence, only the *temperature* may be readily controlled after the incinerator unit is constructed. This may be done by varying the air-to-fuel ratio. If solid carbonaceous waste is to be burned without smoke, a minimum temperature of 760°C (1400°F) must be maintained in the combustion chamber. Upper temperature limits in the incinerator are dictated by the refractory materials available to line the inner wall of the burn chamber. Above 1300°C (2400°F) special refractories are needed.

The degree of *turbulence* of the air for oxidation with the waste fuel will affect the incinerator performance significantly. In general, both mechanical and aerodynamic means are utilized to achieve mixing of the air and fuel. The completeness of combustion and the time required for complete combustion are significantly affected by the amount and the effectiveness of the turbulence.

The third major requirement for good combustion is *time*. Sufficient time must be provided to the combustion process to allow slow-burning particles or droplets to burn completely before they are chilled by contact with cold surfaces or the atmosphere. The amount of time required depends on the temperature, fuel size, and degree of turbulence achieved.

If the waste gas contains organic materials that are combustible, then incineration should be considered as a final method of disposal. When the amount of combustible material in the mixture is below the lower flammable limit, it may be necessary to add small quantities of natural gas or other auxiliary fuel to sustain combustion in the burner. Thus economic considerations are critical in the selection of incinerator systems because of the high costs of these additional fuels.

Boilers for some high-temperature industrial processes may serve as incinerators for toxic or hazardous carbonaceous waste. Cement kilns, which must operate at temperatures in excess of 1400°C (2500°F) to produce cement clinker, can use organic solvents as fuel, and this provides an acceptable method of waste solvent and waste oil disposal.

Incineration is also a possibility for the destruction of liquid wastes. Liquid wastes may be classified into two types from a combustion standpoint: combustible liquids and partially combustible liquids. Combustible liquids include all materials having sufficient calorific value to support combustion in a conventional combustor or burner. Noncombustible liquids cannot be treated by incineration and include materials that would not support combustion without the addition of auxiliary fuel and would have a high percentage of noncombustible constituents such as water. To support combustion in air without the assistance of an auxiliary fuel, the waste must generally have a heat content of 18,500 to 23,000 kJ/kg (8000–10,000 Btu/lb) or higher. Liquid waste having a heating value below 18,500 kJ/kg (8000 Btu/lb) is considered a partially combustible material and requires special treatment.

When starting with a waste in liquid form, it is necessary to supply sufficient heat for vaporization in addition to raising it to its ignition temperature. For a waste to be considered combustible, several rules of thumb should be used. The waste should be plumbable at ambient temperature or capable of being pumped after heating to some reasonable temperature level. Since liquids vaporize and react more rapidly when finely divided in the form of a spray, atomizing nozzles are usually used to inject waste liquids into incineration equipment whenever the viscosity of the waste permits atomization. If the waste cannot be pumped or atomized, it cannot be burned as a liquid but must be handled as a sludge or solid.

Several basic considerations are important in the design of an incinerator for a partially combustible waste. First, the waste material must be atomized as finely as possible to present the greatest surface area for mixing with combustion air. Second, adequate combustion air must be provided to supply all the oxygen required for oxidation or incineration of the organic present. Third, the heat from the auxiliary fuel must be sufficient to raise the temperature of the waste and the combustion air to a point above the ignition temperature of the organic material in the waste.

Incineration of wastes that are not pure liquids but that might be considered sludge or slurries is also an important waste disposal problem. Incinerator types applicable for this kind of waste would be fluidized bed incinerators, rotary kiln incinerators, and multiple-hearth incinerators,[1] all of which increase incineration efficiency.

Incineration is not a total disposal method for many solids and sludge because most of these materials contain noncombustibles and have residual ash. Complications develop with the wide variety of materials that must be burned. Controlling the proper amount of air to give combustion of both solids and sludge is difficult, and with most currently available incinerator designs this is impossible.

Closed incinerators such as rotary kilns and multiple-hearth incinerators are also used to burn solid wastes. Generally, the incinerator design does not have to be limited to a single combustible or partially combustible waste. Often it is both economical and feasible to use a combustible waste, either liquid or gas, as the heat source for the incineration of a partially combustible waste that may be either liquid or gas.

Experience indicates that wastes that contain only carbon, hydrogen, and oxygen and that may be handled in power generation systems may be destroyed in a way that

---

[1] Fluidized bed incinerators burn waste that has been treated to flow almost like a liquid. Rotary kiln incinerators burn waste in a rotating chamber, which exposes new surfaces to the burners as it rotates. Multiple-hearth incinerators have more than one burning chamber.

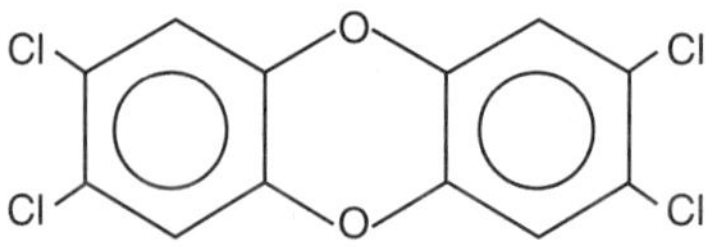

Figure 15-4. 2,3,7,8-Tetrachlorodibenzo-*p*-dioxin.

reclaims some of their energy content. These types of waste may also be judiciously blended with wastes having low energy content, such as the highly chlorinated organic, to minimize the use of purchased fossil fuel. On the other hand, rising energy costs should not be a significant deterrent to the use of thermal destruction methods when they are clearly indicated to be the most desirable method on an environmental basis.

Air emissions from hazardous waste incinerators include the common air pollutants, discussed in Chap. 18. In addition, inadequate incineration may result in emission of some of the hazardous materials that the incineration was intended to destroy. Incomplete combustion, particularly at relatively low temperatures, may also result in production of a class of compounds known collectively as dioxin, including both polychlorinated dibenzodioxins (PCDD) and polychlorinated dibenzofurans (PCDF). The compound in this class that has been identified as a carcinogen and teratogen is 2,3,7,8-tetrachlorodibenzo-*p*-dioxin (2,3,7,8-TCDD), shown in Fig. 15-4.

TCDD was first recognized as an oxidation product of trichlorophenol herbicides (2,4-D and 2,4,5-T, one of the ingredients of Agent Orange) (Tschirley 1986). In 1977, it was one of the PCDDs found present in municipal incinerator fly ash and air emissions, and it has subsequently been found to be a constituent of gaseous emissions from virtually all combustion processes, including trash fires and barbecues. TCDD is degraded by sunlight in the presence of water.

The acute toxicity of TCDD in animals is extremely high (LD_{50} in hamsters of 3.0 mg/kg); carcinogenesis and genetic effects (teratogenesis) have also been observed in chronic exposure to high doses in experimental animals. In humans, the evidence for these adverse effects is mixed. Although acute effects such as skin rashes and digestive difficulties have been observed on high accidental exposure, these are transitory. Public concern has focused on chronic effects, but existing evidence for either carcinogenesis or birth defects in humans from chronic TCDD exposure is inconsistent. Regulations governing incineration are designed to limit TCDD emission to below measurable quantities; these limits may usually be achieved by the proper combination of temperature and residence time in the incinerator. Engineers should understand, however, that public concern about TCDD, and dioxin in general, is disproportionate to the known hazards and is a major factor in opposition to incinerator siting.

Landfills

Landfills must be adequately designed and operated if public health and the environment are to be protected. The general components that go into the design of these

Hazardous Waste 309

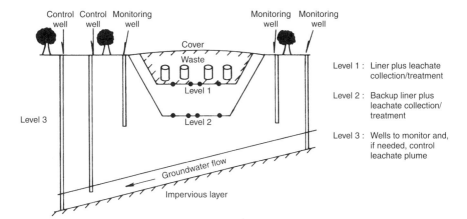

Figure 15-5. Three levels of safeguard in hazardous waste landfills.

facilities, as well as the correct procedure to follow during the operation and postclosure phase of the facility's life, are discussed next.

Design. The three levels of safeguard that must be incorporated into the design of a hazardous landfill are displayed in Fig. 15-5. The primary system is an impermeable liner, either clay or synthetic material, coupled with a leachate collection and treatment system. Infiltration may be minimized with a cap of impervious material overlaying the landfill, and sloped to permit adequate runoff and to discourage pooling of the water. The objectives are to prevent rainwater and snow melt from entering the soil and percolating to the waste containers and, in case water does enter the disposal cells, to collect and treat it as quickly as possible. Side slopes of the landfill should be a maximum of 3:1 to reduce stress on the liner material. Research and testing of the range of synthetic liners must be viewed with respect to a liner's strength, compatibility with wastes, costs, and life expectancy. Rubber, asphalt, concrete, and a variety of plastics are available, and such combinations as polyvinyl chloride overlaying clay may prove useful on a site-specific basis.

A leachate collection system must be designed by contours to promote movement of the waste to pumps for extraction to the surface and subsequent treatment. Plastic pipes, or sand and gravel, similar to systems in municipal landfills and used on golf courses around the country, are adequate to channel the leachate to a pumping station below the landfill. One or more pumps direct the collected leachate to the surface, where a wide range of waste-specific treatment technologies are available, including:

- *Sorbent material*: Carbon and fly ash arranged in a column through which the leachate is passed.
- *Packaged physical–chemical units*: This includes chemical addition and flash mixing, controlled flocculation, sedimentation, pressure filtration, pH adjustment, and reverse osmosis.

The effectiveness of each method is highly waste-specific, and tests must be conducted on a site-by-site basis before a reliable leachate treatment system can be designed. All methods produce waste sludge that must reach ultimate disposal.

A secondary safeguard system consists of another barrier contoured to provide a backup leachate collection system. In the event of failure of the primary system, the secondary collection system conveys the leachate to a pumping station, which in turn relays the wastewater to the surface for treatment.

A final safeguard system is also advisable. This system consists of a series of discharge wells up-gradient and down-gradient to monitor groundwater quality in the area, and to control leachate plumes if the primary and secondary systems fail. Up-gradient wells act to define the background levels of selected chemicals in the groundwater and to serve as a basis for comparing the concentrations of these chemicals in the discharge from the down-gradient wells. This system thus provides an alarm mechanism if the primary and secondary systems fail.

If methane generation is possible in a hazardous waste landfill, a gas collection system must be designed into the landfill. Sufficient vent points must be allowed so that the methane generated may be burned off continuously.

Operation. As waste containers are brought to a landfill site for burial, specific precautions should be taken to ensure the protection of public health, worker safety, and the environment. Wastes should be segregated by physical and chemical characteristics, and buried in the same cells of the landfill. Three-dimensional mapping of the site is useful for future mining of these cells for recovery purposes. Observation wells with continuous monitoring should be maintained, and regular core soil samples should be taken around the perimeter of the site to verify the integrity of the liner materials.

Site Closure. Once a site is closed and does not accept any more waste, the operation and maintenance of the site must continue. The impervious cap on top of the landfill must be inspected and maintained to minimize infiltration. Surface water runoff must be managed, collected, and possibly treated. Continuous monitoring of surface water, groundwater, soil, and air quality is necessary, as ballooning and rupture of the cover material may occur if gases produced or released from the waste rise to the surface. Waste inventories and burial maps must be maintained for future land use and waste reclamation. A major component of postclosure management is maintaining limited access to the area.

CONCLUSION

Hazardous waste is a relatively new concern of environmental engineers. For years, the necessary by-products of an industrialized society were piled "out back" on land that had little value. As time passed and the rains came and went, the migration of harmful chemicals moved hazardous waste to the front page of the newspaper and into the classroom. Engineers employed in all public and private sectors must now face head-on the processing, transportation, and disposal of these wastes. Hazardous waste must be appropriately addressed at the "front end" of the generation process, either by maximizing resource recovery or by detoxification at the site of generation.

In Minnesota, for example, the problem of household hazardous wastes are being handled in permanent, regional collection centers, replacing once-a-day drop-off programs common throughout the United States (Ailbmann 1991). Storage, landfilling in particular, is at best a last resort measure for hazardous waste handling.

PROBLEMS

15.1 Assume you are an engineer working for a hazardous waste processing firm. Your vice president thinks it would be profitable to locate a new regional facility near the state capitol. Given what you know about that region, rank the factors that distinguish a good site from a bad site. Discuss the reasons for this ranking; i.e., why, for example, are hydrologic considerations more critical in that region than, for example, the geology.

15.2 You are a town engineer just informed of a chemical spill on Main Street. Sequence your responses. List and describe the actions your town should take for the next 48 hours if the spill is relatively small (100–500 gal) and is confined to a small plot of land.

15.3 The manifest system, through which hazardous waste must be tracked from generator to disposal site, is expensive for industry. Make these assumptions about a simple electroplating operation: 50 barrels of a waste per day, 1 "trip ticket" per barrel, $25/h labor charge. Assume the generator's technician can identify the contents of each barrel at no additional time or cost to the company because she routinely has done that for years anyway. What is the cost to the generator in person-hours and dollars to comply with the manifest system shown in Fig. 15-1? Document assumptions about the time required to complete each step of each trip ticket.

15.4 Compare and contrast the design considerations of the hazardous waste landfill with the design considerations of a conventional municipal refuse landfill.

15.5 As town engineer, design a system to detect and stop the movement of hazardous wastes into your municipal refuse landfill.

Chapter 16
Radioactive Waste

This chapter presents a background discussion of the interaction of ionizing radiation with matter, as well as a discussion of the environmental effects of nuclear generation of electricity and of radionuclides that are accessible to humans in the environment. The chapter focuses on radioactive waste as an environmental pollutant, discusses the impact of ionizing radiation on environmental and public health, and summarizes engineering options available today for the management and disposal of radioactive waste.

RADIATION

X-rays were discovered by Wilhelm Roentgen toward the end of the year 1895. Almost immediately thereafter, Henri Becquerel observed radiation similar to X-rays emanating from certain uranium salts. In 1898, Marie and Pierre Curie studied radiation from two uranium ores, pitchblende and chalcolite, and isolated two additional elements that exhibited radiation similar to that of uranium but considerably stronger. These two elements were named radium and polonium. The discovery and isolation of these radioactive elements mark the beginning of the "atomic age."

The Curies classified the radiation from radium and polonium into three types, according to the direction of deflection in a magnetic field. These three types of radiation were called alpha (α), beta (β), and gamma (γ) radiation. Becquerel's observation correlated gamma radiation with Roentgen's X-rays. In 1905, Ernest Rutherford identified alpha particles emanating from uranium as ionized helium atoms, and in 1932, Sir James Chadwick characterized as neutrons the highly penetrating radiation that results when beryllium is bombarded with alpha particles. Modern physics has subsequently identified other subatomic particles, including positrons, muons, and pions, but not all of these are of equal concern to the environmental engineer. Management of radioactive waste requires understanding of the sources and effects of alpha, beta, gamma, and neutron emissions.

Radioactive Decay

An atom that is radioactive has an unstable nucleus. The nucleus moves to a more stable condition by emitting an alpha or beta particle or neutron; this emission is frequently accompanied by emission of additional energy in the form of gamma radiation. As a result of these emissions, the radioactive atom is transformed into either an isotope of the same element (neutron or gamma only emission) or into an isotope of a different element (alpha or beta emission). This transformation is known as *radioactive decay*, and the emissions are known as *ionizing radiation*. The rate of radioactive decay, or rate of decrease of the number of radioactive nuclei, can be expressed by a first-order rate equation

$$\frac{dN}{dt} = -K_b Nt, \qquad (16.1)$$

where

N = the number of radioactive nuclei and
K_b = a factor called the *disintegration constant*; K_b has the units of time^{-1}.

Integrating over time, we get the classical equation for radioactive decay:

$$N = N_0 e^{-K_b t}. \qquad (16.2)$$

The data points in Fig. 16-1 correspond to this equation.

After a specific time period $t = t_{1/2}$, the value of N is equal to one-half of N_0, and after each succeeding period of time $t_{1/2}$ the value of N is one-half of the preceding N.

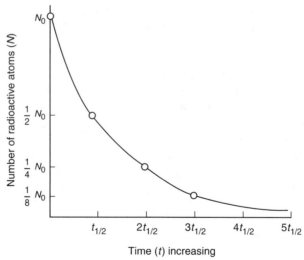

Key:
$t_{1/2}$ = one half-life
$2t_{1/2}$ = two half-lives, etc.

Figure 16-1. General description of radioactive decay.

Radioactive Waste 315

Table 16-1. Some Important Radionuclides

Radionuclide	Type of radiation	Half-life
Krypton-85	Beta and gamma	10 years
Strontium-90	Beta	29 years
Iodine-131	Beta and gamma	8.3 days
Cesium-137	Beta and gamma	30 years
Tritium (Hydrogen-3)	Beta	12 years
Cobalt-60	Beta and gamma	5 years
Carbon-14	Beta	5770 years
Uranium-235	Alpha	7.1×10^8 years
Uranium-238	Alpha	4.9×10^9 years
Plutonium-239	Alpha	24,600 years

That is, one-half of the radioactive atoms have decayed (or disintegrated) during each time period $t_{1/2}$. This time period $t_{1/2}$ is called the *radiological half-life*, or sometimes simply the half-life. Looking at Fig. 16-1, we see that at $t = 2t_{1/2}$, N becomes $\frac{1}{4}N_0$; at $t = 3t_{1/2}$, N becomes $\frac{1}{8}N_0$; and so on. Equation (16.2) is so constructed that N never becomes zero in any finite time period; for every half-life that passes, the number of atoms is halved. If the decay constant K_b is known, the half-life may be determined from Eq. (16.2):

$$\ln 2 = K_b t_{1/2}. \tag{16.3}$$

The radiological half-lives of selected radionuclides are presented in Table 16-1.

EXAMPLE 16.1. An amount of 10.0 g of pure $_6C^{11}$ is prepared. The equation for this nuclear reaction is

$$_6C^{11} \rightarrow {_1e^0} + {_5B^{11}}.$$

The half-life of C-11 is 21 min. How many grams of C-11 will be left 24 h after the preparation? (Note that one atomic mass unit (amu) = 1.66×10^{-24} g.)

Equation (16.2) refers to the number of atoms, so we must calculate the number of atoms in 1.0 g of carbon-11

$$1\,g = \left(\frac{1}{11}\,\text{mole}\right)\left(6.02 \times 10^{23}\,\text{atoms/mole}\right) = 5.47 \times 10^{22}\,\text{atoms} = N_0.$$

Applying Eq. (16.2),

$$24\,\text{h} = (24)(60) = 1440\,\text{min} = t$$

$$K_b = \ln 2 / t_{1/2} = 0.693/21 = 0.033\,\text{min}^{-1}$$

$$N = (5.47 \times 10^{22})e^{-47.52} = (5.47 \times 10^{22})(2.30 \times 10^{-21}) = 126 \text{ atoms.}$$

$$\frac{(116 \text{ atoms})(12 \text{ g/mole})}{6.02 \times 10^{23} \text{ atoms/mole}} = 2.52 \times 10^{-21} \text{ g}$$

Useful figures of merit for radioactive decay are:

- After 10 half-lives, 10^{-3} (or 0.1%) of the original quantity of radioactive material is left.
- After 20 half-lives, 10^{-6} of the original quantity of radioactive material is left.

Alpha, Beta, and Gamma Radiation

Emissions from radioactive nuclei are called, collectively, *ionizing radiation* because collision between these emissions and an atom or molecule ionizes that atom or molecule. Ionizing radiation may be characterized further as alpha, beta, or gamma radiation by its behavior in a magnetic field. Apparatus for such characterization is shown in Fig. 16-2. A beam of radioactively disintegrating atoms is aimed with a lead barrel at a fluorescent screen that is designed to glow when hit by the radiation. Alternately charged probes direct the α and β radiation accordingly. The γ radiation is seen to be "invisible light," a stream of neutral particles that passes undeflected through the electromagnetic field. α and β emissions have some mass, and are considered particles, while γ emissions are photons of electromagnetic radiation.

Alpha radiation has been identified as helium nuclei that have been stripped of their planetary electrons, and each consists of two protons and two neutrons. α particles thus have a mass of about 4 amu (6.642×10^{-4} g) each, and a positive charge of 2.[1]

They are emitted with kinetic energies between 4 and 10 MeV from the nuclei of relatively heavy elements. As these charged particles travel at approximately 10,000 miles/s, they collide with other atoms. Each collision results in a transfer of energy to the electrons of these other atoms, and thus in production of an ion pair: a negatively charged electron and an associated positively charged ion. An alpha particle produces between 30,000 and 100,000 ion pairs per centimeter of air traveled. Its kinetic energy is depleted rapidly and it has a range of between 1 and 8 cm in air and much less in denser media like solid objects or human skin cells. External radiation by α particles presents no direct health hazard, because even the most energetic are stopped by the epidermal layer of skin and rarely reach more sensitive layers. A health hazard occurs when material contaminated with α-emitting radionuclides is eaten or

---

[1] This electric charge is expressed in units relative to an electronic charge of -1.

[2] Mass is designated as energy in the equation $E = mc^2$. One megaelectron volt (MeV) $= 10^6$ electron volts (eV); 1 eV $= 1.603 \times 10^{-19}$ J, and is the energy of an electron accelerated through a field of a million volts.

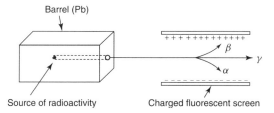

Figure 16-2. Controlled measurement of alpha (α), beta (β), and gamma (γ) radiation.

inhaled, or otherwise absorbed inside the body, so that organs and tissues more sensitive than skin are exposed to α radiation. Collisions between α particles and the atoms and molecules of human tissue may cause disorder of the chemical or biological structure of the tissue.

Beta radiation is a stream of electrons emitted at a velocity approaching the speed of light, with kinetic energy between 0.2 and 3.2 MeV. Given their lower mass of approximately 5.5×10^{-4} amu (9.130×10^{-24} g), interactions between β particles and the atoms of pass-through materials are much less frequent than α particle interactions: fewer than 200 ion pairs are typically formed in each centimeter of passage through air. The slower rate of energy loss enables β particles to travel several meters through air and several centimeters through human tissue. Internal organs are generally protected from external β radiation, but exposed organs such as eyes are sensitive to damage. Damage may also be caused by incorporation of β emitters into the body and resulting in exposure of internal organs and tissue.

Gamma radiation is invisible electromagnetic radiation, composed of photons, much like medical X-rays. γ photons are electrically neutral and collide randomly with the atoms of the material as they pass through. The considerably longer distance that γ rays travel in all media is defined by the *relaxation length*, the distance that the γ photon travels before its energy is decreased by a factor of $1/e$. A typical 0.7-MeV γ photon has a relaxation length of 5, 50, and 10,000 cm in lead, water, and air, respectively — much longer than an α or β particle of the same energy. External doses of γ radiation may have significant human health consequences because the dose is not greatly impacted by passage of the radiation through air. The properties of the more common radioactive emissions are summarized in Table 16-2.

When ionizing radiation is emitted from a nucleus, the nature of that nucleus changes: another element is formed and there is a change in nuclear mass as well. This process may be written as a nuclear reaction, in which both mass and charge must balance for reactants and products. For example, the beta decay of C-14 may be written

$$_6C^{14} = {_{-1}\beta^0} + {_7N^{14}}.$$

That is, C-14 decays to ordinary stable nitrogen (N-14) with emission of a beta particle. The mass balance for this equation is

$$14 = 0 + 14,$$

Table 16-2. Properties of Ionizing Radiation

Particle or photon (Wave)	Mass (amu)	Electric charge
Alpha ($_2He^4$)	4	+2
Beta (electron)	5.5×10^{-4}	-1
Gamma (X-ray)	~ 0	0
Neutron	1	0
Positron (positive electron)	5.5×10^{-4}	+1

and the charge balance is

$$6 = -1 + 7.$$

A typical reaction for α decay, the first step in the U-238 decay chain, is

$$_{92}U^{238} = {_2}He^4 + {_{90}}Th^{234}.$$

When a radionuclide emits a β, the mass number remains unchanged and the atomic number increases by 1 (β decay is thought to be the decay of a neutron in the nucleus to a proton and a β, with subsequent emission of the β). When a nuclide emits an α, the atomic mass decreases by 4 and the atomic number decreases by 2. γ emission does not result in a change of either atomic mass or atomic number.

Nuclear reactions may also be written for bombardment of nuclei with subatomic particles. For example, tritium (H-3) is produced by bombarding a lithium target with neutrons

$$_0n^1 + {_3}Li^6 = {_1}H^3 + {_2}He^4.$$

These reactions tell us nothing about the energy with which ionizing radiation is emitted, or the relative biological damage that can result from transfer of this energy in collisions. These effects are discussed briefly later in this chapter.

Units for Measuring Ionizing Radiation

Damage to living organisms is directly related to the amounts of energy transferred to tissue by collisions with α and β particles, neutrons, and γ radiation. This energy, in the form of ionization and excitation of molecules, results in heat damage to the tissue. Many of the units discussed in this section are thus related to energy transfer.

The International System of Units (SI units) for measuring ionizing radiation, based on the meter–kilogram–second (mks) system, was defined by the General Conference on Weights and Measures in 1960, and adoption of these units has been recommended by the International Atomic Energy Agency. SI units replaced the units that had been in use since about 1930. SI units are used in this chapter, and their relationship to the historical units is also discussed.

A *becquerel* (Bq) is the SI measure of source strength or total radioactivity, and is defined as one disintegration per second; the units of becquerels are second^{-1}. The decay rate dN/dt is measured in becquerels. The historical unit of source strength, the *curie* (Ci), is the radioactivity of 1 g of the element radium, and is equal to 3.7×10^{10} Bq.[3] The source strength in becquerels is not sufficient for a complete characterization of a source; the nature of the radionuclide (e.g., Pu-239, Sr-90) and the energy and type of emission (e.g., 0.7 MeV, γ) are also necessary.

The relationship between activity and mass of a radionuclide is given by

$$Q = \frac{(\ln 2)(N^0)\left(\frac{M}{W}\right)}{t_{1/2}} = K_b N^0 \left(\frac{M}{W}\right), \quad (16.4)$$

where

Q = number of becquerels,
K_b = the disintegration constant = $0.693/t_{1/2}$,
$t_{1/2}$ = half-life of the radionuclide, in seconds,
M = mass of the radionuclide, in grams,
N^0 = Avogadro's number, 6.02×10^{23} atoms/gram–atom, and
W = atomic weight of the radionuclide, in grams/gram–atom.

The *gray* (Gy) is the SI measure of the quantity of ionizing radiation that results in absorption of one joule of energy per kilogram of absorbing material. One gray is the equivalent of 100 *rad*, the historical unit of "radiation absorbed dose" equal to 100 ergs/g of absorbing material. The gray is a unit of *absorbed radiation dose*.

The *sievert* (Sv) is the SI measure of *absorbed radiation dose equivalent*. That is, 1 Sv of absorbed ionizing radiation does the same amount of biological damage to living tissue as 1 Gy of absorbed X-ray or γ radiation. The standard for comparison is γ radiation having a linear energy transfer in water of 3.5 keV/μm and a dose rate of 0.1 Gy/min. As previously stated, all ionizing radiation does not produce the same biological effect, even for a given amount of energy delivered to human tissue. The dose equivalent is the product of the absorbed dose in grays and a *quality factor* (QF) (sometimes called the *relative biological effectiveness*, or RBE). The historical dose equivalent unit is the *rem*, for "roentgen equivalent man." One sievert is equivalent to 100 rem. These units are related as follows:

$$QF = Sv/Gy = rem/rad.$$

Quality factors take into account chronic low-level exposures as distinct from acute high-level exposures and the pathway of the radionuclide into the human body (inhalation, ingestion, dermal absorption, external radiation) as well as the nature of the

---

[3] Marie Curie is said to have rejected a unit of one disintegration/second as too small and insignificant. The much larger unit of the curie was then established.

Table 16-3. Sample Quality Factors for Internal Radiation

Type of radiation	Quality factor
Internal alpha	10
Plutonium alpha in bone	20
Neutrons (atom bomb survivor dose)	20
Fission spectrum neutrons	2–100
Beta and Gamma	
$\quad E_{max}{}^a > 0.03$ MeV	1.0
$\quad E_{max} < 0.03$ MeV	1.7

$^a E_{max}$ refers to the maximum energy of emissions from the β or γ source.

ionizing radiation. Table 16-3 gives sample QFs for the internal dose from radionuclides incorporated into human tissue.

Dose equivalents are often expressed in terms of *population dose equivalent*, which is measured in *person–Sv*. The population dose equivalent is the product of the number of people affected and the average dose equivalent in sieverts. That is, if a population of 100,000 persons receives an average dose equivalent of 0.05 Sv, the population dose equivalent will be 5000 person–Sv. The utility of this concept will become evident in the discussion on health effects.

Table 16-4 gives 1996 estimated average radiation dose equivalents in the United States. The dose equivalent is product of the absorbed dose and the QF. The *effective dose equivalent* (EDE) is the risk-weighted sum of the dose equivalents to the individually irradiated tissues or organs. The *committed effective dose equivalent* (CEDE) is the total effective dose equivalent to the individual over a 50-year period. The EDE and CEDE are usually cited by the U.S. Environmental Protection Agency (EPA) and other regulatory agencies. Note that 55% of the "background" effective dose equivalent is from radon exposure and 82% is from natural (nonanthropogenic) sources. The effective dose equivalent is the number usually cited as average annual background exposure for an individual living in the United States. Different radionuclides exhibit different EDEs and CEDEs. These values may be found for each isotope in the *Handbook of Health Physics and Radiological Health* (Schleien et al. 1998).

Measuring Ionizing Radiation

The particle counter, the ionization chamber, photographic film, and the thermoluminescent detector are four methods widely used to measure radiation dose, dose rate, and the quantity of radioactive material present.

Particle counters are designed to detect the movement of single particles through a defined volume. Gas-filled counters collect the ionization produced by the radiation as it passes through the gas and amplify it to produce an audible pulse or other signal. Counters are used to determine radioactivity by measuring the number of particles emitted by radioactive material in a given time.

Table 16-4. Average Annual Dose Equivalent of Ionizing Radiation to a Member of the U.S. Population[a]

Source of radiation	Dose equivalent, mSv	(mrem)	Effective dose equivalent, mSv
Natural			
Radon	24	(2400)	2.0
Cosmic radiation	0.27	(27)	0.27
Terrestrial	0.28	(28)	0.28
Internal	0.39	(39)	0.39
Total	24.94		3.0
Anthropogenic			
Medical: diagnostic X-ray	0.39	(39)	0.39
Medical: nuclear medicine	0.14	(14)	0.14
Medical: consumer products	0.10	(10)	0.1
Occupational	0.009	(0.9)	<0.1
Nuclear fuel cycle	<0.01	(<1.0)	<0.1
Fallout	<0.01	(<1.0)	<0.1
Miscellaneous	<0.01	(<1.0)	<0.1
Total	0.67		0.67

[a] From A. Upton *et al.* (Eds.), National Academy of Sciences, *Health Effects of Exposure to Low Levels of Ionizing Radiation: BEIR V*, National Academy Press, Washington, DC, 1990.

Ionization chambers consist of a pair of charged electrodes that collect ions formed within their respective electric fields. Ionization chambers can measure dose or dose rate because they provide an indirect representation of the energy deposited in the chamber.

Photographic film darkens on exposure to ionizing radiation and is an indicator of the presence of radioactivity. Film is often used for determining personnel exposure and making other dose measurements for which a record of dose accumulated over a period of time is necessary, or for which a permanent record is required.

Thermoluminescent detectors (TLD) are crystals, such as NaI, that can be excited to high electronic energy levels by ionizing radiation. The excitation energy is then released as a short burst or flash of light, which can be detected by a photocell or photomultiplier. TLD systems are replacing photographic film because they are more sensitive and consistent. Liquid scintillators, organic phosphors operating on the same principle as TLDs, are used in biochemical applications.

HEALTH EFFECTS

When α and β particles, and γ radiation, interact with living tissue they transfer energy to the receiving material through a series of collisions with its atoms or nuclei.

Table 16-5. Representative LET Values

Radiation	Kinetic energy (MeV)	Average LET (keV/μm)	Dose equivalent
X-ray	0.01–0.2	3.0	1.00[a]
γ	1.25	0.3	0.7
β	0.1	0.42	1.0
	1.0	0.25	1.4
α	0.1	260	
	5.0	95	10
Neutrons	Thermal		4–5
	1.0	20	2–10
Protons	2.0	16	2
	5.0	8	2

[a]These equivalents are in terms of X-ray dose, which is defined as 1.00.

Molecules along the path of the ionizing radiation are damaged in the process, as chemical bonds are broken and electrons are ejected (ionization). Resulting biological effects are due mainly to the interactions of these electrons with molecules of tissue. The energy transferred through these collisions and interactions per unit path length through the tissue is called *linear energy transfer* (LET) of the radiation. The more ionization observed along the particle's path, the more intense the biological damage. LET can serve as a qualitative index for ranking ionizing radiation with respect to biological effect. Table 16-5 gives some typical LET values.

Biological effects of ionizing radiation may be grouped as *somatic* and *genetic*. Somatic effects are impacts on individuals who are directly exposed to the radiation. Radiation sickness (circulatory system breakdown, nausea, hair loss, and sometimes death) is an *acute* somatic effect occurring after very high exposure, as from a nuclear bomb, intense radiation therapy, or a catastrophic nuclear accident. Such an accident occurred at the Chernobyl nuclear electric generating plant (located near Kiev, Ukraine) in April 1986. Forty-five people received whole-body doses between 4 and 16 Gy, and died during the 50 days following the accident. An additional 158 people who received doses between 0.8 and 4 Gy suffered acute radiation sickness; all but one of these individuals recovered after treatment with red blood cell replacement.

A very tragic accident at a uranium processing facility at Tokaimura, Japan, in September of 1999 provided an opportunity to study these effects in detail, as well as developing treatment for acute radiation sickness. Three workers were transferring a uranium solution from one container to another without proper precautions, when the solution went critical, releasing a burst of neutrons. It is estimated that two of the workers received about 12 Sv and a third about 7 Sv. The two most heavily exposed individuals lived about three and nine months, respectively. The third exposed worker recovered. Treatment with tissue transplants and a variety of enzymes, blood cell replacement, and hormone treatment prolonged both the life and comfort of the exposed workers.

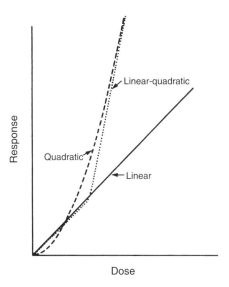

Figure 16-3. Linear-quadratic dose–response relationship.

Chronic effects resulting from long-term exposure to low doses of ionizing radiation may include both somatic and genetic effects that may occur because ionizing radiation damages the genetic material of the cell. Our knowledge of both somatic and genetic effects of low-dose ionizing radiation is based on animal studies and a very limited number of human epidemiological studies: studies of occupational exposure, the Japanese Atomic Bomb Survivors' Life Study, and studies of effects of therapeutic radiation treatment. Based on extrapolation from these studies, the dose–response relationship has been estimated to be linear at low doses and quadratic at higher doses (Fig. 16-3). Much of the low-dose part of the curve is below any range of experiment. As with all carcinogens, the absence of a threshold has been assumed. The extrapolated linear dose–response relationship is often referred to as the linear-nonthreshold (LNT) theory of radiation health effect.

Reevaluation of human epidemiological studies indicates that there may be a threshold below which there is no chronic effect. That is, there is growing evidence that the LNT theory may be incorrect. A study of cancer incidence in Japanese atom bomb survivors showed that excess cancer incidence among survivors exposed to less than 0.02 Gy (2 rad) was actually lower than in a population that had no exposure above background.[4] Similarly, a study of oral and laryngeal cancers in radium watch dial painters showed a clear threshold below which no effects due to radium exposure were observed.[5] There has been no similar verification of the LNT theory. Revision of

---

[4] S. Hattori, "State of Research and Radiation Hormesis in Japan," *J. Occupational Med. Toxicol.* **3**, 203–217, 1994.

[5] R. G. Thomas, "The US Radium Luminisers: A Case for a Policy of 'Below Regulatory Concern,'" *J. Radiol. Protection*, **14**, 141–153, 1994.

Table 16-6. Yields of DNA Damage Necessary to Kill 63% of the Cells Exposed[a]

Agent	Number of DNA lesions per damaged cell
Ionizing radiation	1000
UV light	400,000
Hydrogen peroxide	2,600,000
Benzo(a)pyrene 4,5-oxide	100,000
Aflatoxin	10,000

[a] From National Academy of Sciences (1990).

Table 16-7. Average Fatal Somatic Risk from Lifetime Exposure to 1 mSv/yr Low LET Radiation

Year	Dose model	Effects per 10^5 persons
1977	Quadratic	75–175
1980	Linear	403
1980	Quadratic	169
1985	Linear	280
1990	Linear-quadratic for males	520
1990	Linear-quadratic for females	600
1996	Linear (general public)	500
1996	Linear (occupational)	400

the dose–response relationship for ionizing radiation would result in drastic changes in regulatory standards and limits for radioactive pollutants.

Although we do not know the precise mechanism by which ionizing radiation produces somatic and genetic effects, we understand that it involves damage to the DNA of the cell nucleus. Table 16-6 compares DNA damage caused by a variety of carcinogens. If there is a threshold for effects of ionizing radiation, it would be related to the high frequency of DNA repair that occurs naturally in human and other organisms.

Somatic effects include decrease in organ function and carcinogenesis. A variety of estimates of average fatal somatic risk, using a variety of dose–response models, have been developed over the years. These estimates are summarized in Table 16-7. The 1996 estimates given in Table 16-7 are from Publication 60 of the International Commission on Radiation Protection and are usually used in environmental impact assessments.

Genetic effects result from radiation damage to chromosomes and have been demonstrated to be inheritable in animals, but not in the human population. The human genetic risk is estimated to be between 1 and 45 additional genetic abnormalities per 10 mSv (1 rem) per million liveborn offspring in the first generation affected, and between 10 and 200 per 10 mSv per million liveborn at equilibrium. The spontaneous rate of human genetic abnormality is presently estimated to be about 50,000 per million liveborn in the first generation.

Health risk from ionizing radiation may be summarized as follows: there is a documented risk from ionizing radiation, but it is apparently small, clearly uncertain, and depends on a number of factors. These include:

- the magnitude of the absorbed dose,
- the type of ionizing radiation,
- penetrating power of the radiation,
- the sensitivity of the receiving cells and organs,
- the rate at which the dose is delivered,
- the proportion of the target organ or organism exposed, and
- the possibility of a threshold below which there is no discernible damage.

The risk from various radionuclides affecting health in various ways is summarized in an excellent reference, the *Handbook of Health Physics and Radiological Health*. To protect public health, the environmental engineer must do what he or she can to minimize unnecessary radioactive exposure of both the general public and those who work with ionizing radiation, including handling and disposal of radioactive waste.

SOURCES OF RADIOACTIVE WASTE

The nuclear fuel cycle, radiopharmaceutical manufacture and use, biomedical research and application, nuclear weapons manufacture, and a number of industrial uses generate radioactive waste. The behavior of the radionuclides in waste (as in any other form) is determined by their physical and chemical properties; radionuclides may exist as gases, liquids, or solids and may be soluble or insoluble in water or other solvents.

Until 1980, there was no classification of radioactive wastes. The U.S. Nuclear Regulatory Commission (NRC) has classified radioactive wastes and other materials into the following categories:

• *High-level waste* (HLW). HLW includes spent nuclear fuel from commercial nuclear reactors and the solid and liquid waste from reprocessing spent or irradiated fuel.[6] The NRC reserves the right to classify additional materials as HLW as necessary.

• *Uranium mining and mill tailings*. The pulverized rock and leachate from uranium mining and milling operations.

• *Transuranic waste* (TRU). Radioactive waste that is not HLW but contains more than 3700 Bq (100 nCi) per gram of elements heavier than uranium (the elements with atomic number higher than 92). Most TRU waste in the United States is the product of defense reprocessing and plutonium production.

---

[6]"Reprocessing" is the chemical treatment of irradiated fuel to recover plutonium and fissile uranium.

- *By-product material.* Any radioactive material, except fissile nuclides, produced as waste during plutonium production or fabrication.
- *Low-level waste* (LLW). LLW includes everything not included in one of the other categories. LLW is not necessarily less radioactive than HLW and may have higher specific activity in becquerels per gram. The distinguishing feature of LLW is that it contains virtually no alpha emitters. To ensure appropriate disposal, the NRC has designated several classes of LLW:

Class A contains only short-lived radionuclides or extremely low concentrations of longer-lived radionuclides, and must be chemically stable. Class A waste may be disposed of in designated LLW landfills as long as it is not mixed with hazardous or flammable waste.

Class B contains higher levels of radioactivity and must be physically stabilized before transportation or disposal. It must not contain free liquid.

Class C is waste that will not decay to acceptable levels in 100 years and must be isolated for 300 years or more. Power plant LLW is in this category.

Greater than Class C (GTCC) will not decay to acceptable levels in 300 years. A small fraction of power plant Class C waste is in this category. Some nations like Sweden treat GTCC waste like HLW, and the United States may do the same.

Mixed Low-Level Waste (MLLW) is LLW that contains hazardous waste as defined under the Resource Conservation and Recovery Act (RCRA).

Formerly used sites Remedial Action Program (FUSRAP) waste is contaminated soil from radium and World War II uranium refining and atomic bomb development. Little was known at that time about the long-term effects of ionizing radiation. Consequently, there was widespread contamination in the structures in which this work was done and of the land surrounding those structures. FUSRAP waste contains very low concentrations of radionuclides, but there is a large amount of such waste.

The Nuclear Fuel Cycle

The nuclear fuel cycle (shown in Fig. 16-4) generates radioactive waste at every stage. Uranium mining and milling generate the same sort of waste that mining and milling operations generate, including acid mine drainage, as well as radioactive uranium daughter elements, including a considerable amount of Rn-222. Mining and milling dust must be stabilized to prevent both windborne dispersion and leaching into ground and surface water.

Partially refined uranium ore, called "yellowcake" because of its bright yellow color, must be enriched in the fissile isotope U-235 before nuclear fuel (as used in the United States) can be fabricated from it. Mined uranium is more than 99% U-238, which is not fissile, and only 0.711% U-235. The concentration of U-235 is increased to about 3% to 5% by converting to UF_6 and concentrating the lighter isotope by gas diffusion or gas centrifugation; the UF_6 enriched in the lighter isotope is then converted to UO_2 and fabricated into fuel. Enrichment and fabrication both produce low-level waste. In addition, enrichment also produces depleted uranium

Radioactive Waste 327

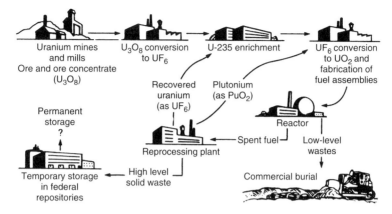

Figure 16-4. The nuclear fuel cycle.

(any uranium that contains less than 0.711% U-235 by weight), an exceedingly dense and pyrophoric material used in radiation shielding and in armor-piercing weapons.

Nuclear fuel is inserted into a reactor core, where a controlled fission reaction produces heat, which in turn produces pressurized steam for electric power generation. The steam system, turbines, and generators in a nuclear power plant are essentially the same as those in any thermal (fossil fuel burning) electric generating plant. The difference between nuclear and fossil fuel electric generation is in the evolution of the heat that drives the plant.

Figure 16-5 is a diagram of a typical pressurized water nuclear reactor. Commercial reactors in the United States are either pressurized water reactors, in which the water that removes heat from the nuclear reactor core (the "primary coolant") is under pressure and does not boil, or boiling water reactors, in which the primary coolant is permitted to boil.[7] The primary coolant transfers heat from the core to the steam system (the "secondary coolant") by a heat exchange system that ensures complete physical isolation of the primary from the secondary coolant. A third cooling system provides water from external sources to condense the spent steam in the steam system.

All thermal electric power generation produces large quantities of waste heat. Fossil fuel electric generating plants have, typically, a thermal efficiency of about 42%; that is, 42% of the heat generated by combustion of the fuel is converted to electricity, and 58% is dissipated in the environment. By comparison, nuclear plants are about 33% thermally efficient.

The nuclear reactor is perhaps the key radioactive waste producer in the nuclear fuel cycle. The production of HLW is a direct consequence of fission reactions in

---

[7]The Fort St. Vrain Plant reactor in Colorado, now being decommissioned, used helium gas as the primary coolant.

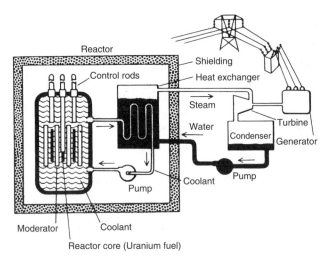

Figure 16-5. Typical pressurized water nuclear reactor.

the core. One such reaction is

$$_{92}U^{235} + {_0}n^1 = {_{42}}Mo^{95} + {_{57}}La^{139} + 2\,{_0}n^1 + 204 \text{ MeV}.$$

Several features of this reaction merit discussion.

- A large amount of energy is released in the reaction: 204 MeV per uranium atom, or 80 million Btu per gram of uranium. Uranium fission released about 100,000 times as much heat per gram of fuel as natural gas combustion. The development of commercial nuclear power is based on this phenomenon.
- One neutron is required for the fission reaction, but the reaction itself produces two neutrons, each of which can initiate another fission reaction that will produce two neutrons, and so on, resulting in a *fission chain reaction*. However, the concentration of fissile material (U-235 in commercial power reactors) must be high enough so that neutrons produced are likely to collide with U-235 nuclei. The mass of fissile material needed to sustain a fission chain reaction is called the *critical mass*.
- Neutron flow in a reactor may be interrupted, and the fission reaction stopped, by inserting control rods that absorb neutrons (Fig. 16-5). Although inserting control rods stops the fission reaction, the insertion does not stop heat generation in the core, since the many radioactive fission products continue to emit energy. Therefore, continued coolant flow is critical.
- In this particular reaction, Mo-95 and La-139 are the products of the fission. However, a fissile nucleus can split apart in about 40 different ways, yielding approximately 80 different fission fragments. Although many of these have very short half-lives and decay very quickly, some have long half-lives. These long half-life fission products

Table 16-8. Long-Lived Reactor Fission Products

Isotope	Type of emission	Half-life	Ci produced/MW$_t$–yr
Kr-85	Beta/gamma	10.6 years	500
Sr-89	Beta	53 days	23,000
Sr-90	Beta	29 years	2600
Y-91	Beta/gamma	58 days	33,500
Zr-95	Beta	65 days	50,000
Nb-95	Beta/gamma	35 days	52,000
Ru-103	Beta	40 days	40,000
Ru-106	Beta	1 year	19,000
Ag-111	Beta	7.6 days	1300
Sn-125	Beta	9.4 days	420
Sb-125	Beta	2 years	200
Te-127	Gamma	105 days	580
Te-129	Beta	33.5 days	3000
I-129	Beta	1.6×10^7 years	<200
I-131	Beta	8.3 days	28,000
Xe-131	Beta	12 days	270
Xe-133	Beta/gamma	5.3 days	56,500
Cs-134	Beta	2.1 years	2000
Cs-137	Beta	30 years	3600
Ba-140	Beta	13 days	45,000
Ce-141	Beta	33 days	48,000
Pr-143	Beta	13.7 days	46,000
Ce-144	Beta	286 days	38,000
Nd-147	Beta	11 days	19,000
Pm-147	Beta	2.6 years	7600
Pm-148	Gamma	42 days	620
Eu-156	Beta	14 days	4800

make up most of the radioactivity of HLW. Table 16–8 lists the relatively long-lived fission products from a typical reactor.

As can be seen from the Table 16-8, this mixture of fission products has a very high specific radioactivity. Fission products are beta or gamma emitters, because they are too small to emit alpha particles. The longest-lived fission products, Cs-137 and Sr-90, have half-lives of 30 and 29 years, respectively. A large quantity of fission products containing these two radionuclides could thus be a significant source of radioactivity for 600 years.

About one fission reaction in 10,000 yields three fission fragments instead of two; the third fragment is tritium (H-3), which has a 12.3-year half-life. Since tritium is chemically virtually identical to hydrogen, it exchanges freely with nonradioactive hydrogen gas in a reactor and with H$^+$ ions in the reactor cooling water. Containment of tritium is thus difficult.

In a fission reaction, not all of the neutrons will collide with a fissile nucleus to initiate fission. Some neutrons will collide, and react with, the fuel container and the reactor vessel itself, producing neutron activation products. Some of these are relatively long-lived, notably Co-60 (half-life, 5.2 years) and Fe-59 (half-life, 45 days). Other neutrons react with nonfissile isotopes, like U-238, as in the reaction

$$_{92}U^{238} + _{0}n^{1} = _{92}U^{239} = _{93}Np^{239} + _{-1}e^{0}$$

$$_{93}Np^{239} = _{94}Pu^{239} + _{-1}e^{0}.$$

Pu-239, "weapons-grade" plutonium, has a half-life of 24,600 years. Other isotopes of plutonium are formed also. Radionuclides like U-238, from which fissile material like plutonium can be produced by nuclear reaction, are known as *fertile* isotopes. Another isotope of plutonium, Pu-241, decays by beta emission to Am-241 (half-life, 433 years). Am-241 is commonly used as an ion producer in smoke detectors.

Each uranium isotope is part of a decay series; when U-238 decays, the daughter element is also radioactive and decays, producing another radioactive daughter,

Table 16-9. U-238 Decay Series

Isotope	Type of emission	Half-life
U-238	Alpha	4.5×10^9 years
Th-234	Beta	24.1 days
Pa-234	Beta	1.18 min
U-234	Alpha	248,000 years
Th-230	Alpha	80,000 years
Ra-226	Alpha	1620 years
Rn-222	Alpha	3.28 days
Po-218	Alpha/beta	3.05 min
At-218	Alpha	2 s
Pb-214	Beta	26.8 min
Bi-214	Alpha/beta	19.7 min
Po-214	Alpha	0.00016 s
Tl-210	Beta	1.32 mins
Pb-210	Beta	19.4 years
Bi-210	Alpha/beta	5.0 days
Po-210	Alpha	138.4 days
Tl-206	Beta/gamma	4.2 min
Pb-206		Stable

Note. The chain branches where two emissions are listed; e.g., Pb-214 and At-218 are both daughters of Po-218.

and so on, until a stable element (usually a lead isotope) is reached. Table 16-9 shows the U-238 decay series. The plutonium isotopes also decay in a series of radionuclides. Spent nuclear fuel (HLW) thus contains fission products, tritium, neutron activation products, plutonium, and plutonium and uranium daughter elements in a very radioactive, very long-lived mixture that is chemically difficult to segregate.

Reprocessing Waste and Other Reactor Waste

In the United States, plutonium is used only for nuclear warheads, which were manufactured from the end of World War II until 1989. Plutonium was produced by irradiating U-238 with neutrons (in the reaction given above) in military breeder reactors. The irradiated fuel was then completely dissolved in nitric acid, and plutonium along with fissile uranium and neptunium were extracted with tributyl phosphate. Further partition and selective precipitation resulted in recovery of plutonium, uranium, neptunium, strontium, and cesium.

The fissile isotopes of plutonium and uranium are categorized as *special nuclear material*; the other nuclides are considered by-product material. The term *reprocessing* refers to the entire process of extracting fissile material from irradiated fuel. Although the production and extraction of special nuclear material have ceased, the very large quantities of neutralized acid solvent sludge and the organic extraction solvents used contain high concentrations of radionuclides and are classified as HLW. The process also yielded TRU waste and LLW.[8]

Commercial fuel is not currently reprocessed in the United States. In France, plutonium for nuclear power generation is produced from fertile material in the Superphenix breeder reactor. Plutonium for nuclear weapons has not been produced in the United States since 1993, and surplus plutonium from the U.S. weapons program is now being fabricated into mixed uranium/plutonium oxide (MOX) fuel. Great Britain and France also produce MOX fuel. Although no MOX fuel is presently burned in U.S. commercial nuclear power plants, it is being used in Europe and Canada.

The primary and secondary coolants in a nuclear generating plant pick up considerable radioactive contamination through controlled leaks. Contaminants are removed from the cooling water by ion exchange and the loaded ion-exchange columns are Class C or GTCC waste. Class A and Class B wastes are also produced in routine cleanup activities in nuclear reactors.

After 30 to 40 years of operation, the reactor core and the structures immediately surrounding it will have become very radioactive, primarily by neutron activation, and the reactor must be shut down and decommissioned. At present, 10 commercial reactors in the United States have either been decommissioned or are undergoing decommissioning.

---

[8]Information about the exact quantities of particular constituents of reprocessing waste is not made available to the public for reasons of national security.

Additional Sources of Reactor Waste

The increasingly widespread use of radionuclides in research, medicine, and industry has created a lengthy list of potential sources of radioactive waste. This list ranges from laboratories using a few isotopes, to medical and research laboratories where large quantities of many different radioisotopes are used (and often wasted), to a growing list of industrial uses like well logging.

Liquid scintillation counting has become an important biomedical tool and produces large volumes of mixed hazardous and radioactive waste (MLLW): organic solvents like toluene that are contaminated, usually with tritium or C-14, but have a low specific radioactivity. The RCRA prohibits landfill disposal of the solvents in an LLW site, and there is considerable public opposition to incineration of any radioactive material, no matter how low the specific activity. MLLW incinerators are operated at two U.S. Department of Energy sites.

Naturally occurring radionuclides like those listed in Table 16-10 may pose threats to public health. As may be seen in Table 16-4, by far the largest fraction of background ionizing radiation comes from radon (Rn-222). Rn-222 is a member of the uranium decay series and is thus found ubiquitously in rock. Chemically, Rn-222 is an inert gas, like helium, and does not become chemically bound. When uranium-bearing minerals or rock are crushed or machined, Rn-222 is released; there is even a steady release of

Table 16-10. Some Radionuclides of Concern

Isotope	Type of emission	Half-life
Cosmic ray production		
H-3	Beta	12 years
Be-10	Beta	2.5×10^6 years
C-14	Beta	5730 years
Si-32	Beta	650 years
Cl-36	Beta	310,000 years
Rock and building materials		
Rn-222	Alpha	3.8 days
Uranium + daughters	Alpha/beta/gamma	
Thorium + daughters	Alpha/beta/gamma	
K-40	Beta	1.3×10^9 years
Atmospheric fallout and radioactive waste		
C-14	Beta	5730 years
Cs-137	Beta	30 years
Sr-90	Beta	29 years
I-131	Beta	8.3 days
H-3	Beta	12 years

Rn-222 from undisturbed rock outcrops. Buildings that are insulated to prevent excessive heat loss often have too little air circulation to keep the interior purged of Rn-222; thus concentrations of Rn-222 in homes and commercial buildings with restricted circulation can become quite high. Rn-222 itself has a short half-life, but decays to shorter and longer-lived metallic radionuclides (see Table 16-9). The combined dose from radon and its radioactive progeny can be significant when radon is inhaled.

Coal combustion, copper mining, and phosphate mining release isotopes of uranium and thorium into the environment. K-40, C-14, and H-3 are found in many foodstuffs. Although atmospheric nuclear testing was discontinued about 30 years ago, radioactive fallout from past tests continues to enter the terrestrial environment.

MOVEMENT OF RADIONUCLIDES THROUGH THE ENVIRONMENT

A radionuclide is an environmental pollutant because of a combination of the following properties:

- Half-life
- Chemical nature and properties: a radioactive isotope will behave chemically and biochemically like stable (nonradioactive) isotopes of the same element.
- Abundance
- Nature of the radioactive emissions

Like any other waste material, radioactive wastes may contaminate air, water, soil, and vegetation, and this contamination may affect public health adversely. Figure 16-6 shows the pathways of contamination from a hypothetical source. Doses of radiation from environmental contamination are usually classified by pathway, that is, as *inhalation dose, ingestion dose*, and *immersion dose*.

Radionuclides released and transported through the air that people breathe enter the human body by the atmospheric dispersion and inhalation pathway. The NRC prescribes strict containment of almost all airborne radionuclides, although there can be accidental releases. However, atmospheric releases of Kr-85, Xe-134, radioiodine, and tritium from boiling water reactors are planned, since these are gases and cannot be completely trapped, as are some gaseous radionuclide releases from fuel reprocessing plants. During the period 1945–1955 there were deliberate releases of I-131 from defense reprocessing facilities. Today, the greatest amount of radioactivity from planned releases is the venting of Rn-222 to the air.

The atmospheric pathway has received considerable attention because of the analogies between radioactive emissions and the more ordinary gaseous air pollutants generated by industries, automobiles, diesel trucks, etc. Chapters 18 and 19 discuss air pollution dispersion and meteorology; most of that discussion is equally applicable to airborne radioactive gases and particles. Airborne radionuclides can also enter the ingestion or food-chain pathway through deposition on the soil and on vegetation. The accident at the Chernobyl nuclear reactor, discussed previously, which released a very large quantity of radioactive material into the air, resulted in radioactively contaminated

334 ENVIRONMENTAL ENGINEERING

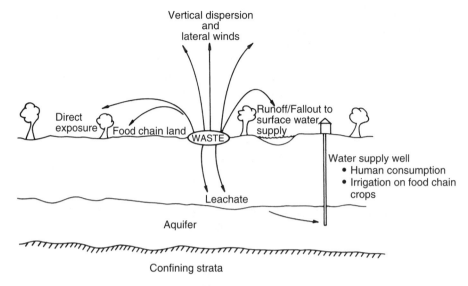

Figure 16-6. Potential movement of radioactive materials from waste storage and "disposal" areas to the accessible environment.

food in several Eastern and Northern European countries. Close to the Chernobyl plant, enough radioactive material was deposited from the air to produce an external, or immersion, dose; this sort of external dose is sometimes referred to as *groundshine*.

Water transport occurs whenever radionuclides in surface or subsurface soil, or on vegetation surfaces, erode or leach into a watercourse, as well as when airborne radionuclides fall out on surface waters. Leaching into waterways, and transport of radionuclides in groundwater, can take enough time so that shorter half-life nuclides may have decayed to negligible levels before entering the human food chain. These processes remain problematical for long half-life nuclides, however. Chapters 6 and 11 address water transport considerations that are somewhat applicable to the problems associated with radioactive waste.

Radionuclides also move from deposition or contamination of ground surfaces into the human food chain. Plant uptake mechanisms cannot distinguish between radioactive and stable isotopes of the same element, and will take up radioactive isotopes of stable nutrient elements. Animals eating the plants will absorb radioisotopes in the same way. For example, a cow eating a plant contaminated with Sr-90 will metabolize the radionuclide into her milk, and thus pass it along in the food chain.

RADIOACTIVE WASTE MANAGEMENT

The objective of the environmental engineer is to prevent the introduction of radioactive materials into the biosphere, and particularly into the environment accessible to human activity, for the effective lifetime (about 20 half-lives) of these materials. Control of the potential direct impact on the human environment is necessary but not sufficient, because radionuclides can be transmitted through water, air, and land pathways for many years and even for generations. Some radioactive waste may be retrieved and recycled by future reprocessing, but reprocessing creates its own radioactive waste stream. Most radioactive waste can be treated only by isolating it from the accessible environment until its radioactivity no longer poses a threat. Isolation requirements differ for different classes of radioactive waste.

Engineers charged with radioactive waste control must, because of technical, political, and economic factors outside their control, focus on long-term storage technologies, i.e., disposal. Some radionuclides, particularly those that make up HLW, have half-lives of tens of thousands of years, or even hundreds of thousands of years. It is difficult to imagine a technology that truly offers ultimate disposal for these wastes; thus we think in terms of very long-term storage. Many of the issues discussed in Chapter 15 are applicable to the radioactive waste problem.

High-Level Radioactive Waste

A number of options were considered for HLW disposal, and it is agreed internationally that mined geologic disposal offers the best option for isolation, although research on transmutation (of long-lived into shorter-lived radionuclides) is continuing.

In 1979, the U.S. Geological Survey (USGS) recommended a two-stage barrier system for the geologic disposal of HLW, in order to minimize the likelihood of leakage from the repository and dispersal in the accessible environment. The waste form itself — radioactive material dissolved and dispersed in a glass matrix — would provide the first barrier. The second barrier would be the geologic rock formation itself. This double-barrier scheme is being implemented for defense reprocessing HLW. Commercial spent nuclear fuel, however, is not reprocessed and will be stored in the geologic repository in the form in which it leaves the reactor core: spent fuel rods. These rods will be sealed in heavy casks of steel and depleted uranium.

When the fissile uranium in a bundle of fuel rods has been used to a point where the fission rate is too slow for efficient power generation, the rods are ejected into a very large pool of water, where they remain until the short-lived radionuclides have decayed away and until the rods are thermally cool enough to handle with ordinary machinery. This takes about six months, but the lack of any other storage facility for most spent fuel in the United States has resulted in on-site pool storage for as long as 10 years. In 1998, the U.S. Nuclear Regulatory Commission approved casks for dry surface storage for sufficiently cooled spent nuclear fuel, and several nuclear power plants now have dry surface storage capability.

The United States is bound by nonproliferation agreements to accept and store spent fuel that this nation has supplied to nuclear reactors in other countries. At present, foreign spent fuel is stored at two U.S. Department of Energy facilities: the Savannah River Site in South Carolina and the Idaho National Engineering and Environmental Laboratory in Idaho. When a repository is available, the aged and cooled spent fuel will be loaded into casks for disposal in the repository.

Investigations for a mined geologic repository began in the United States in 1972 with Project Salt Vault, a study of a mined-out salt mine near Lyons, KS. Salt mines are still under investigation as HLW repositories; two German sites, at Gorleben and Asse, are being investigated (or, in the terminology of the nuclear waste industry, *characterized*). Granite, frozen clay, and basalt have been investigated as repository sites in various parts of the world. In the United States, investigations have shifted to hard rock formations. The U.S. Nuclear Waste Policy Act, as amended in 1987, mandates that only the volcanic tuff at Yucca Mountain, NV (about 100 miles NNW of Las Vegas, NV) be characterized, and that no other site be considered unless the Yucca Mountain is found unsuitable.

The U.S. HLW repository was scheduled to begin receiving waste in 1998 and to close, sealing the waste permanently and irretrievably, in 2098. Delays in characterization have delayed the opening of the repository, which is now projected to occur in 2010. A temporary storage site for commercial spent fuel, from which the spent fuel could be retrieved at some future time for placement in a repository, is also being considered.

Transuranic (TRU) Waste

A salt formation in southeastern New Mexico, under investigation since 1978 as a possible radioactive waste repository, began receiving TRU waste during the spring of 1999. The Waste Isolation Pilot Project (WIPP) was found to be better suited as a repository for TRU defense waste than for the more thermally stressful HLW. The WIPP was completed and certified by the U.S. Environmental Protection Agency and received its first shipment of TRU waste in March 1999. The facility — the only operating geologic radioactive waste repository in the world — is now receiving about 100 shipments of TRU waste each month.

Low-Level Radioactive Waste

Unlike HLW, commercial LLW has been the responsibility of the private sector since 1960. It has been disposed of, without any treatment at all, in shallow burial trenches about 60 ft long with a floor area of about 25,000 ft^2. Three of the existing commercial sites were shut down between 1975 and 1978 because of leaks of radionuclides into surface water and drinking water. Three of the original sites remain open today, at Hanford, WA, Barnwell, SC, and Beatty, NV. The last two of these are nearing capacity and are due to be closed. An additional commercial site in north-central Utah is now also receiving waste.

Before 1980 there were no uniform disposal regulations or practices for handling LLW; in fact, there was no regulatory definition of LLW. Passage of the Low-Level Radioactive Waste Policy Act of 1980 (and its subsequent amendments) and the promulgation of 10 CFR Part 61 of the U.S. Code of Federal Regulations, the regulations for shallow land disposal of LLW, have improved both the method of selecting LLW disposal sites and the environmental safety of LLW handling.

Techniques now being applied to LLW include impervious packaging, compaction, incineration, and stabilization in an asphalt or cement matrix. Compaction can achieve volume reduction of about 8:1 and incineration can reduce volumes by factors of 30 or more. Incineration appears to be an excellent treatment option for mixed radioactive and organic chemical wastes like liquid scintillators. Incineration emissions require careful monitoring and decontamination factors of 10,000 or more. Air emission controls are discussed further in Chap. 20.

The Low-Level Radioactive Waste Policy Act of 1980 as amended requires states to form regional interstate compacts and site regional LLW disposal facilities. Siting and operating these facilities will require careful environmental engineering.

TRANSPORTATION OF RADIOACTIVE WASTE

Some radioactive waste, particularly HLW, has such high activity that blocking all external radiation during transportation would make the transport container far too heavy for heavy truck transport or even for rail transport. As a result, transported radioactive material is allowed a certain external dose by Nuclear Regulatory Commission regulation. This dose, called the Transport Index, is the dose rate in mrem/hour at a perpendicular distance of 1 m from the outside edge of the vehicle or trailer. Transport containers must also meet certain other regulatory requirements. For most LLW, the container must be tested and certified to withstand the rigors and physical stresses of normal transportation. For HLW, TRU, and some LLW, the container must be able to withstand certain accident conditions, including combinations of fire and mechanical stress.

CONCLUSION

Energy can be transferred from one place to another by direct radiation. Radiated energy travels in straight lines and can be visualized either as a stream of particles or as waves. In its wave formulation, energy forms a continuous spectrum from X-rays at the very-short-wavelength end to radar and heat at the very-long-wavelength end. Very-short-wavelength radiation can ionize atoms and molecules. Its source can be the spontaneous process in which unstable atoms of an element emit excess energy from their nuclei, as well as anthropogenic chain reactions in nuclear power facilities and nuclear detonations.

This ionizing energy flux may have adverse impacts on biological matter. Radiation sickness, cancer, shortened life, or immediate death may result from varying exposures. Radiation doses to living tissue are measured as grays or rads, units of absorbed energy, and sieverts or rems, units of relative biological damage. These units take into account the effects on living tissue of alpha, beta, gamma, and neutron radiation.

The general concern of the environmental engineer is radiation from anthropogenic sources, particularly the wastes from nuclear power plants, radiation from natural

338   ENVIRONMENTAL ENGINEERING

sources like mine tailings, and radon, because it is ubiquitous. Wastes are handled in much the same manner as the hazardous wastes discussed in Chap. 15.

The United States no longer operates plutonium production reactors, and complete shutdown of the nuclear industry is suggested from time to time by various groups. Even if the U.S. nuclear industry were shut down, the existing waste would have to be dealt with. Moreover, other countries throughout the world are not simply maintaining their nuclear power production capability, but expanding it. Like virtually every other pollution problem, radioactive waste is a worldwide concern, and not limited to the United States or the Western Hemisphere.

In the late 1990s, the shutdown of the U.S. nuclear industry, or its decline by attrition, was seriously considered. Localized electricity shortages, like the shortage that forced "rolling blackouts" in California in 2000 and 2001, has led to a rethinking of the future role of nuclear energy in the U.S. energy picture. At present, nuclear power supplies about 20% of the electricity in the United States. In a few states the share is about 40%, and in one state, Illinois, it is better than 60%. In the event of shutdown, no more than about 25% of this energy could be saved by conservation. So-called "alternative" energy sources like wind power, solar thermal generation, geothermal energy, and small-scale hydropower have not yet been implemented on a large scale. The adverse environmental effects of large-scale implementation are as yet unknown, but cannot arbitrarily be assumed to be small or negligible.

Other methods of thermal generation of electricity, combustion of fossil fuels, have quite well-understood adverse environmental effects. The adverse effects of burning coal are particularly severe. In addition, fuel combustion contributes to the concentration of CO_2 in the earth's atmosphere, while nuclear electric generation does not. Thorough and objective comparisons of electric generating methods should guide decisions about power generation. Generating electricity, no matter how it is done, causes irrevocable environmental damage on a scale directly proportional to the power produced.

During the past half-century, the pendulum of public opinion swung from the "pro-nuclear" extreme of the 1950s and 1960s to the "anti-nuclear" extreme of the 1980s and 1990s. Politics tends to follow public opinion, but political decisions, including those about nuclear power and radioactive waste, are emotion-laden and not always scientifically or environmentally sound. The role of engineers, now more than ever, is to inform and influence political decisions.

PROBLEMS

16.1 Show that after 10 half-lives, about 0.1% of the initial radioactive material is left, and that after 20 half-lives, 10^{-4}% is left.

16.2 Fe-55 has a half-life of 2.4 years. Calculate the disintegration constant. If a reactor core vessel contains 16,000 Ci of Fe-55, how many curies will be left after 100 years? How many becquerels will be left? How many grams?

16.3 I-131 is used in the diagnosis and treatment of thyroid disease, and a typical diagnostic amount is 1.9×10^6 Bq injected intravenously. If the *physiological* half-life of iodine is 15 days and a patient injected with I-131 produces about 3.5 L of urine a day, what will be the specific activity (in Bq/L) of the patient's urine on the 10th day after injection? What will be the specific activity in curies per liter (Ci/L)?

16.4 Write nuclear reactions for the following, identifying the product elements and particles:

   a. Fusion of two deuterium atoms to form He-3,
   b. Beta decay of Sr-90,
   c. Beta decay of Kr-85,
   d. Neutron emission from Kr-87, and
   e. Decay of Th-230 to Ra-226.

16.5 The dose in Problem 16.3 is adjusted not to exceed 0.5 Gy in 24 h. Calculate the somatic risk in LCF from this dose. If there were a threshold of adverse effect at 0.2 Gy, would there still be somatic risk?

16.6 How many grams of Cs-137 are produced during each 24-h day of operation by a nuclear power plant that has an electrical output capacity of 750 MW_e? How many grams would be left if this amount were allowed to decay for 100 years?

16.7 From Table 16-4, calculate the energy absorbed by a 130-lb female in one year from natural (nonanthropogenic) sources of ionizing radiation. Repeat the calculation for a 175-lb male. What assumptions did you make?

16.8 The EPA considers a risk of 10^{-6} (one in a million) per source of radioactivity per year acceptable. Calculate the dose, in sieverts of low-level gamma radiation that corresponds to this risk, assuming the linear nonthreshold theory.

16.9 The EPA has set a standard for the WIPP TRU waste repository as follows: in the first 10,000 years after closure, the probability that 1/10,000 of the plutonium activity in the repository can leak out is 0.1, and the probability that 10 times this amount of plutonium activity can leak out is 0.001. Show this standard in a two-dimensional graph. What leak rate would have a probability of 1? (Hint: plot probability on the vertical axis and leak rate on the horizontal axis.)

16.10 The first three steps in the ^{238}Pu decay chain are

$$^{238}_{94}Pu \rightarrow ^{234}_{92}U \rightarrow ^{230}_{90}Th.$$

The half-lives of these three radionuclides are Pu-238 87 years, U-234 248,000 years, and Th-230 73,400 years. Solving a series of equations like Eq. (16.1), and starting with 10 g of Pu-238, write an equation for the amount of Th-230 formed as a function of time. This is not a trivial problem! If you cannot solve the problem analytically, try solving by iteration using a spreadsheet program, and plot the results.

16.11 Analysis indicates that 99% of the risk from spent nuclear fuel is due to three radionuclides: Sr-90, Cs-137, and Co-60. As much as 99.7% is due to these three plus Pu-238. What analyses would have been done to arrive at these estimates? What do these estimates suggest about long-term disposition of spent nuclear fuel? (There is no "right answer" to this question.)

Chapter 17

Solid and Hazardous Waste Law

Laws controlling environmental pollution are discussed in this text in terms of their evolution from the courtroom through Congressional committees to administrative agencies. The gaps in common law are filled by statutory laws adopted by Congress and state legislatures and implemented by administrative agencies such as the U.S. Environmental Protection Agency (EPA) and state departments of natural resources. For several reasons, this evolutionary process was particularly rapid in the area of solid waste law.

For decades, and in fact centuries, solid waste was disposed of on land no one really cared about. Municipal refuse was historically taken to a landfill in the middle of a woodland, and industrial waste, which was often hazardous, was generally "piled out back" on land owned by the industry itself. In both cases, environmental protection and public health were not perceived as issues. Solid wastes were definitely out of sight and conveniently out of mind.

Only relatively recently has public interest in solid waste disposal sites reached the level of concern that equals that for water and air pollution. Public interest was triggered in part by the rate at which land was being filled with waste and by the increasing cost of land for disposal. Some local courtrooms and zoning commissions have dealt with siting selected disposal facilities over the years, but most decisions simply resulted in the city or industry hauling the waste a little farther away from the complaining public. At these remote locations, solid waste was not visible like the smokestacks that emitted pollutants into the atmosphere and the pipes discharging wastewater into the rivers. Pollution of the land and underground aquifers from solid waste disposal facilities was and is much more subtle.

Federal and state environmental statutes did not initially address subtle effects. The highly visible problem of air pollution was addressed in a series of clean air acts, and the other obvious pollution was confronted in the water pollution control laws. Finally, as researchers dug deeper into environmental and public health concerns, they realized that even obscure landfills and holding lagoons had the potential to pollute the land significantly and to impact public health. Subsurface and surface water supplies, and even local air quality, are threatened by solid waste.

This chapter addresses solid waste law in two major sections: nonhazardous waste and hazardous waste. The division reflects the regulatory philosophy for dealing with these two very distinct problems. Because of the general lack of common law in this area, we move directly into statutory controls of solid waste.

NONHAZARDOUS SOLID WASTE

The most significant solid waste disposal regulations were developed under the Resource Conservation and Recovery Act (RCRA) of 1976. This federal statute, which amended the elementary Solid Waste Disposal Act of 1965, reflected the concerns of the public in general and Congress in particular about: (1) protecting public health and the environment from solid waste disposal, (2) filling the loopholes in existing surface water and air quality laws, (3) ensuring adequate land disposal of residues from air pollution technologies and sludge from wastewater treatment processes, and (4) most importantly, promoting resource conservation and recovery.

The EPA implemented RCRA in a manner that reflected these concerns. Disposal sites were cataloged as either landfills, lagoons, or land spreading operations, and the adverse effects of improper disposal were grouped into seven categories:

- *Floodplains* were historically prime locations for industrial disposal facilities because many industries elected to locate along rivers for water supply, power generation, or transportation of process inputs or production outputs. When the rivers flooded, the disposed wastes were washed downstream, with immediate adverse impact on water quality.
- *Endangered and threatened species* may be affected by habitat destruction as the disposal site is being developed and operated, or harmed by toxic or hazardous materials leaking from the site. Animals may be poisoned as they wander onto unfenced sites.
- *Surface water quality* may also be affected by certain disposal practices. Without proper controls of runoff and leachate, rainwater has the potential to transport pollutants from the disposal site to nearby lakes and streams.
- *Groundwater quality* is of great concern because about half of the nation's population depends on groundwater for water supply. Soluble and partially soluble substances are leached from the waste site by rainwater; the leachate can contaminate underground aquifers.
- *Food-chain crops* may be adversely affected by landspreading or solid waste that may adversely affect both public health and agricultural productivity. Food-chain crops, especially leafy vegetables like lettuce and animal feed crops like alfalfa, often bioconcentrate heavy metals and other trace chemicals.
- *Air quality* may be degraded by pollutants emitted from waste decomposition, such as methane, and may cause serious pollution problems downwind from a disposal site. Uncovered waste in landfills can sustain spontaneous combustion, and landfill fires ("dump fires") can degrade air quality severely.
- *Health* and *safety* of on-site workers and the nearby public may be compromised by dump fires, as well as explosions of gas generated at the site. Uncovered waste attracts flocks of birds that pose a severe hazard to aircraft.

The EPA guidelines spell out the operational and performance requirements to mitigate or eliminate these seven types of impact from solid waste disposal.

Under operational standards, technologies, designs, or operating methods are specified to a degree that theoretically ensures the protection of public health and the environment. Any or all of a long list of operational considerations could go into a plan to build and operate a disposal facility: the type of waste to be handled, the facility location, facility design, operating parameters, and monitoring and testing procedures. The advantage of operational standards is that the best practical technologies can be used for solid waste disposal, and the state agency mandated with environmental protection can determine compliance with a specified operating standard. The major drawback is that compliance is generally not measured by monitoring actual effects on the environment, but rather as an either/or situation in which the facility either does or does not meet the required operational requirement.

On the other hand, performance standards are developed to provide given levels of protection to the land and the air and water quality around the disposal site. Determining compliance with performance standards is not easy because the actual monitoring and testing of groundwater, surface water, and land and air quality are costly, complex undertakings. Because solid wastes are heterogeneous and because site-specific considerations are important in their adequate disposal, the federal regulatory effort realized the necessity of allowing state and local discretion in protecting the environmental quality and public health. Therefore, the EPA developed both operational and performance standards to minimize the effect of each of the eight potential impacts of solid waste disposal.

Floodplains. The protection for floodplain focuses on limiting disposal facilities in such areas unless the local area has been protected against being washed out by the base flood. A base flood (sometimes called a "100-year flood") is defined as a flood that has a 1% or greater chance of occurring in any year, or a flood equaled or exceeded only once in every 100 years on the average.

Endangered and Threatened Species. Solid waste disposal facilities must be constructed and operated so they will not contribute to the taking of endangered and threatened species. *Taking* means the harming, pursuing, hunting, wounding, killing, trapping, or capturing of species so listed by the U.S. Department of the Interior. Operators of a solid waste disposal facility may be required to modify operation to comply with these rules if nearby areas are designated as critical habitats.

Surface Water. Solid waste disposal leads to the pollution of surface waters whenever rainwater percolates through the refuse and runoff occurs, or whenever spills take place during solid waste shipments. Point source discharges from disposal sites are regulated by the NPDES permit system discussed in Chap. 10. Nonpoint source, or diffused water, movement is regulated by the requirements implementing an area-wide or statewide water quality management plan approved by the EPA. Generally, these plans regulate the degradation of surface water quality by facility design, operation, and maintenance. Artificial and natural runoff barriers such as liners, levees, and dikes are often required by states. If runoff waters are collected, the site becomes, by definition, a point source of pollution, and the facility must have an NPDES permit if this new point source discharges into surface watercourses.

Groundwater. To prevent contamination of groundwater supplies, the owners and operators of a solid waste disposal facility must comply with one or more of five design and operational requirements: (1) utilize natural hydrogeologic conditions like underlying confining strata to block flow to aquifers, (2) collect and properly dispose of leachate by installing natural or synthetic liners, (3) reduce the infiltration of rainwater into the solid waste with correct cover material, (4) divert contaminated groundwater or leachate from groundwater supplies, and (5) conduct groundwater monitoring and testing procedures.

Land Used for Food or Forage Crops. Federal guidelines attempt to regulate the potential movement of heavy metals and synthetic organics, principally polychlorinated biphenyl (PCB), from solid waste into crops that are part of the human food chain. Typically, the solid waste of concern here is sludge from municipal wastewater treatment facilities. To control heavy metals, the regulations give the operator of such disposal facilities two options to follow: the controlled application approach and the dedicated facility approach. The operator electing the controlled application approach must conform with an annual application rate of cadmium from solid waste. Cadmium is controlled basically as a function of time and the type of crop to be grown. A cadmium addition of 2.0 kg/ha (1.8 lb/acre) is allowed for all food-chain crops other than tobacco, leafy vegetables, and root crops grown for human consumption. An application rate of 0.5 kg/ha (0.45 lb/acre) is in force for these accumulator crops. Under this option, the operator must conform to a cumulative limit on the amount of cadmium that may be disposed of on food-chain land. This cumulative limit is a function of the cation exchange capacity (CEC) of the soil and the background soil pH. At low pH and low CEC, a maximum cumulative application rate of 5.0 kg/ha (4.50 lb/acre) is allowed, whereas at high CEC and soils with high or near neutral pH levels, a 20 kg/ha (18 lb/acre) amount is allowed. The theory is that in soils of high CEC and high pH, the cadmium remains bound in the soil and is not taken up by the plants. All operators electing this option must ensure that the pH of the solid waste and soil mixture is 6.5 or greater at the time of application.

The second option open to food-chain landspreaders is the dedicated facility approach. This option is distinguished from the first option by its reliance on output control, or crop management as opposed to input control, or limiting the amounts of cadmium that may be applied to the soil. This second cadmium control option is designed specifically for facilities with the resources and the capabilities to closely manage and monitor the performance of their respective operations. The requirements under this option include: (1) only growing animal feed crops, (2) the pH of the solid waste and soil mixture is 6.5 or greater at the time of application, (3) there is a facility operating plan that demonstrates how the animal feed will be distributed to preclude direct ingestion by humans, and (4) there are deed notifications for future property owners that the property has received solid wastes at high cadmium application rates and that food-chain crops should not be grown owing to a possible health hazard.

Solid waste with PCB concentrations greater than or equal to 10 mg/kg of dry weight must be incorporated in the soil when applied to land used for producing animal feed, unless the PCB concentration is found to be less than 1.5 mg/kg fat basis in the produced milk. By requiring soil incorporation, the solid waste regulations attempt to provide that the Food and Drug Administration levels for PCB in milk and animal feed will not be exceeded.

Air Quality. A solid waste disposal facility must comply with the Clean Air Act and the relevant state Air Quality Implementation Plans developed by state and local air emission control boards. Open burning of residential, commercial, institutional, and industrial solid waste is generally prohibited. However, several special wastes are excluded from this prohibition: diseased trees, land-clearing debris, debris from emergency cleanup operations, and wastes from silviculture and agriculture operations. Open burning is defined as the combustion of solid waste without: (1) control of combustion in the air to maintain adequate temperatures for efficient combustion, (2) containment of the combustion reaction in an enclosed device to provide adequate residence time, or (3) control of the emissions from the combustion process.

Health. The federal rules require that the operation of a solid waste disposal facility protect the public health from disease vectors. Disease vectors are any routes by which disease is transmitted to humans (birds, rodents, flies, and mosquitoes are good examples). This protection must be achieved by minimizing the availability of food for the vectors. At landfills, an effective means to control vectors, especially rodents, is the application of cover material at the end of each day of operation. Other techniques include poisons, repellents, and natural controls such as the supply of predators. Treating sewage sludge with pathogen reduction processes serves to control the spread of disease from landscaping activities.

Safety. Fires at solid waste disposal facilities are a constant threat to public safety. Death, injury, and vast property damage have resulted in the past from fires breaking out in open dumps. The ban on open burning also reduces the chance of accidental fires. In terms of aircraft safety, facilities should not be located between airports and bird feeding, roosting, or watering sites. Flocks of birds continually fly around the working faces of even properly operated sanitary landfills, and pose a severe hazard to nearby aircraft flying landing/takeoff patterns.

HAZARDOUS WASTE

Common law, in particular, is not well developed in the hazardous waste area. The issues discussed in Chap. 15 are newly recognized by society, and little case law has had a chance to develop. Since the public health impacts of solid waste disposal in general and hazardous waste disposal in particular were so poorly understood, few plaintiffs ever bothered to take a defendant into the common law courtroom to seek payment for damages or injunctions against such activities. Hazardous waste law is therefore discussed here as statutory law, specifically in terms of the compensation for victims of improper hazardous waste disposal and efforts to regulate the generation, transport, and disposal of hazardous waste. Historically, federal statutory law was generally lacking in describing how victims should be compensated for improper hazardous waste disposal. A complex, repetitious, confusing list of federal statutes was the only recourse for victims.

The federal Clean Water Act only covers wastes discharged to navigable waterways. Only surface water and ocean waters within 200 miles from the coast are considered, and a revolving fund is set up by the act and administered by the Coast Guard. Fines and charges are deposited into the fund to compensate victims of discharges, but the funds are only available if the discharge of the waste is clearly identifiable. The fund is used most often for compensating states for cleanup of spill from large tanker ships. The federal Drinking Water Act includes groundwater protection, and directs EPA to set "maximum contaminant levels" (MCL) for groundwater contaminants.

The Outer Continental Shelf (OCS) Lands Act sets up two funds to help pay for hazardous waste cleanup and to compensate victims. Under the act, OCS leaseholders are required to report spills from petroleum-producing sites, and the Offshore Oil Spill Pollution Fund exists to finance cleanup costs and compensate injured parties for loss of the use of property, natural resources, profits, and state or local government's loss of tax revenue. The U.S. Department of Transportation (DOT) is responsible for the administration of this fund, provided by a tax on oil produced at the OSC sites. If the operator of the site cooperates with the DOT after a spill, his or her liability is limited.

The Fisherman's Contingency Fund also exists under the OCS Lands Act to repay fishermen for loss of profit and equipment owing to oil and gas exploration, development, and production. The U.S. Department of Commerce is responsible for the administration of this fund. Assessments are collected by the U.S. Department of the Interior from holders of permits and pipeline easements. If a fisherman cannot pinpoint the site responsible for the discharge of hazardous waste, his or her claim against the fund may still be acceptable if the boat was operated within the vicinity of OCS activity and if a claim is filed within 5 days. Two agencies are involved in the administration of the fund, making the implementation of the fund especially confusing.

The Price–Anderson Act provides compensation to victims of accidental discharge from nuclear facilities. Radioactive material/toxic spills and emissions are covered, as are explosions. The licensees of a nuclear facility are required by the Nuclear Regulatory Commission (NRC) to obtain insurance protection. If damages from an accident exceed this insurance coverage, the federal government indemnifies the licensee up to $500 million. In no case does financial liability exceed $560 million. If an accident occurs such that liability exceeds $560 million, federal disaster relief is called upon. The Patriot Act of 2002 has added maximum indemnity of $90 billion in the event of deliberate attack.

Under the Deepwater Ports Act, the Coast Guard acts to remove oil from deepwater ports. The DOT administers a liability fund that helps pay for the cleanup and that compensates injured parties. Financed by a 2¢ per barrel tax on oil loaded or unloaded at such a facility, the fund takes effect once the required insurance coverage is exceeded. The port itself must hold insurance for $50 million for claims against its waste discharges, and vessels are insured for $20 million for claims against their waste spills. Once these limits are exceeded, the fund established by the Deepwater Ports Act takes over. Again, the usefulness of the fund is dependent on how well two agencies work together and how well insurance claims are administered, and administration is confusing at best. After the 1990 oil spill at Valdez, AK, the Exxon Corporation was required to provide additional compensation for loss of fisheries.

Statutes at the state level have generally paralleled these federal efforts. New Jersey has a Spill Compensation and Control Act for hazardous wastes listed by the EPA. The fund covers cleanup costs, loss of income, loss of tax revenues, and the restoration of damaged property and natural resources. A tax of 1¢ per barrel of hazardous substance finances a fund that pays injured parties if they file a claim within six years of the hazardous waste discharge and within one year of the discovery of the damage.

The New York Environmental Protection and Spill Compensation Fund is similar to the New Jersey fund, but differs in two respects. In New York, only petroleum discharges are covered, and the generator may not blame a third party for a spill or discharge incident. Thus, if a trucker or handler accidentally spills hazardous waste, the generator of the waste may still be held liable for damages.

Other states have limited efforts in the compensation of victims of hazardous waste incidents. Florida has a Coastal Protection Trust Fund of $35 million that compensates victims for spills, leaks, and dumping of waste. However, the coverage is limited to injury resulting from oil, pesticides, ammonia, and chlorine, which provides a loophole for many types of hazardous waste discussed in Chap. 15. Other states that do have compensation funds generally spell out limited compensationable injuries and provide limited funds, often less than $100,000.

In the past, the value of these federal and state statutes has been questioned. Even when taken as a group, they do not provide for complete compensation strategy for dealing with hazardous waste. Few funds address personal injuries, and abandoned disposal sites are not considered. Huge administrative problems also exist as several agencies attempt to respond to a hazardous waste spill. Questions of which fund applies, which agency has jurisdiction, and what injuries may receive compensation linger.

In 1980, the federal Comprehensive Environmental Response, Compensation, and Liability Act (CERCLA), commonly called "Superfund," was enacted. Originally, the Superfund Act was to serve two purposes: to provide money for the cleanup of abandoned hazardous waste disposal sites, and to establish liability so dischargers could be required to pay for injuries and damages. Three considerations went into the development of this act:

- *The types of incidents to be covered* by Superfund include accidental spills and abandoned sites as well as on-site toxic pollutants in harbors and rivers. Fires and explosions caused by hazardous materials are also a focus of this consideration as the Superfund concept has evolved.
- *The types of damages to be compensated* include three categories: environmental cleanup costs; economic losses associated with property use, income, and tax revenues; and personal injury in the form of medical costs for acute injuries, chronic illness, death, and general pain and suffering.
- *The sources of payments for the compensation* include federal appropriations into the fund, industry contributions for sales tax, and income tax surcharges. Federal and state cost sharing is also possible, as is the establishment of fees for disposal of hazardous waste at permitted disposal facilities.

A fund of between $1 billion and $4 billion is established, which is financed 87% by a tax on the chemical industry and 13% by general revenues of the federal government. Small payments are permitted out of the fund, but only for out-of-pocket medical costs and partial payment for diagnostic services. The Superfund Act does not set strict liability for spills and abandoned hazardous wastes. The Superfund Amendments and Reauthorization Act (SARA) of 1986 reauthorized and strengthened CERCLA (Martin and Koszynsk 1991).

Compensation for damages is one concern; regulations that control generators, transporters, and disposers of hazardous waste are the other side of the hazardous waste law coin. The federal RCRA, which is discussed in this chapter as it relates to solid waste, also is the principal statute that deals with hazardous waste. The RCRA requires the EPA to establish a comprehensive regulatory program to control hazardous waste. This solid and hazardous waste act offers a good example of the types of reporting and record-keeping requirements mandated in federal environmental laws. The Clean Water Act and Clean Air Act places requirements on water and air polluters similar to the type of requirements for generators of hazardous waste.

A *generator* of hazardous waste must meet the following EPA requirements. This manifest system of "cradle-to-grave" tracking (see Fig. 15-3) is the key to the regulatory system.

1. A determination that the waste is hazardous, either as defined by an EPA listing of hazardous waste, or by EPA testing procedures, or as indicated by the materials and processes used in production, must be made. These questions of definition are not trivial. A waste must be mobile in the environment and its constituents must be toxic, corrosive, flammable, or ignitable. In 1990, the EPA replaced the EP (Extraction Procedure) Toxicity test with the TCLP (Toxicity Characteristic Leaching Procedure) to determine which, if any, constituents are mobile (Henriches 1991). The TCLP is troublesome at best to follow in the laboratory.

2. An EPA general identification number must be obtained.

3. A facility permit must be obtained if hazardous waste is stored at the generating site for 90 days or longer.

4. Appropriate containers and labels must be used before shipment off-site.

5. A manifest for tracking the waste shipment off-site must be prepared, as described in Chap. 15 (see Fig. 15-3).

6. Arrival of the waste at the disposal site must be ensured.

7. Annual summaries of activities must be submitted to federal or state regulatory agencies or both.

A *transporter* of hazardous waste must meet several requirements under RCRA:

1. An EPA transporter identification number must be obtained.
2. Compliance with provisions of the manifest system is required.
3. The entire quantity of waste must be delivered to the disposal or processing site.
4. A copy of the manifest must be retained for 3 years.
5. DOT rules for responding to spills of hazardous waste must be followed.

RCRA also places requirements on *facilities* that treat, store, or dispose of hazardous waste:

1. The owner of the facility must apply for an operator's permit and supply data on the proposed site and waste to be handled. This permit will spell out the terms of compliance: construction and operating schedules, as well as monitoring and record-keeping procedures.

2. As permits are granted by the appropriate federal or state agency, minimum operating standards will be placed on the facility. Design and engineering standards for containing, neutralizing, and destroying the wastes are addressed in these permits, as are safety and emergency measures in the event of an accident. Personnel training is included in each permit's requirements.

3. The financial responsibility of the facility is defined, and a trust fund generally must be established at site closure. This fund enables the monitoring of groundwater and surface discharges to continue after the site is closed, and enables the executor of the trust to maintain the site for years to come.

The responsibilities for administering RCRA are passed from the EPA to state agencies as these agencies show they have the authority and expertise to regulate hazardous waste effectively.

In addition to the CERCLA and RCRA control of hazardous waste, the Toxic Substances Control Act (TOSCA) regulates hazardous substances before they become hazardous waste. Under one facet of this regulatory program, the Premanufacturing Notification (PMN) System, all manufacturers must notify the EPA before marketing any substance not included in the EPA's 1979 inventory of toxic substances. In this notification, the manufacturer must analyze the predicted effect of the substance on workers, on the environment, and on consumers. This analysis must be based on test data and all relevant literature. A compliance checklist is presented in Table 17-1 in

Table 17-1. Checklist to Comply with TOSCA's PMN System

Step	Requirement
1.	Inspect EPA inventory of toxic substances to see whether the new product is listed.
2.	Assess EPA rules for testing procedures to determine whether substances not listed by the EPA are, in fact, "toxic."
3.	Prepare and submit the PMN form at least 90 days before manufacture.
4.	Obtain one of the following EPA rulings:
	• No ruling: Manufacturing may begin at the end of the 90 days.
	• EPA order: An EPA order for more data requires a response before manufacturing may proceed.
	• EPA ruling: EPA may prohibit or limit the proposed manufacturing, processing, use, or disposal of the substance.

350   ENVIRONMENTAL ENGINEERING

which specific steps are spelled out for manufacturers who wish to produce a new product.

CONCLUSION

Solid waste statutory law developed rather rapidly once the public, along with scientists and engineers, realized the real and potential impacts of improper disposal. Air, land, and water quality from such disposal became a key concern to local health officials and federal and state regulators. Hazardous and nonhazardous solid wastes are now regulated under a complex system of federal and state statutes that place operating requirements on facility operators.

The effectiveness of these statutes is somewhat in doubt. How will the EPA implement RCRA in all cases? If states are given the responsibility to control hazardous and nonhazardous waste within their boundaries, how will they respond? Will the "joint and several liability" provision of Superfund, which makes any and all waste disposers at a site liable irrespective of the amount of waste for which they are responsible, stalemate its implementation? Since September 11, 2001, we also ask who is responsible if a deliberate attack creates a Superfund site. In the end, what will be the public health, environmental quality, and economic impacts of RCRA and the Superfund?

PROBLEMS

17.1 Local land use ordinances often play key roles in limiting the number of sites available for a solid or hazardous waste disposal facility. Discuss the types of these zoning restrictions that apply in your home town.

17.2 Assume you work for a firm that contracts to clean the inside and outside of factories and office buildings in the state capitol. Your boss thinks she could make more money by expanding her business to include the handling and transportation of hazardous wastes generated by her current clients. Outline the types of data you must collect to advise her to expand or not expand.

17.3 Laws that govern refuse collection often begin with controls on the generator, i.e., rules that each household must follow if city trucks are to pick up their solid waste. If you are asked by a town council to develop a set of such "household rules," what controls would you include? Emphasize public health concerns, minimizing the cost of collection, and even resource recovery considerations.

17.4 The oceans have long been viewed as bottomless pits into which the solid wastes of the world may be dumped. Engineers, scientists, and politicians are divided on the issue. Should ocean disposal be banned? Develop arguments pro and con.

Chapter 18

Meteorology and Air Pollution

Given that the earth's atmosphere is about 100 miles deep, that thickness and volume sometimes are suggested to be enough to dilute all of the chemicals and particles thrown into it. However, 95% of this air mass is within 12 miles of the earth's surface and it is this 12-mile depth that contains the air we breathe as well as the pollutants we emit. This layer, called the *troposphere*, is where we have our weather and air pollution problems.

Weather patterns determine how air contaminants are dispersed and move through the troposphere, and thus determine the concentration of a particular pollutant that is breathed or the amount deposited on vegetation. An air pollution problem involves three parts: the pollution source, the movement or dispersion of the pollutant, and the recipient (Fig. 18-1). Chapter 19 discusses the source and effects of pollutants; this chapter concerns itself with the transport mechanism: how the pollutants travel through the atmosphere. The environmental engineer should be conversant enough with some basic meteorology to be able to predict, approximately, the dispersion of air pollutants.

BASIC METEOROLOGY

Pollutants circulate the same way the air in the troposphere circulates. Air movement is caused by solar radiation and the irregular shape of the earth and its surface, which causes unequal absorption of heat by the earth's surface and atmosphere. This differential heating and unequal absorption creates a dynamic system.

The dynamic thermal system of the earth's atmosphere also yields differences in barometric pressure. We associate low-pressure systems with both hot and cold weather fronts. Air movement around low-pressure fronts in the Northern Hemisphere is counterclockwise and vertical winds are upward, where condensation and precipitation take place. High-pressure systems bring sunny and calm weather — stable atmospheric conditions — with winds (in the Northern Hemisphere) spiraling clockwise and downward. Low- and high-pressure systems, commonly called *cyclones* and *anticyclones*, are illustrated in Fig. 18-2. Anticyclones are weather patterns of high stability, in which dispersion of pollutants is poor, and are often precursors to air pollution episodes.

352   ENVIRONMENTAL ENGINEERING

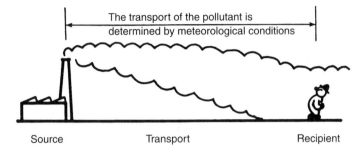

Figure 18-1. Meteorology of air pollutants.

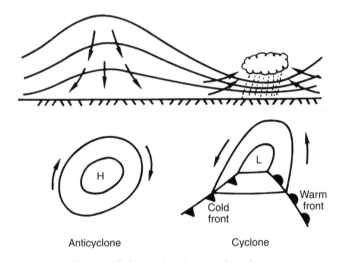

Figure 18-2. Anticyclone and cyclone.

Figure 18-3 shows a weather map for the eastern United States for January 1968. The high-pressure area around the Chesapeake Bay indicates a region of stable air, where pollutants build up and do not disperse. An air pollution episode was avoided by the front that moved in from the west, dispersing the pollutants. Air quality management involves both control of air pollution sources and effective dispersion of pollutants in the atmosphere.

HORIZONTAL DISPERSION OF POLLUTANTS

The earth receives light energy at high frequency from the sun and converts this to heat energy at low frequency, which is then radiated back into space. Heat is transferred from the earth's surface by radiation, conduction, and convection. Radiation is direct transfer of energy and has little effect on the atmosphere; conduction is the transfer of

Meteorology and Air Pollution   353

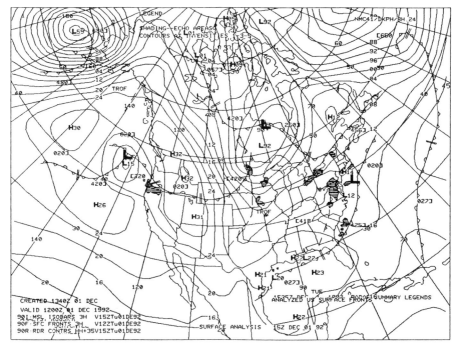

Figure 18-3. Weather map for 1 December 1992.

heat by physical contact (the atmosphere is a poor conductor since the air molecules are relatively far apart); convection is transfer of heat by movement of warm air masses. Solar radiation warms the earth and thus the air above it. This heating is most effective at the equator and least at the poles. The warmer, less dense air rises at the equator and cools, becomes more dense, and sinks at the poles. If the earth did not rotate then the surface wind pattern would be from the poles to the equator. However, the rotation of the earth continually presents new surfaces to be warmed, so that a horizontal air pressure gradient exists as well as the vertical pressure gradient. The resulting motion of the air creates a pattern of winds around the globe, as shown by Fig. 18-4.

Seasonal and local temperature, pressure and cloud conditions, and local topography complicate the picture. Land masses heat and cool faster than water so that shoreline winds blow out to sea at night and inland during the day. Valley winds result from cooling of air high on mountain slopes. In cities, brick and concrete buildings absorb heat during the day and radiate it at night, creating a *heat island* (Fig. 18-5), which sets up a self-contained circulation called a *haze hood* from which pollutants cannot escape.

Horizontal wind motion is measured as wind velocity. Wind velocity data are plotted as a *wind rose*, a graphic picture of wind velocities and the direction *from which the wind came*. The wind rose in Fig. 18-6 shows that the prevailing winds were

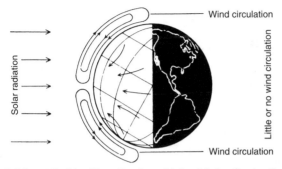

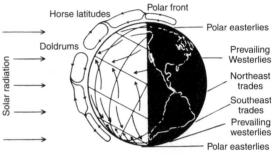

Figure 18-4. Global wind patterns. (Courtesy of the American Lung Association.)

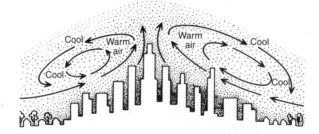

Figure 18-5. Heat island formed over a city.

from the southwest. The three features of a wind rose are:

- the *orientation* of each segment, which shows the direction from which the wind came,
- the *width* of each segment, which is proportional to the wind speed, and
- the *length* of each segment, which is proportional to the percent of time that wind at that particular speed was coming from that particular direction.

Air pollution enforcement engineers sometimes use a *pollution rose*, a variation of a wind rose in which winds are plotted only on days when the air contamination level

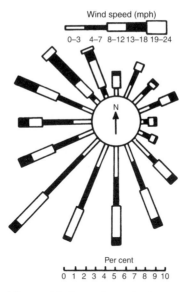

Figure 18-6. Typical wind rose.

exceeds a given amount. Figure 18-7 shows pollution roses at three points plotted only for days when the SO_2 level exceeded 250 $\mu g/m^3$. Note that because the roses indicate the directions *from* which the wind is coming, two of the roses point *to* Plant 3, the apparent primary pollution source. Wind is probably the most important meteorological factor in the movement and dispersion of air pollutants, or, in simple terms, pollutants move predominantly downwind.

VERTICAL DISPERSION OF POLLUTANTS

As a parcel of air in the earth's atmosphere rises through the atmosphere, it experiences decreasing pressure and thus expands. This expansion lowers the temperature of the air parcel, and therefore the air cools as it rises. The rate at which dry air cools as it rises is called the *dry adiabatic lapse rate* and is independent of the ambient air temperature. The term "adiabatic" means that there is no heat exchange between the rising parcel of air under consideration and the surrounding air. The dry adiabatic lapse rate may be calculated from basic physical principles.

$$dT/dz|_{\text{dry-adiabatic}} = -9.8°\text{C/km}, \tag{18.1}$$

where T = temperature and z = altitude.

The actual measured rate at which air cools as it rises is called the *ambient* or *prevailing lapse rate*. The relationships between the ambient lapse rate and the dry adiabatic lapse rate essentially determine the stability of the air and the speed with which pollutants will disperse. These relationships are shown in Fig. 18-8.

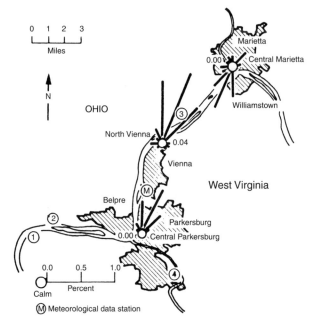

Figure 18-7. Pollution roses for SO_2 concentrations greater than $250\,\mu g/m^3$. The four chemical plants are the suspected sources, but the roses clearly point to Plant 3, identifying it as the primary culprit.

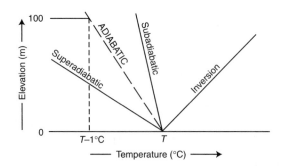

Figure 18-8. Ambient lapse rates and the dry adiabatic lapse rate.

When the ambient lapse rate is exactly the same as the dry adiabatic lapse rate, the atmosphere has *neutral* stability. *Superadiabatic* conditions prevail when the air temperature drops more than 9.8°C/km (1°C/100 m). *Subadiabatic* conditions prevail when the air temperature drops at a rate less than 9.8°C/km. A special case of subadiabatic conditions is the *temperature inversion*, when the air temperature actually increases with altitude and a layer of warm air exists over a layer of cold air. Superadiabatic atmospheric conditions are unstable and favor dispersion; subadiabatic

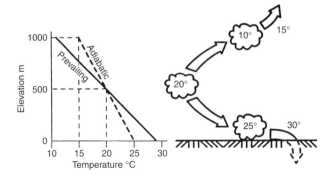

A. Super-adiabatic conditions (unstable)

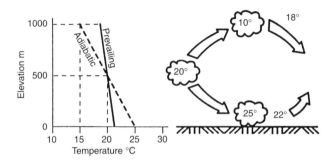

B. Sub-adiabatic conditions (stable)

Figure 18-9. Stability and vertical air movement.

conditions are stable and result in poor dispersion; inversions are extremely stable and trap pollutants, inhibiting dispersion. These conditions may be illustrated by the following example, illustrated in Fig. 18-9:

The air temperature at an elevation of 500 m is 20°C, and the atmosphere is super-adiabatic: the ground level temperature is 30°C and the temperature at an elevation of 1 km is 10°C. The (superadiabatic) ambient lapse rate is −20°C/km. If a parcel of air at 500 m moves up adiabatically to 1 km, what will be its temperature? According to the dry adiabatic lapse rate of −9.8°C/km the air parcel would cool by 4.9°C to about 15°C.

However, the temperature at 1 km is not 15°C but 10°C. Our air parcel is 5°C warmer than the surrounding air and will continue to rise. In short, under subadiabatic conditions, a rising parcel of air keeps right on going up. Similarly, if our parcel were displaced downward to, say, 250 m, its temperature would increase by 2.5°C to 22.5°C. The ambient temperature at 250 m, however, is 25°C, so that our parcel of air is now *cooler* than the surrounding air and keeps on sinking. There is no tendency to stabilize; conditions favor instability.

Now let us suppose that the ground level temperature is 22°C, and the temperature at an elevation of 1 km is 15°C. The (subadiabatic) ambient lapse rate is now −7°C/km. If our parcel of air at 500 m moves up adiabatically to 1 km, its temperature would again drop by 4.9°C to about 15°C, the same as the temperature of the surrounding air at 1 km. Our air parcel would cease rising, since it would be at the same density as the surrounding air.

If the parcel were to sink to 250 m, its temperature would again be 22.5°C, and the ambient temperature would be a little more than 20°C. The air parcel is slightly warmer than the surrounding air and tends to rise back to where it was. In other words, its vertical motion is damped, and it tends to become stabilized; subadiabatic conditions favor stability and limit vertical mixing.

Figure 18-10 is an actual temperature sounding for Los Angeles. Note the beginning of an inversion at about 1000 ft that puts an effective cap on the city and holds in the air pollution. This type of inversion is called a *subsidence inversion*, caused by a large mass of warm air subsiding over a city.

A more common type is the *radiation inversion*, caused by radiation of heat from the earth at night. As heat is radiated, the earth and the air closest to it cool, and this cold air is trapped under the warm air above it (Fig. 18-11). Pollution emitted during the night is caught under the "inversion lid."

Atmospheric stability may often be recognized by the shapes of plumes emitted from smokestacks as seen in Figs. 18-12 and 18-13. Neutral stability conditions usually

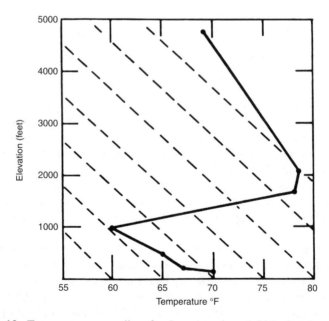

Figure 18-10. Temperature sounding for Los Angeles, 4 PM, October 1962. The dotted lines show the dry adiabatic lapse rate.

Meteorology and Air Pollution   359

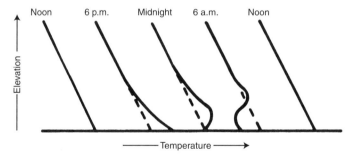

Figure 18-11. Typical ambient lapse rates during a sunny day and clear night.

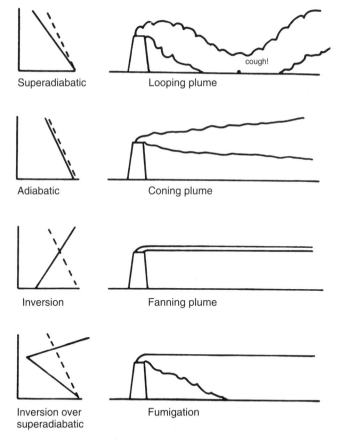

Figure 18-12. Plume shapes and atmospheric stability.

Figure 18-13. Iron oxide dust looping plume from a steel mill.

result in *coning* plumes, while unstable (superadiabatic) conditions result in a highly dispersive *looping* plume. Under stable (subadiabatic) conditions, the *fanning* plume tends to spread out in a single flat layer. One potentially serious condition is called *fumigation*, in which pollutants are caught under an inversion and are mixed owing to strong lapse rate. A looping plume also produces high ground-level concentrations as the plume touches the ground.

Assuming adiabatic conditions in a plume allows estimation of how far it will rise or sink, and what type of plume it will be during any given atmospheric temperature condition, as illustrated by Example 18.1.

EXAMPLE 18.1. A stack 100 m tall emits a plume whose temperature is 20°C. The temperature at the ground is 19°C. The ambient lapse rate is −4.5°C/km up to an altitude of 200 m. Above this the ambient lapse rate is +20°C/km. Assuming perfectly adiabatic conditions, how high will the plume rise and what type of plume will it be?

Figure 18-14 shows the various lapse rates and temperatures. The plume is assumed to cool at the dry adiabatic lapse rate 10°C/km. The ambient lapse rate below 200 m is subadiabatic, the surrounding air is cooler than the plume, so it rises, and cools as it rises. At 225 m, the plume has cooled to 18.7°C, but the ambient air is at this temperature also, and the plume ceases to rise. Below 225 m, the plume would have been slightly coning. It would not have penetrated 225 m.

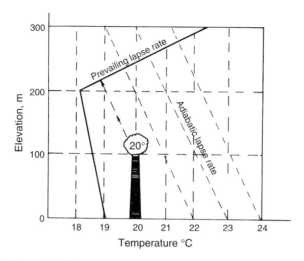

Figure 18-14. Atmospheric conditions in Example 18.1.

Effect of Water in the Atmosphere

The *dry* adiabatic lapse rate is characteristic of dry air. Water in the air will condense or evaporate, and in doing so will release or absorb heat, respectively, making calculations of the lapse rate and atmospheric stability complicated. In general, as a parcel of air rises, the water vapor in that parcel will condense and heat will be released. The rising air will therefore cool more slowly as it rises; the *wet* adiabatic lapse rate will in general be less negative than the *dry* adiabatic lapse rate. The wet adiabatic lapse rate has been observed to vary between -6.5 and $-3.5°C/km$. Water in the atmosphere affects air quality in other ways as well. Fogs are formed when moist air cools and the moisture condenses. Aerosols provide the condensation nuclei, so that fogs tend to occur more frequently in urban areas. Serious air pollution episodes are almost always accompanied by fog (remember that the roots of the word "smog" are "smoke" and "fog"). The tiny water droplets in fog participate in the conversion of SO_3 to H_2SO_4. Fog sits in valleys and stabilizes inversions by preventing the sun from warming the valley floor, thus often prolonging air pollution episodes.

ATMOSPHERIC DISPERSION

Dispersion is the process by which contaminants move through the air and a plume spreads over a large area, thus reducing the concentration of the pollutants it contains. The plume spreads both horizontally and vertically. If it is a gaseous plume, the motion of the molecules follows the laws of gaseous diffusion.

The most commonly used model for the dispersion of gaseous air pollutants is the Gaussian model developed by Pasquill, in which gases dispersed in the atmosphere

362 ENVIRONMENTAL ENGINEERING

are assumed to exhibit ideal gas behavior. Rigorous derivation of the model is beyond the scope of this text, but the principles on which the model is based are:

- The predominant force in pollution transport is the wind; pollutants move predominantly downwind.
- The greatest concentration of pollutant molecules is along the plume center line.
- Molecules diffuse spontaneously from regions of higher concentration to regions of lower concentration.
- The pollutant is emitted continuously, and the emission and dispersion process is steady state.

Figure 18-15 shows the fundamental features of the Gaussian dispersion model, with the geometric arrangement of source, wind, and plume. We can construct a Cartesian coordinate system with the the emission source at the origin and the wind direction along the x axis. Lateral and vertical dispersions are along the y and z axes, respectively. As the plume moves downwind, it spreads both laterally and vertically away from the plume centerline as the gas molecules move from higher to lower concentrations. Cross sections of the pollutant concentration along both the y and the z axes thus have the shape of Gaussian curves, as shown in Fig. 18-15.

Since stack gases are generally emitted at temperatures higher than ambient, the buoyant plume will rise some distance before beginning to travel downwind. The sum of this vertical travel distance and the geometric stack height is H, the *effective* stack height. The source of the pollutant plume is, in effect, a source elevated above the ground at elevation

$$z = H,$$

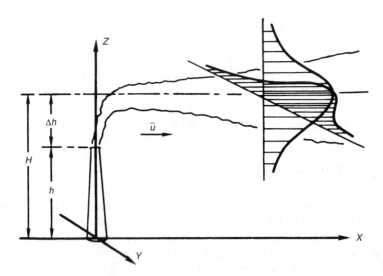

Figure 18-15. Gaussian dispersion model.

and the downwind concentration emanating from this elevated source may be written

$$C(x, y, z) = \frac{Q}{2\pi u \sigma_y \sigma_z} \exp\left(-\frac{y^2}{2\sigma_y^2}\right) \left(\exp\left(-\frac{(z+H)^2}{2\sigma_z^2}\right) + \exp\left(-\frac{(z-H)^2}{2\sigma_z^2}\right)\right), \tag{18.2}$$

where

$C(x, y, z)$ is the concentration at some point in space with coordinates x, y, z, and
$Q =$ the emission rate of the pollution source (in g/s),
$u =$ the average wind speed in (m/s),
$\sigma_y =$ the standard deviation of the plume in the y direction (m), and
$\sigma_z =$ the standard deviation of the plume in the z direction (m).

The units of concentration are grams per cubic meter (g/m³). Since pollution concentrations are usually measured at ground level, that is, for $z = 0$, Eq. (18.2) usually reduces to

$$C(x, y, 0) = \frac{Q}{\pi u \sigma_y \sigma_z} \exp\left(-\frac{y^2}{2\sigma_y^2}\right) \left(\exp\left(-\frac{(H)^2}{2\sigma_z^2}\right)\right). \tag{18.3}$$

This equation takes into account the reflection of gaseous pollutants from the surface of the ground.

We are usually interested in the greatest value of the ground level concentration in any direction, and this is the concentration along the plume centerline; that is, for $y = 0$. In this case, Eq. (18.3) reduces to

$$C(x, 0, 0) = \frac{Q}{\pi u \sigma_y \sigma_z} \exp\left(\frac{-H^2}{2\sigma_z^2}\right). \tag{18.4}$$

Finally, for a *source of emission* at ground level, $H = 0$, and the ground level concentration of pollutant downwind along the plume centerline is given by

$$C(x, 0, 0) = \frac{Q}{\pi u \sigma_y \sigma_z}. \tag{18.5}$$

For a release above ground level the maximum downwind ground level concentration occurs along the plume centerline when the following condition is satisfied:

$$\sigma_z = \frac{H}{\sqrt{2}}. \tag{18.6}$$

The standard deviations σ_y and σ_z are measures of the plume spread in the crosswind (lateral) and vertical directions, respectively. They depend on atmospheric stability and on distance from the source. Atmospheric stability is classified in categories A through F, called stability classes. Table 18-1 shows the relationship between stability

Table 18-1. Atmospheric Stability under Various Conditions

Wind speed at 10 m (m/s)	Day Incoming solar radiation			Night Thin overcast	
	Strong	Moderate	Slight	1/2 low cloud	3/8 cloud
<2	A	A–B	B		
2–3	A–B	B	C	E	F
3–5	B	B–C	C	D	E
5–6	C	C–D	D	D	E
>6	C	D	D	D	D

class, wind speed, and sunshine conditions. Class A is the least stable; Class F is the most stable. In terms of ambient lapse rates, Classes A, B, and C are associated with superadiabatic conditions; Class D with neutral conditions; and Classes E and F with subadiabatic conditions. A seventh, Class G, indicates conditions of extremely severe temperature inversion, but in considering frequency of occurrence is usually combined with Class F. Urban and suburban populated areas rarely achieve stability greater than Class D, because of the heat island effect; stability classes E and F are found in rural and unpopulated areas. Values for the lateral and vertical dispersion constants, σ_y and σ_z, are given in Figs. 18-16 and 18-17. Use of the figures is illustrated in Examples 18.2 and 18.3.

EXAMPLE 18.2. An oil pipeline leak results in emission of 100 g/h of H_2S. On a very sunny summer day, with a wind speed of 3.0 m/s, what will be the concentration of H_2S 1.5 km directly downwind from the leak?

From Table 18-1, we may assume Class B stability. Then, from Fig. 18-16, at $x = 1.5$ km, σ_y is approximately 210 m and, from Fig. 18-17, σ_z is approximately 160 m, and

$$Q = 100 \text{ g/h} = 0.0278 \text{ g/s}.$$

Applying Eq. (18.5), we have

$$C(1500, 0, 0) = \frac{0.0278 \text{ g/s}}{\pi (3.0 \text{ m/s})(210 \text{ m})(160 \text{ m})} = 8.77 \times 10^{-8} \text{ g/m}^3 = 0.088 \text{ } \mu\text{g/m}^3.$$

EXAMPLE 18.3. A coal-burning electric generating plant emits 1.1 kg/min of SO_2 from a stack with an effective height of 60 m. On a thinly overcast evening, with a wind speed of 5.0 m/s, what will be the ground level concentration of SO_2 500 m directly downwind from the stack?

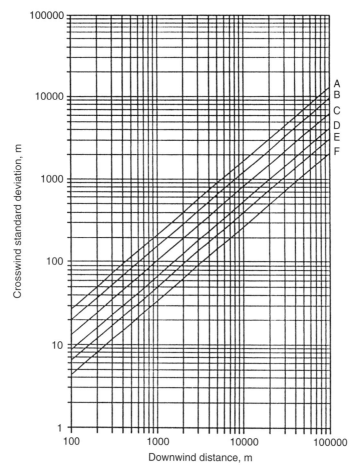

Figure 18-16. Standard deviation or dispersion coefficient, σ_y, in the crosswind direction as a function of downwind distance (Wark and Warner 1986).

From Table 18-1, we may assume Class D stability. Then, from Fig. 18-16, at $x = 0.5$ km, σ_y is approximately 35 m and σ_z is approximately 19 m, and

$$Q = 1.1 \text{ kg/min} = 18 \text{ g/s}.$$

In this problem, the release is elevated, and $H = 60$ m.

Applying Eq. (18.4), we have

$$C(0.5, 0, 0) = \frac{18 \text{ g/s}}{\pi (5 \text{ m/s})(35 \text{ m})(19 \text{ m})} \exp\left(\frac{-(60)^2}{2(19)^2}\right) = 11.8 \times 10^{-6} \text{ g/m}^3,$$

or $11.8 \, \mu\text{g/m}^3$.

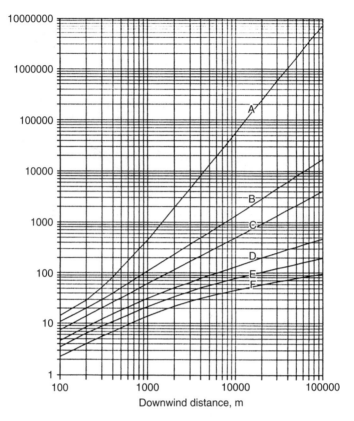

Figure 18-17. Standard deviation or dispersion coefficient, σ_z, in the vertical direction as a function of downwind distance (Wark and Warner 1986).

Variation of Wind Speed With Elevation

The model used so far assumes that the wind is uniform and unidirectional, and that its velocity can be estimated accurately. These assumptions are not realistic: wind direction shifts and wind speed varies with time as well as with elevation. The variation of wind speed with elevation can be approximated by a parabolic wind velocity profile. That is, the wind speed u at an elevation h may be calculated from the measured wind speed u_0 at a given elevation h_0 using the relationship

$$u = u_0 \left(\frac{h}{h_0}\right)^n. \tag{18.7}$$

The exponent n, called the stability parameter, is an empirically determined function of the atmospheric stability, and is given in Table 18-2. Wind is often measured in weather stations at an elevation of 10 m above ground level.

Table 18-2. Relationship between the Stability Parameter and Atmospheric Stability[a]

Stability condition	n
Large lapse rate (Classes A, B, C)	0.20
Zero or small lapse rate (Class D)	0.25
Moderate inversion (Class E)	0.33
Strong inversion (Classes F and G)	0.50

[a] From Wark and Warner (1986).

Effective Stack Height

The effective stack height is the height above ground at which the plume begins to travel downwind: the effective release point of the pollutant and the origin of its dispersion. A number of empirical models exist for calculating the plume rise h — the height above the stack to which the plume rises before dispersing downwind. Three equations that give a reasonably accurate estimate of plume rise have been developed by Carson and Moses (1969) for different stability conditions. For superadiabatic conditions

$$\Delta h = 3.47 \frac{V_s d}{u} + 5.15 \frac{Q_h^{0.5}}{u}; \tag{18.8a}$$

for neutral stability

$$\Delta h = 0.35 \frac{V_s d}{u} + 2.64 \frac{Q_h^{0.5}}{u}; \tag{18.8b}$$

and for subadiabatic conditions

$$\Delta h = -1.04 \frac{V_s d}{u} + 2.24 \frac{Q_h^{0.5}}{u}, \tag{18.8c}$$

where

V_s = stack gas exit speed (in m/s),
d = stack diameter (in m), and
Q_h = heat emission rate from the stack (in kJ/s).

As before, length is in meters and time is in seconds, and the heat emission rate is measured in kilojoules per second.

EXAMPLE 18.4. A power plant has a stack with a diameter of 2 m and emits gases with a stack exit velocity of 15 m/s and a heat emission rate of 4,800 kJ/s. The wind speed is 5 m/s. Stability is neutral. Estimate the plume rise. If the stack has a geometric height of 40 m, what is the effective stack height?

Using Eq. (18.8b),

$$\Delta h = 0.35 \frac{(15)(2)}{5} + 2.64 \frac{\sqrt{4800}}{5} = 38.7 \text{ m}$$

$$H = h_g + h$$

$$H = 40 \text{ m} + 38.7 \text{ m} = 78.7 \text{ m}.$$

The accuracy of plume rise and dispersion analysis is not very good. Uncalibrated models predict ambient concentrations to within an order of magnitude at best. To ensure reasonable validity and reliability, the model should be calibrated with measured ground level concentrations.

The model discussed applies only to a continuous, steady point source of emission. Discrete discontinuous emissions or puffs, larger areas that act as sources, like parking lots, and line sources, like highways, are modeled using variants of the Gaussian approach, but the actual representation used in each case is quite different.

Computer Models for Assessing Atmospheric Dispersion

A number of computer models that run on a desktop PC exist for assessing atmospheric dispersion of pollutants. These are essentially codifications of the Gaussian dispersion equations that solve the equations many times and output an isopleth plot. Some models are:

- DEPOSITION 2.0 (U.S. Nuclear Regulatory Commission NUREG/GR-0006, 1993)
- CAP88-PC (U.S. Department of Energy, ER 8.2, GTN, 1992)
- RISKIND (Yuan et al. 1993)
- HAZCON (Sandia National Laboratories 1991)
- TRANSAT (Sandia National Laboratories 1991)
- HOTSPOT (Lawrence Livermore National Laboratory, 1996)
- MACCS 2 (Sandia National Laboratories, 1993)

CLEANSING THE ATMOSPHERE

Processes by which the atmosphere cleans itself do exist, and include the effect of gravity, contact with the earth's surface, and removal by precipitation.

Gravity

Particles in the air, if they are larger than about a millimeter in diameter, are observed to settle out under the influence of gravity; the carbon particles from elevated diesel truck exhaust are a very good example of such settling. However, most particles of air pollutants are small enough that their settling velocity is a function of atmospheric turbulence, viscosity, and friction, as well as of gravitational acceleration, and settling can be exceedingly slow. Particles smaller than 20 μm in diameter will seldom settle out by gravity alone. Gases are removed by gravitational settling only if they are adsorbed onto particles or if they condense into particulate matter. Sulfur trioxide, for example, condenses with water and other airborne particulates to form sulfate particles.

Particles small enough to stay in the air for appreciable periods of time are dispersed in the air, but in a slightly different way than gaseous pollutants are dispersed. The dispersion equation must be modified by considering the settling velocity of these small particles. For particles between 1 and 100 μm in diameter, the settling velocity follows Stokes' law

$$V_t = gd^2 \frac{\rho}{18\mu} \tag{18.9}$$

where

V_t = settling or terminal velocity,
g = acceleration due to gravity,
d = particle diameter,
ρ = particle density, and
μ = viscosity of air.

The settling velocity modifies the Gaussian dispersion equation, Eq. (18.3), to give the analogous equation

$$C(x, y, 0) = \frac{Q}{2\pi u \sigma_y \sigma_z} \exp\left(\frac{-y^2}{2\sigma_y^2}\right) \exp\left(-\frac{(H - (V_t x/u))^2}{2\sigma_z^2}\right) \tag{18.10}$$

for dispersion of small particles.

The factor of $\frac{1}{2}$ in the first term arises because falling particles are not reflected at the ground surface.

The *rate*, ω, at which particulate matter is being deposited on the ground, is related to the ambient concentration as shown in

$$\omega = V_t C(x, y, 0) = \frac{Q}{2\pi u \sigma_y \sigma_z} \exp\left(\frac{-y^2}{2\sigma_y^2}\right) \exp\left(-\frac{(H - (V_t x/u))^2}{2\sigma_z^2}\right), \tag{18.11}$$

where ω = the deposition rate (in g/s–m^2).

EXAMPLE 18.5. Using the data of Example 18.3, and assuming that the emission consists of particles 10 μm in diameter and having a density of 1 g/cm^3, calculate (1) the ambient ground level concentration at 200 m downwind along the plume centerline, and (2) the deposition rate at that point. The viscosity of the air is 0.0185 g/m-s at 25°C.
From Eq. (18.9), the settling velocity is

$$V_t = (9.8 \text{ m/s}^2)(10^{-5} \text{ m})^2 \frac{(1 \text{ g/cm}^3)}{(10^{-6} \text{ m}^3/\text{cm}^3)(18)(0.0185 \text{ g/m-s})} = 0.0029 \text{ m/s},$$

or 0.29 cm/s. From Example 18.3

$$Q = 18 \text{ g/s}$$

$$C(0.2, 0, 0) = \frac{18 \text{ g/s}}{2\pi(5 \text{ m/s})(35 \text{ m})(19 \text{ m})}$$

$$\times \exp\left\{-\frac{1}{2}\left[\frac{60 \text{ m} - \left(\frac{(.0029 \text{ m/s})(200 \text{ m})}{5 \text{ m/s}}\right)}{19 \text{ m}}\right]^2\right\}$$

$$= 6.03 \times 10^{-6} \text{ g/m}^3$$

or 6.03 μg/m^3. The deposition rate is then

$$\omega = (0.0029 \text{ m/s})(6.03 \times 10^{-6} \text{ g/m}^3\text{-s})$$

$$= 1.75 \times 10^{-8} \text{ g/m}^2\text{-s}$$

or 17.5 ng/m^2-s.

Surface Sink Absorption

Many atmospheric gases are absorbed by the features of the earth's surface, including stone, soil, vegetation, bodies of water, and other materials. Soluble gases like SO_2 dissolve readily in surface waters, and such dissolution can result in measurable acidification.

Precipitation

Precipitation removes contaminants from the air by two methods. *Rainout* is an "in-cloud" process in which very small pollutant particles become nuclei for the formation of rain droplets that grow and eventually fall as precipitation. *Washout* is a "below-cloud" process in which rain falls through the pollutant particles and molecules, which are entrained by the impinging rain droplets or which actually dissolve in the rainwater.

The relative importance of these removal mechanisms was illustrated by a study of SO_2 emissions in Great Britain, where the surface sink accounted for 60% of the SO_2,

15% was removed by precipitation, and 25% blew away from Great Britain, heading northwest toward Norway and Sweden.

CONCLUSION

Polluted air results from both emissions into the air and meteorological conditions that control the dispersion of those emissions. Pollutants are moved predominantly by wind, so that very light wind results in poor dispersion. Other conditions conducive to poor dispersion are:

- little lateral wind movement across the prevailing wind direction,
- stable meteorological conditions, resulting in limited vertical air movement,
- large differences between day and night air temperatures, and the trapping of cold air in valleys, resulting in stable conditions,
- fog, which promotes the formation of secondary pollutants and hinders the sun from warming the ground and breaking inversions, and
- high-pressure areas resulting in downward vertical air movement and absence of rain for washing the atmosphere.

Air pollution episodes can now be predicted, to some extent, on the basis of meteorological data. The EPA and many state and local air pollution control agencies are implementing early warning systems, and acting to curtail emissions and provide emergency services in the event of a predicted episode.

PROBLEMS

18.1 Consider the following temperature soundings: ground level, 21°C; 500 m, 20°C; 600 m, 19°C; 1000 m, 20°C. If we release a parcel of air at 600 m, will it rise or sink or remain at 600 m? If a 70-m stack releases a plume with a temperature of 30°C, what type of plume results? How high does the plume rise?

18.2 A weather station located at an elevation of 10 m above ground measures wind speed at 1.5 m/s. Plot the wind speed vs elevation for the meteorologic conditions given in Table 18-2.

18.3 Plot the ambient lapse rate given the following temperature sounding:

Elevation (m)	Temperature (°C)
0	20
50	15
100	10
150	15
200	20
250	15
300	20

372   ENVIRONMENTAL ENGINEERING

What type of plume would you expect if the exit temperature of the plume at the stack were 15°C and the smokestack were 40 m tall? 120 m tall? 240 m tall?

18.4 Three industrial emitters and an air sampling station are located at the following coordinates on a map (the Y axis points North; the X axis East):

Industrial Plant A at $X = 3, Y = 3$
Industrial Plant B at $X = 3, Y = 1$
Industrial Plant C at $X = 8, Y = 8$
Air sampling station at $X = 5, Y = 5$.

The data at the air sampling station are:

Day	Wind direction	Particulates ($\mu g/m^3$)	SO_2 ($\mu g/m^3$)
1	N	80	80
2	NE	120	20
3	NW	30	30
4	N	90	40
5	NE	130	20
6	SW	20	180
7	S	30	100
8	SW	40	200
9	E	100	60
10	W	10	100

Draw wind roses and pollution roses, and determine which plant is most guilty of the measured air pollution.

18.5 Using Eq. (18.7), derive an expression for the maximum ground level concentration in terms of the wind speed u and the dispersion constants σ_y and σ_z.

18.6 A power plant burns 1000 tons of coal per day, 2% of which is sulfur. All of the sulfur is burned completely and emitted into the air from a stack with an effective height of 100 m. For a wind speed of 6 m/s, calculate (a) the ground level SO_2 concentration along the plume center line 10 km downwind, (b) the maximum ground level SO_2 concentration for B stability, using a conservative value for the wind speed, and (c) the downwind distance of maximum ground level SO_2 for B stability.

18.7 The power plant in Problem 18.6 uses coal with a 14% by weight ash content, half of which is fly ash (entrained in the off-gas). If the ash particles (which we assume to be uniform!) have a diameter of 15 μm and a density of 1.5 g/cm^3, calculate the ash deposition rate along the plume center line (a) 5 km downwind on a bright sunny day and (b) 10 km downwind on a night when the weather is overcast. Use the minimum compatible wind speeds.

18.8 The odor threshold of hydrogen sulfide is about 0.7 $\mu g/m^3$. If 0.08 g of H_2S is emitted per second out of a stack with an effective height of 40 m during an overcast night with a wind speed of 3 m/s, calculate the maximum ground level concentration of H_2S and estimate how far downwind this concentration occurs. Estimate the area (in terms of x and y coordinates) where H_2S could be detected by its odor. A programmable calculator or a computer spreadsheet program may facilitate this calculation, or you can write a computational program.

18.9 A facility emits 100 g SO_2/s. On relatively clear nights (slight cloud cover) and a wind speed of 2 m/s, the maximum ground level SO_2 concentration cannot exceed 60 $\mu g/m^3$. If the only control on SO_2 emission is stack height, what would the effective stack height have to be?

18.10 A copper smelter processes 1000 metric tons of ore per day, principally $CuFeS_2$, producing SO_2 as a pollutant. The stack is 80 m tall and 2 m in exit diameter, and heat loss through the stack is 3000 kJ/s. For an overcast night with a wind speed (measured at 5 m) of 4 m/s:

a. Calculate the effective stack height.
b. Calculate the maximum ground level concentration of SO_2.
c. Estimate the percent SO_2 control needed to meet the EPA 24-h secondary ambient air quality standard (see Chap. 20).

18.11 A power plant has a 90-m stack with an outlet diameter of 10 m. The plant burns 15,500 metric tons of coal per day, with a sulfur content of 3.5% by weight, and all the sulfur is converted to SO_2. The coal has a heat content of 12,500 Btu/lb, and about 6% of the heat generated is lost out the stack. The temperature of the plume emerging from the stack is 200°C. Wind measurements at 3 m from the ground on an overcast winter afternoon, when the ground temperature is 0°C, show a wind from due north at 4 m/s.

a. Estimate the *effective* stack height by two methods.
b. Using the more conservative value of the effective stack height, plot a graph of the ground level SO_2 concentration with downwind distance. At the 10-km point, plot the SO_2 at right angles to the wind (as in Fig. 18-16) in both the y and z directions.

18.12 If an air sampling site were located on the ground 8 km SE of the plant in Problem 18.11, what SO_2 concentration would the sampler expect to measure?

18.13 The concentration of H_2S 200 m downwind from an abandoned oil well is 3.2/m³. What is the emission rate on a partially overcast afternoon when the wind speed is 2.5 m/s at ground level?

18.14 A power plant burns 2% sulfur coal and has a 70-m stack with a diameter of 1.5 m. The exit velocity and temperature are 15 m/s and 340°C, respectively. The atmospheric conditions are 28°C and a barometric pressure of 90 kPa, and the weather is a clear summer afternoon with a wind speed of 2.8 m/s measured at 10 m. At a point 2.5 km downwind and 300 m off the plume centerline, the ground level SO_2 concentration was measured at 143 $\mu g/m^3$. How much coal does the plant burn?

18.15 The power plant in Problem 18.14 burns coal with a 10% ash content, and 20% of the ash (by weight) is emitted into the air. The ash has an average density of 1.0 g/cm³ and an average diameter of 2 μm. Calculate the deposition rate at the point indicated in Problem 18.14.

374 ENVIRONMENTAL ENGINEERING

18.16 The national average frequency of each of the Pasquill stability classes is:

Class A: 1.1%
Class B: 6.8%
Class C: 11.4%
Class D: 47.2%
Class E: 12.1%
Class F/G: 21.4%

a. What is the minimum wind speed consistent with each of the stability classes? Why?
b. Select a minimum wind speed and construct isopleths for the power plant in Problem 18.14. You will need to use a spreadsheet program for this problem.

Chapter 19

Measurement of Air Quality

Air quality measurements are designed to determine the levels of all types of contaminants in the atmosphere we breathe with no attempt made to differentiate between naturally occurring contaminants and those that result from human activity. Measurements of air quality fall into three classes:

- Measurement of emissions. This is called stack sampling when a stationary source is analyzed. Samples are drawn out through a hole or vent in the stack for on-the-spot analyses. Mobile sources like automobiles are tested by sampling exhaust emissions while the engine is running and working against a load.
- Meteorological measurements. The measurement of meteorological factors: wind speed, wind direction, lapse rates, etc., is necessary to determine how pollutants travel from source to recipient.
- Measurement of ambient air quality. Ambient air quality is measured by a variety of monitors discussed in this chapter. Almost all evidence of health effects of air pollution are based on correlation of these effects with measured ambient air quality.

Modern monitoring and measurement equipment uses power-driven pumps and other collection devices, and can sample a larger volume of air in a relatively shorter time than older equipment could. Gas measurement is often by wet chemistry, in that collected gas is either dissolved into or reacted with a collecting fluid. Modern monitors provide continuous readout. The measurement of pollutant concentrations is almost instantly translated by a readout device, so that the pollution may be measured while it is happening.

MEASUREMENT OF PARTICULATE MATTER

Both total suspended particulate matter (TSP) and particles $10\,\mu\text{m}$ in diameter or less (PM_{10}) are measured when monitoring ambient air. Measurement devices used are variants of high-volume, or hi-vol, samplers. A hi-vol sampler (see Fig. 19-1) operates much like a vacuum cleaner by pumping air at a high rate through a filter. About $2000\,\text{m}^3$ ($70,000\,\text{ft}^3$) can be pumped in 24 h, so that sampling time can be cut to between 6 and 24 h. Analysis is gravimetric; the filter is weighed before and after the sampling period. The weight of particles collected is then the difference between these two weights.

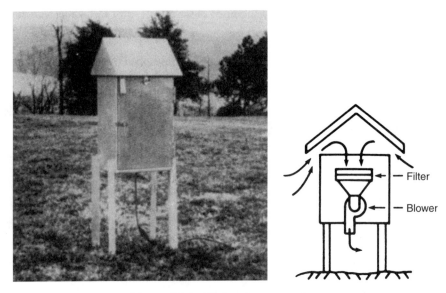

Figure 19-1. High-volume sampler.

Air flow through the filter is measured with a flow meter, usually calibrated in cubic feet of air per minute. Because the filter collects dirt during its hours of operation, less air passes through it during the latter part of the test than in the beginning, and the air flow must therefore be measured at both the start and end of the test period and the values averaged.

EXAMPLE 19.1. A clean filter is found to weigh 10.00 g. After 24 h in a hi-vol sampler, the filter plus dust weighs 10.10 g. The air flow at start and end of the test was 60 and 40 ft^3/min, respectively. What is the concentration of particulate matter?

Weight of the particulates (dust) = $(10.10 - 10.00)$ g $\times$ 10^6 µg/g = 0.1×10^6 µg.

Average air flow = $(60 + 40)/2 = 50$ ft^3/min.

Total air through the filter = 50 ft^3/min $\times$ 60 min/h $\times$ 24 h/day $\times$ 1 day

$= 72{,}000$ ft^3 $\times$ 0.0283 m^3/ft^3 = 2038 m^3.

Total suspended particulate matter = $(0.1 \times 10^6$ µg$)/2038$ m^3 = 49 µg/m^3.

The high-volume sampler operated in this way measures TSP. Hi-vol samplers can be fitted with a variety of filters and used to measure smaller particles and particles of a particular size. Measurement of 10-µm particles requires about 10 times the air flow in a hi-vol sampler: 16.7 L/min as compared to 1.4 L/min for TSP. The ambient standard is more stringent than it is for TSP (see Chap. 21).

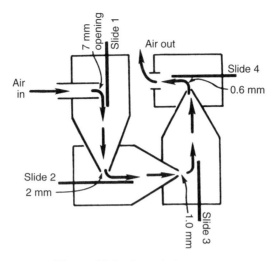

Figure 19-2. Cascade impactor.

The cascade impactor, shown in Fig. 19-2, may also be used to measure fine particles including respirable particles less than 1.0 μm in diameter — small enough to penetrate the lung. The impactor consists of four tubes, each with a progressively smaller opening, thus forcing progressively higher throughput velocities. A particle entering the device may be small enough to follow the streamline of flow through the first nozzle without hitting the microscope slide. At the next nozzle, however, the velocity may be sufficiently high to prevent the particle from negotiating the turn, and it will impinge on the slide.

Particle-measuring devices are usually fitted with an automatic computer input and recording arrangement. The hi-vol sampler/computer recorder combination is often referred to as a CAPS, or computer-assisted particle monitor.

The nephelometer (Fig. 19-3) can make continuous measurements of real-time data. A nephelometer measures the intensity of light scattered by fine particles in the air, and the scattered light intensity is proportional to the concentration of smoke or very fine particulate matter in the air. Fine particles interfere with visibility by scattering light; this scattering is what we know as haze. In a nephelometer, the scattered light intensity is measured at a 90° angle from the incident light. The instrument can be calibrated either in units of percent visibility decrease or in units of micrograms per cubic meter ($\mu g/m^3$).

MEASUREMENT OF GASES

Ambient concentrations of gaseous air pollutants are usually measured by reacting the gas chemically with a colorimetric reagent and measuring the intensity of the reaction product color.

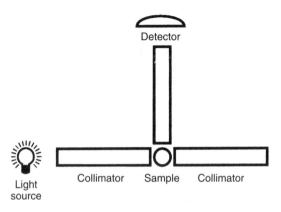

Figure 19-3. Diagram of a nephelometer.

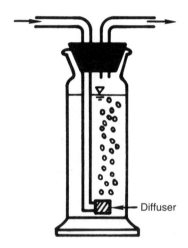

Figure 19-4. A typical bubbler used for measurement of gaseous air pollutants.

Sulfur dioxide may also be measured by impregnating filter papers with chemicals that react with SO_2 and change color. For example, lead peroxide reacts to form dark lead sulfate, by the reaction

$$PbO_2 + SO_2 \rightarrow PbSO_4.$$

The extent of dark areas on the filter and the depth of color may be related to the SO_2 concentration. Like the ozone measurement, this type of SO_2 measurement takes several days of exposure to yield measurable results.

Use of a gas bubbler, shown in Fig. 19-4, shortens the time required for sampling considerably. As shown in Fig. 19-4, the air sample is bubbled through a solution that reacts chemically with the particular gaseous pollutant being measured. The concentration is then measured with further wet chemical techniques. For example,

Measurement of Air Quality 379

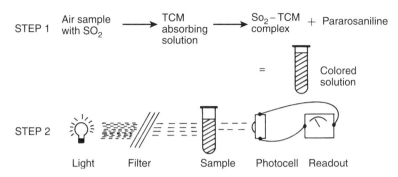

Figure 19-5. Schematic diagram of the pararosaniline method for measuring SO_2.

SO_2 may be measured by bubbling air through hydrogen peroxide, so that the following reaction occurs:

$$SO_2 + H_2O_2 \rightarrow H_2SO_4.$$

The amount of sulfuric acid formed can then be determined by titrating against a base of known concentration.

Figure 19-5 is a schematic diagram of the pararosaniline technique for measuring SO_2, which is a standard method for measuring atmospheric SO_2. In this method, air is bubbled through a solution of tetrachloromercurate (TCM). The SO_2 and TCM combine to form a complex that then is reacted with pararosaniline to form a colored solution. The intensity of the color is proportional to the SO_2 concentration, absorbs light at wavelength $560 \, \mu m$, and can be measured with a colorimeter or spectrophotometer. A similar colorimetric technique, for measuring ammonia concentration, is described in Chap. 5.

Most bubblers are not 100% efficient; not all of the gas bubbled through the liquid will be absorbed, and some will escape. The quantitative efficiency of a bubbler is established by testing and calibrating with known concentrations of various gases in air. Gas chromatography is a newer and very useful second-generation measurement method, particularly since trapping the pollutant in a bubbler is not necessary. Long sampling times are not needed, and the air sample can usually be introduced directly into the gas chromatograph.

One widely used third-generation device is nondispersive infrared spectrophotometry (nondispersive IR), used for measurement of CO, including CO measurement for routine automobile inspection and maintenance. Like all asymmetric gas molecules, CO absorbs at the specific infrared frequencies that correspond to molecular vibrational and rotational energy levels. As shown in Fig. 19-6, the air sample is pumped into one of two chambers in the detector. The other chamber contains a reference gas like nitrogen. Infrared lamps shine through both the sample cell and the reference cell. CO in the sample will absorb IR in direct proportion to its concentration in the sample. After passing through the two cells, the radiant energy is absorbed by the gas in the two detector cells, both of which contain CO. Absorption of radiant

380 ENVIRONMENTAL ENGINEERING

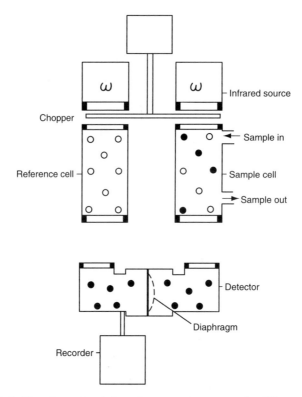

Figure 19-6. Nondispersive infrared spectrophotometer for CO measurement.

energy by the CO causes the gas in the detector cells to expand, but the detector under the reference cell will receive more energy (since the reference gas did not absorb) and expand, moving the diaphragm separating the two detector cells. The diaphragm movement is detected electronically, and the continuously transmitted signal is read out on a recorder.

REFERENCE METHODS

Many methods are available for measuring air pollutant concentration, and new, more rapid, and more accurate and precise methods are always being developed. The U.S. Environmental Protection Agency (EPA) has designated a series of reference methods with which results from all other methods can be compared. Although the EPA standard reference methods may not always be the most sensitive, they have been standardized and have a history of independently duplicated results. Some reference methods change with each annual edition of the Code of Federal Regulations. The reader is encouraged to access the EPA webpage at http://www.epa.gov/ttn/emc/ for the most recent standard measuring and monitoring methods. The existence of reference methods is especially

Measurement of Air Quality 381

important when there is a question of compliance with air quality standards and when analyzing for hazardous air pollutants.

GRAB SAMPLES

It is often necessary to obtain a sample of a gas for analysis in the laboratory, and obtaining this sort of grab sample of air presents some difficulties. Grab samples of exhaust gases, as from automobiles, are relatively straightforward to collect; care must be taken only that the container can withstand the temperature of the exhaust gas being sampled. Plastic or aluminum-coated bags are often used. The gas is pumped or exhausted with some positive pressure into the bag, and allowed to escape through a small hole. By displacing several volumes of gas in this way, contamination problems can be avoided.

Evacuated containers can be used for grab samples. The air to be sampled is allowed to be drawn into a previously evacuated container, usually attached to a vacuum system. Some contamination is always possible, since no container can be completely evacuated and the concentration of air pollutants being sampled is usually small. If gases being sampled are not soluble in water, allowing gas to displace water in a container is a useful sampling method. Unfortunately, most air pollutants measured in grab samples are water soluble.

STACK SAMPLES

Stack sampling is an art worthy of individual attention. Sampling of gas directly from a stack is necessary to evaluate compliance with emission standards and to determine the efficiency of air pollution control equipment. In moderate- to large-diameter industrial stacks, when the gas is exhausted at relatively high temperatures, the concentration of exhaust gas constituents is not uniform across the gas stream, and the sampler must take care in getting a representative sample. A thorough survey is made of the flow, temperature, and pollutant concentration both across the effluent stream and at various locations within the stack. A train of instruments, as shown in Fig. 19-7, is often used for stack sampling, so that a number of measurements may be determined at each positioning of the intake nozzle.

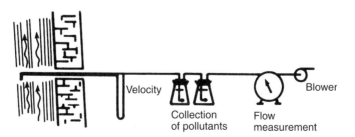

Figure 19-7. A stack sampling train.

SMOKE AND OPACITY

Visible smoke from a stack is often the only immediate evidence of a pollution violation external to an industrial source. Effluent gases cannot be sampled and analyzed without a complicated system that always involves notifying the emitter, but smoke issuing from a stack can be seen. The opacity of a smoke plume is thus still the only enforcement method that may be used without the emitter's knowledge (and, often, cooperation). Therefore, many regulations are still written on the basis of visually estimated smoke density.

The density of black or gray smoke is measured by visual observation and estimation of smoke opacity — the amount of light blocked by the smoke. Although this method might appear to have very large inherent uncertainty, it has been used enough so that the reproducibility of observations is fairly reliable. Some data on reproducibility, developed by EPA, are:

- For black plumes (133 sets at a smoke generator), 100% of the sets were read with a positive error of less than 7.5% opacity; 99% were read with a positive error of less than 5% opacity. (Note: For a set, positive error = average opacity determined by observers' 25 observations − average opacity determined from transmissometer's 25 recordings.)
- For white plumes (170 sets at a smoke generator, 168 sets at a coal-fired power plant, 298 sets at a sulfuric acid plant), 99% of the sets were read with a positive error of less than 7.5% opacity; 95% were read with a positive error of less than 5% opacity.

A positive observational error associated with an average of 25 readings may thus be established at 5%.

A typical opacity standard is 20% opacity, which is a barely visible plume, with allowances for 40% opacity for very short periods of time. Modern practice consists of training enforcement agents to recognize opacities by repeated observation of smoke of predetermined opacity.

CONCLUSION

As with water pollution, the analytical tests of air quality can be only as good as the samples or sampling techniques used. Moreover, the prevailing analytical techniques leave much to be desired in both precision and accuracy. Most measurements of environmental quality, including these, are reasonable estimates at best.

PROBLEMS

19.1 An empty 6-in. diameter dustfall jar weighs 1560 g. After sitting outside for the prescribed amount of time, the jar weighs 1570 g. Report the dustfall in the usual manner.

19.2 A hi-vol clean filter weighs 20.0 g. After exposure in the sampler for 24 h, the filter (now dirty) weighs 20.5 g. The initial and final air flows through the hi-vol sampler were 70 and 50 ft^3/min, respectively. What was the concentration of particulate matter in the air? Were any National Ambient Air Quality Standards exceeded? (See Chap. 21.)

19.3 A hi-vol sampler draws air at an average rate of 70 ft^3/min. If the measured particle concentration is 200 µg/m^3 and the hi-vol sampler was operated for 12 h, what was the weight of dust on the filter?

Chapter 20

Air Pollution Control

Limiting emissions into the air is both technically difficult and expensive. However, since rain and deposition are the only air cleaning mechanisms available, and neither is very efficient, good air quality depends on pollution prevention and on limiting what goes into the air. The control of air emissions may be realized in a number of ways. Figure 20-1 shows five separate possibilities for control. Dispersion is discussed in Chap. 18, and the four remaining control points are discussed individually in this chapter.

SOURCE CORRECTION

Changing or eliminating a process that produces a polluting air effluent is often easier than trying to trap the effluent. A process or product may be needed or necessary, but could be changed to control emissions. For example, automobile exhaust has caused high lead levels in urban air. Elimination of lead from gasoline, which was needed for proper catalytic converter operation, has also resulted in reducing lead in urban air. Similarly, removal of sulfur from coal and oil before the fuel is burned reduces the amount of SO_2 emitted into the air. In these cases, the source of air pollution has been corrected.

Processes may also be modified in order to reduce air pollution. Odors from municipal incinerators may be controlled by operating the incinerator at a high enough temperature to effect more complete oxidation of odor-producing organic compounds. The 1990 Clean Air Act mandates the use of oxygenated fuels in urban areas in order to limit CO emissions from automobiles.

Strictly speaking, such measures as process change, raw material substitution, and equipment modification to meet emission standards are known as *controls*. In contrast, *abatement* is the term used for all devices and methods for decreasing the quantity of pollutant reaching the atmosphere, once it has been generated by the source. For simplicity, however, we will refer to all of the procedures as controls.

COLLECTION OF POLLUTANTS

Collection of pollutants for treatment is often the most serious problem in air pollution control. Automobile exhaust is a notorious polluter mainly because the effluent

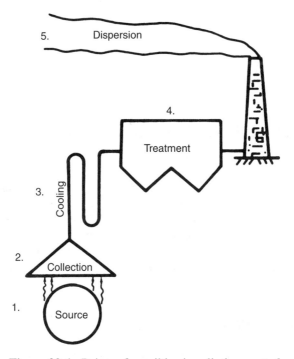

Figure 20-1. Points of possible air pollution control.

(exhaust) is so difficult to trap and treat. If automobile exhaust could be channeled to a central treatment facility, treatment could be more efficient in controlling each individual car.

Recycling of exhaust gases is one means of control. Although automobiles could not meet 1990 exhaust emission standards only by recycling exhaust and blow-by gases, this method proved a valuable start to automobile emission controls. Many stationary industries recycle exhaust gases, usually CO and volatile organic compounds, as fuel for the process, since even CO releases heat when burned to CO_2.

Process exhaust gases channeled through one or more stacks are relatively easy to collect, but fugitive emissions from windows and doors, cracks in the walls, and dust raised during on-site transportation of partially processed materials pose a difficult collection problem. Some industries must overhaul the entire plant air flow system to provide adequate control.

COOLING

The exhaust gases to be treated are sometimes too hot for the control equipment, and must first be cooled. Cooling may also drop the temperature below the condensation point of some pollutants, so that they may be collected as liquids. Dilution, quenching,

Air Pollution Control 387

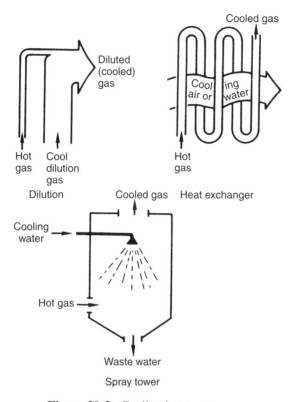

Figure 20-2. Cooling hot waste gases.

and heat exchange, shown in Fig. 20-2, are all acceptable cooling methods. Quenching has the added advantage of scrubbing out some gases and particulate matter, but may yield a dirty, hot liquid that itself requires disposal. Cooling coils are probably the most widely used cooling method and are especially appropriate where heat can be conserved.

TREATMENT

Selection of the correct treatment device requires matching the characteristics of the pollutant with features of the control device. Pollutant particles vary in size over many orders of magnitude, from ideal gas molecules to macroscopic particles several millimeters in diameter. One device will not be effective and efficient for all pollutants, or even for all pollutants coming from the same stack. The chemical behavior of pollutants may also dictate selection of a control process. The various air pollution control devices are conveniently divided into those that control particulate matter and those that control gaseous pollutants.

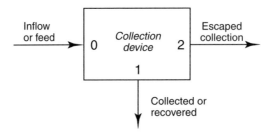

Figure 20-3. Diagram of a collection device.

The effectiveness of a collection device is evaluated according to the recovery of the pollutant, as in (see Fig. 20-3)

$$R = \frac{x_1}{x_0} \times 100, \quad (20.1)$$

where

R = recovery of pollutant,
x_1 = mass flow of pollutant recovered (in kg/s), and
x_0 = mass flow of pollutant produced (in kg/s).

If the mass flow of pollutant that escapes capture is designated x_2, then

$$x_0 = x_1 + x_2,$$

and

$$R = \frac{x_1}{x_1 + x_2} \times 100. \quad (20.2)$$

If the concentration of the pollutants and the rate of air flow are known, the recovery may be expressed as

$$R = \frac{Q_1 C_1}{Q_0 C_0} \times 100, \quad (20.3)$$

where Q = air flow rate (in m³/s) and C = concentration of pollutant (in kg/m³), and the subscripts refer to inflow and captured pollutant (see Fig. 20-3). When air flow rates are not known, and only concentrations are measured, the recovery may be shown to be

$$R = \frac{C_1(C_0 - C_2)}{C_0(C_1 - C_2)} \times 100. \quad (20.4)$$

In air pollution terms, since the pollutants generated are to be separated from the air, the recovery is called *separation efficiency* or *collection efficiency*.

Cyclones

The cyclone is a popular, economical, and effective means of controlling particulates. Cyclones alone are generally not adequate to meet stringent air pollution control regulations, but serve as precleaners for control devices like fabric filters or electrostatic precipitators. Figure 20-4 shows a simple diagram of a cyclone, and Fig. 20-5 a more detailed picture. Dirty air enters the cyclone off-center at the bottom; a violent swirl of air is thus created in the cone of the cyclone and particles are accelerated centrifugally outward toward the cyclone wall. Friction at the wall slows the particles and they slide to the bottom, where they can be collected. Clean air exits at the center of the top of the cone. As may be seen in Fig. 20-6, cyclones are reasonably efficient for large particle collection, and are widely used as the first stage of dust removal.

Cyclones are sized on the basis of desired centrifugal acceleration. Consider a particle moving radially, from the central axis of the cyclone to the outside wall. The velocity of this movement, assuming laminar flow and spherical particles, is

$$v_R = \frac{(\rho_s - \rho)r\omega^2 d^2}{18\mu}, \qquad (20.5)$$

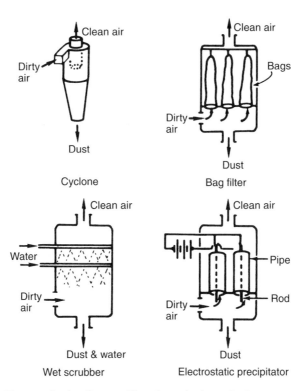

Figure 20-4. Four methods of controlling (trapping) particulate matter from stationary sources.

390 ENVIRONMENTAL ENGINEERING

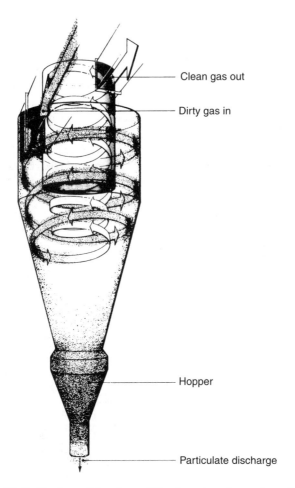

Figure 20-5. Cyclone. (Courtesy of the American Lung Association.)

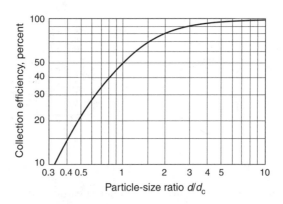

Figure 20-6. Cyclone efficiency.

where

v_R = radial velocity (in m/s),
ρ_s = density of particle (in kg/m³),
ρ = density of air (in kg/m³),
r = radial distance (in m),
ω = rotational velocity (in rad/s),
d = particle diameter (in m), and
μ = viscosity of air (in kg/m–s).

The centrifugal acceleration is equal to the square of the tangential velocity divided by the radius, or

$$r\omega^2 = \frac{v_{\text{tan}}^2}{r}.$$

Substituting for $r\omega^2$ in Eq. (20.5) yields

$$v_R = \frac{(\rho_s - \rho)d^2}{18\mu} \frac{v_{\text{tan}}^2}{r}. \tag{20.6}$$

However,

$$\frac{(\rho_s - \rho)d^2}{18\mu} = \frac{v}{g},$$

where v is the settling or gravitational velocity, so that

$$v_R = \frac{v}{g} \frac{v_{\text{tan}}^2}{r}. \tag{20.7}$$

The *separation factor* S for a cyclone is defined as

$$S = \frac{v_R}{v} = \frac{v_{\text{tan}}^2}{rg}. \tag{20.8}$$

S is dimensionless, and may take values from 5 to as high as 2500. A large separation factor requires high tangential velocities, thus a small diameter and high velocity, leading to large pressure drops across the cyclone. A large-diameter cyclone, on the other hand, would have a small separation factor.

The pressure drop through a cyclone can be calculated by an empirical relationship

$$\Delta P = \frac{3950 K Q^2 P \rho}{T}, \tag{20.9}$$

where

ΔP = pressure drop, in meters of water
Q = m³/s of gas,

Table 20-1. Values of K for Calculating Cyclone Pressure Drop

Cyclone diameter (in.)	K
29	10^{-4}
16	10^{-3}
8	10^{-2}
4	0.1

P = absolute pressure (in atm),
ρ = gas density (in kg/m³),
T = temperature (in K), and
K = proportionality factor, a function of cyclone diameter.

Approximate values of K are given in Table 20-1. Typical pressure drops in cyclones are between 1 and 8 in. of water.

The collection efficiency of a cyclone can be estimated by using the concept of a *cut diameter*, the particle diameter at which 50% of the particles are removed from the gas stream by the cyclone. The cut diameter is

$$d_c = \left[\frac{9\mu b}{2\pi N v_i (\rho - \rho_s)}\right]^{1/2}, \qquad (20.10)$$

where

μ = gas viscosity (in kg/m–h),
b = cyclone inlet width (in m),
N = effective number of outer turns in cyclone (usually about four),
v_i = inlet gas velocity (in m/s),
ρ_s = particle density (in kg/m³), and
ρ = gas density (in kg/m³), usually negligible compared to ρ_s.

Figure 20-6 shows how the cut diameter can be used to establish the collection efficiency for particles of any diameter d. The number of turns N can also be approximated as

$$N = \frac{\pi}{H}(2L_2 - L_1), \qquad (20.11)$$

where

H = height of the inlet (in m),
L_1 = length of cylinder (in m), and
L_2 = length of cone (in m).

EXAMPLE 20.1. A cyclone has an inlet width of 10 cm and four effective turns ($N = 4$). The gas temperature is 350 K and the inlet velocity is 10 m/s. The average particle diameter is 8 µm and the average density is 1.5 g/cm³. What is the collection efficiency?

The viscosity of air at 350 K is 0.0748 kg/m–h. We can assume that ρ is negligible compared to ρ_s:

$$d_c = \left[\frac{9(0.0748 \text{ kg/m–h})(0.1 \text{ m})}{2\pi(4)(10 \text{ m/s})(3600 \text{ s/h})(1500 \text{ kg/m}^3)} \right]^{1/2}$$
$$= 7.04 \times 10^{-6} \text{m} = 7.04 \, \mu\text{m}.$$

Then, using Eq. (20.10)

$$\frac{d}{d_c} = \frac{8}{7.04} = 1.14,$$

and from Fig. 20-6, the expected removal efficiency is about 55%.

Fabric Filters

Fabric filters used for controlling particulate matter (Figs. 20-4 and 20-7) operate like a vacuum cleaner. Dirty gas is blown or sucked through a fabric filter bag. The fabric bag collects the dust, which is removed periodically by shaking the bag. Fabric filters can be very efficient collectors for even submicrometer-sized particles and are widely

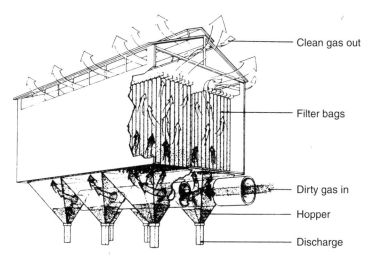

Figure 20-7. Industrial fabric filter apparatus. (Courtesy of American Lung Association.)

used in industrial applications, although they may be sensitive to high temperatures and humidity.

The basic mechanism of dust removal in fabric filters is thought to be similar to the action of sand filters in water quality management, as discussed in Chap. 7. Dust particles adhere to the fabric because of surface force that result in entrapment. Particles are brought into contact with the fabric by impingement or Brownian diffusion. The removal mechanism cannot be simply sieving, since fabric filters commonly have an air space-to-fiber ratio of 1-to-1.

As particles adhere to the fabric, the removal efficiency increases, but so does the pressure drop. The pressure drop is the sum of the pressure drop owing to the fabric and the drop owing to the caked or adhered particles, and is expressed as

$$\Delta P = v\mu \left\{ \frac{x_f}{K_f} + \frac{x_p}{K_p} \right\}, \qquad (20.12)$$

where

ΔP = overall pressure drop in meters of water,
v = superficial gas velocity through the fabric (in m/s),
μ = gas viscosity (in P),
x = thickness of filter (f) and particle layer (p) (in m), and
K = filter (f) and particle layer (p) permeability.

The value of K must be obtained by experiment for each type of fabric and type of dust to be collected.

Wet Collectors

The spray tower or scrubber pictured in Figs. 20-4 and 20-8 can remove larger particles effectively. More efficient scrubbers promote the contact between air and water by violent action in a narrow throat section into which the water is introduced. Generally, the more violent the encounter, hence the smaller the gas bubbles or water droplets, the more effective the scrubbing. A *venturi scrubber*, as shown in Fig. 20-9, is a frequently used high-energy wet collector. Gas flow is constricted through a venturi throat section and water is introduced as high-pressure streams perpendicular to the gas flow. The venturi scrubber is essentially 100% efficient in removing particles >5 μm in diameter. The pressure drop is estimated by an empirical equation

$$\Delta P = v^2 L \times 10^{-6}, \qquad (20.13)$$

where

ΔP = pressure drop across the venturi in centimeters of water,
v = gas velocity in the throat (in cm/s), and
L = water-to-gas volume ratio (in L/m^3).

Air Pollution Control 395

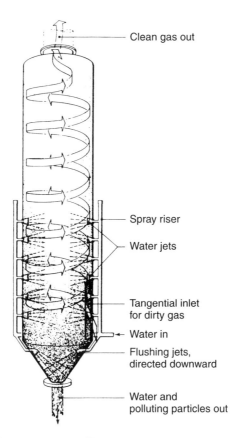

Figure 20-8. Spray tower. (Courtesy of American Lung Association.)

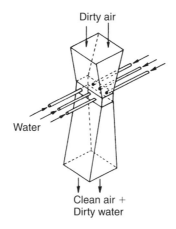

Figure 20-9. Venturi scrubber.

396   ENVIRONMENTAL ENGINEERING

Although wet scrubbers are very efficient, and can trap gaseous pollutants as well as very fine particulate matter, they have disadvantages as well. Scrubbers use a great deal of water that either requires further treatment or has limited use after being used to scrub dirty gas. In places like the Colorado Basin, where water supplies are limited, use in a scrubber may have a very low priority among other uses for available water. Moreover, scrubbers use energy and are expensive to construct as well as to operate. Finally, scrubbers usually produce a visible plume of water vapor.

Electrostatic Precipitators

Electrostatic precipitators are widely used to trap fine particulate matter in applications where a large amount of gas needs treatment and where use of a wet scrubber is not appropriate. Coal-burning electric generating plants, primary and secondary smelters, and incinerators often use electrostatic precipitators. In an electrostatic precipitator, particles are removed when the dirty gas stream passes across high-voltage wires, usually carrying a large negative DC voltage. The particles are electrically charged on passage past these electrodes and then migrate through the electrostatic field to a grounded collection electrode. The collection electrode can be either a cylindrical pipe surrounding the high-voltage charging wire or a flat plate, like that shown in Fig. 20-10. In either case, the collection electrode must be periodically rapped with small hammer-heads to loosen the collected particles from its surface.

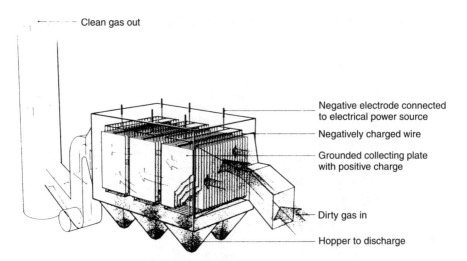

Figure 20-10. Flat-plate electrostatic precipitator. (Courtesy of the American Lung Association.)

The collection efficiency of an electrostatic precipitator may be estimated by an empirical equation

$$R = 1 - \exp\left(\frac{-Av_d}{Q}\right), \qquad (20.14)$$

where

- A = total area of the collecting surface of the collection electrodes (m^2),
- Q = flow rate of gas through the pipe (m^3/s), and
- v_d = drift velocity (m/s).

The drift velocity is the velocity of the particles toward the collecting electrode, and may be calculated theoretically by equating the electrostatic force on the charged particle in the electrical field with the drag force as the particle moves through the gas. The drift velocity is analogous to the terminal settling velocity, as given, for example, in Eq. (18.9), except that in the latter case the force acting in opposition to the drag force is gravitational rather than electrostatic. Drift velocity may be estimated by

$$v_d = 0.5d, \qquad (20.15)$$

where d is the particle size (in μm). Drift velocities are usually between 0.03 and 0.2 m/s.

As the dust layer builds up on the collecting electrode, the collection efficiency may decrease, particularly if the collection electrode is the inside of a cylindrical pipe. Moreover, some dust has a highly resistive surface, does not discharge against the collection electrode, and sticks to the electrode. Heated or water-flushed electrodes may solve this difficulty. Electrostatic precipitators are very efficient collectors of very fine particles. However, the amount of dust collected is directly proportional to the current drawn, so that the electrical energy used by an electrostatic precipitator can be substantial, with resulting high operating cost.

Figure 20-11 shows the effectiveness of an electrostatic precipitator in controlling emissions from a power plant. The large white boxes in the foreground are the electrostatic precipitators. The one on the right, leading to the two stacks on the right, has been turned off to show the effectiveness by comparison with the almost undetectable emission from the stacks on the left.

Comparison of Particulate Control Devices

Figure 20-12 shows the approximate collection efficiencies, as functions of particle size, for the devices discussed. Costs of collection also vary widely.

Figure 20-11. Effectiveness of electrostatic precipitators on a coal-fired electric generating plant. The electrostatic precipitator on the right has been turned off.

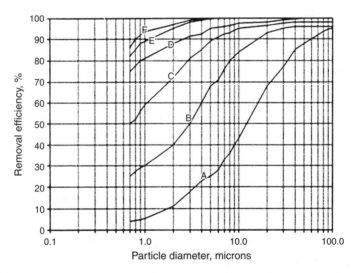

Figure 20-12. Comparison of removal efficiencies: (A) simple and (B) high-efficiency cyclone; (C) electrostatic precipitator; (D) spray tower; (E) Venturi scrubber; and (F) bag filter.

HEPA Filters

Very hazardous or toxic particulate matter of diameter less than 1 μm must sometimes be controlled to better than 99.9%. A single stage of *h*igh *e*fficiency *p*article *a*ttenuation (HEPA) micropore or glass frit filters, through which the precleaned gas is forced or sucked by vacuum, can achieve this level of control, and four to six HEPA filter stages in series can achieve 99.9999% control. HEPA filters are used to control emission of radioactive particles.

CONTROL OF GASEOUS POLLUTANTS

Gaseous pollutants may be removed from the effluent stream by trapping them from the stream, by changing them chemically, or by changing the process that produces the pollutants.

The *wet scrubbers* discussed can remove pollutants by dissolving them in the scrubber solution. SO_2 and NO_2 in power plant off-gases are often controlled in this way. *Packed scrubbers*, spray towers packed with glass platelets or glass frit, carry out such solution processes more efficiently than ordinary wet scrubbers; removal of fluoride from aluminum smelter exhaust gases is an example of packed scrubber use. *Adsorption*, or chemisorption, is the removal of organic compounds with an adsorbent like activated charcoal (see Fig. 20-13).

Incineration (Fig.20-14), or flaring, is used when an organic pollutant can be oxidized to CO_2 and water, or in oxidizing H_2S to SO_2. *Catalytic combustion* is a variant

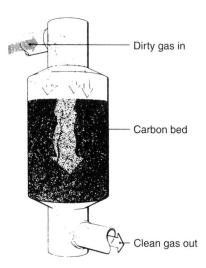

Figure 20-13. Adsorber for control of gaseous pollutants. (Courtesy of American Lung Association.)

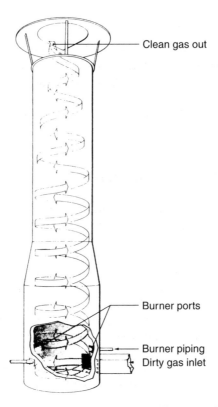

Figure 20-14. Incinerator for controlling gaseous pollutants. (Courtesy of American Lung Association.)

of incineration in which the reaction is facilitated energetically and carried out at a lower temperature by surface catalysis, and will be discussed further in connection with mobile source control. Figure 20-15 compares incineration and catalytic combustion.

Control of Sulfur Dioxide

Sulfur dioxide (SO_2) is both ubiquitous and a serious pollution hazard. The largest single source in the United States, and probably in the industrialized world, is generation of electricity by burning oil or coal, both of which contain sulfur. Increasingly strict standards for SO_2 control have prompted the development of a number of options and techniques for reducing SO_2 emissions. Among these options are:

- *Change to low-sulfur fuel.* Natural gas is exceedingly low in sulfur, while oil burned for industrial heat and electric power generation contains between 0.5 and 3% sulfur, and coal between 0.3 and 4%. Low-sulfur fuel, however, is expensive and the supply is uncertain.

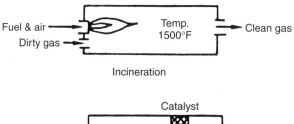
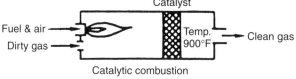

Figure 20-15. Comparison between incineration and catalytic combustion.

- *Desulfurization.* Sulfur may be removed from heavy industrial oil by a number of chemical methods similar to those used to lower the H_2S content of crude oil. In coal, sulfur may be either inorganically bound, as pyrite (FeS_2), or organically bound. Pyrite can be removed by pulverizing the coal and washing with a detergent solution. Organically bound sulfur can be removed by washing with concentrated acid. Preferred methods are coal gasification, which produces pipeline-quality gas, or solvent extraction, which produces low-sulfur liquid fuel.
- *Tall stacks.* Although the use of a tall stack does result in reduced ground-level concentration of gaseous pollutants (see Eqs. 18.5 and 18.6), it is not considered to be a pollution control measure. U.S. federal regulations do not sanction pollution control by use of tall stacks alone.
- *Flue-gas desulfurization.* The off-gases from combustion or other SO_2-producing processes are called flue gases. SO_2 may be cleaned from flue gas by chemical processes, some of which are discussed later in this chapter. The reaction of SO_2 to form sulfuric acid or other sulfates is the most frequently used flue-gas desulfurization method. The method has two limitations: (1) the flue gas must be cleaned of particulate matter before entering an acid-producing plant and (2) acid formation is energetically favorable only for a fairly concentrated gas stream (about 3% — 30,000 ppm — SO_2). The reactions for acid formation are

$$SO_2 + \tfrac{1}{2}O_2 \rightarrow SO_3$$
$$SO_3 + H_2O \rightarrow H_2SO_4.$$

A double-contact acid plant can produce industrial grade 98% sulfuric acid. The nonferrous smelting and refining industry has made the most use of this control method. Analogous reactions can be carried out to produce $(NH_4)_2SO_4$, a fertilizer, and gypsum, $CaSO_4$.

The SO_2 in flue gas from fossil-fuel combustion is too dilute to permit trapping as acid or commercial fertilizer or gypsum. Coal combustion off-gases are also too dirty.

SO$_2$ from fossil-fuel burning off-gas was first controlled by trapping as calcium sulfate in a lime–limestone mixture. This mixture was produced by calcining limestone and injecting lime into the scrubber and adding limestone to the boiler. Although the SO$_2$ was absorbed as CaSO$_3$ and CaSO$_4$, the lime–limestone absorbing material created a solid waste disposal problem of staggering proportions.

Flue-gas desulfurization methods that trap SO$_2$ as sulfite instead of sulfate allow regeneration of the absorbing material, and thus mitigate the solid waste disposal problem. A typical method is single-alkali scrubbing, for which the reactions are

$$SO_2 + Na_2SO_3 + H_2O \rightarrow 2NaHSO_3$$
$$2NaHSO_3 \rightarrow Na_2SO_3 + H_2O + SO_2.$$

The concentrated SO$_2$ that is recovered from this process can be used industrially in pulp and paper manufacture and sulfuric acid manufacture. Figure 20-16 is a diagram for single-alkali scrubbing.

A modification of this system that is both more efficient and less wasteful is double-alkali scrubbing, in which a primary scrubbing cycle dissolves the SO$_2$ in a sodium hydroxide solution, forming sodium bisulfite by the reaction

$$SO_2 + NaOH \rightarrow NaHSO_3.$$

Addition of lime in the second step regenerates the NaOH and produces calcium sulfite.

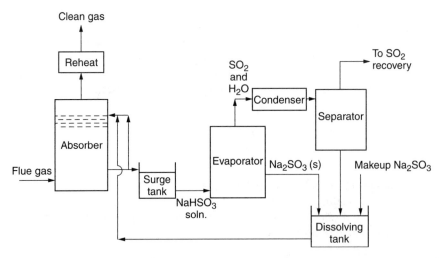

Figure 20-16. Simplified diagram for single-alkali scrubbing of flue gas with regeneration.

Air Pollution Control

SO$_2$ may also be removed from flue gas by dissolution in an aqueous solution of sodium citrate by the reaction

$$SO_2(g) + H_2O(l) \rightarrow HSO_3^- + H^+.$$

Citrate does not itself enter into the reaction but buffers the solution at about pH 4.5, which the reaction requires. The citrate buffer is readily regenerated. Removal efficiency of the citrate process is between 80 and 99%: much better than the 75% achievable by alkali scrubbing.

Control of Nitrogen Oxides

Wet scrubbers absorb NO$_2$ as well as SO$_2$, but are usually not installed primarily for NO$_2$ control. An effective method often used on fossil fuel burning power plants is off-stoichiometric burning. This method controls NO formation by limiting the amount of air (or oxygen) in the combustion process to just a bit more than is needed to burn the hydrocarbon fuel in question. For example, the reaction for burning natural gas

$$CH_4 + O_2 \rightarrow 2O_2 + CO_2$$

competes favorably with the high-temperature combination of nitrogen in the air with oxygen in the air to form NO (which eventually is oxidized to NO$_2$)

$$N_2 + O_2 \rightarrow 2NO.$$

The stoichiometric ratio of oxygen needed in natural gas combustion is

$$32 \text{ g of } O_2 : 16 \text{ g of } CH_4.$$

A slight excess of oxygen in the combustion air will cause virtually all of the oxygen to combine with fuel rather than with nitrogen. In practice, off-stoichiometric combustion is achieved by adjusting the air flow to the combustion chamber until any visible plume disappears.

Control of Volatile Organic Compounds and Odors

Volatile organic compounds and odors are controlled by thorough destructive oxidation, either by incineration or by catalytic combustion, since they are only slightly soluble in aqueous scrubbing media. "Catalytic incinerators," which use a metal oxide or mixed metal oxide catalyst and operate at 450°C and sometimes higher temperatures, can achieve 95 to 99% destruction of volatile organic compounds, like chlorinated hydrocarbons. These very efficient incinerators depend on catalysts that can withstand and function at high temperatures. Such catalysts were generally not available before 1996.

Control of Carbon Dioxide

Carbon dioxide is produced by oxidation of carbon compounds; that is, all combustion, all respiration, and all slow oxidative decay of vegetable matter produce CO_2. The world's oceans absorb CO_2 as carbonate, and plant photosynthesis removes CO_2 from the air. However, these natural phenomena have not kept pace with the steadily increasing concentration of CO_2 in the air, even though increasing CO_2 concentrations increases the rate of photosynthesis somewhat in accordance with the Law of Mass Action. Fossil fuel combustion for electrical production and for transportation appear to be the greatest contributors to increased CO_2 concentration.

CO_2 could be scrubbed from power plant effluent gas by alkaline solution and fixed as carbonate, but this relatively inefficient process would require large quantities of scrubber solution and would produce very large amounts of carbonate. The current approach to CO_2 emission reduction is substitution of other electrical generating sources for fossil fuel combustion. Nuclear, hydroelectric, solar, and wind generation do not produce CO_2, although, like all energy conversion methods, all have some adverse environmental impacts. Hydroelectric and wind generation are limited by the finite number of physical locations where they can be implemented. Nuclear power generation produces radioactive waste, even with reprocessing of fissile materials, and solar power is relatively inefficient and requires a very large land area. Biomass combustion produces CO_2.

Energy conservation is an obvious method for reducing CO_2. It produces no effluent at all, but significantly limiting CO_2 emission by energy conservation alone would require more than voluntary conservation and could result in considerable lifestyle and social changes.

A discussion of non-fossil-fuel energy conversion methods is beyond the scope of this text. However, we may expect to see increased utilization of such methods in the coming century.

CONTROL OF MOVING SOURCES

Mobile sources pose special pollution control problems, and one such source, the automobile, has received particular attention in air pollution control. Pollution control for other mobile sources, such as light-duty trucks, heavy trucks, and diesel engine-driven vehicles, requires controls similar to those used for control of automobile emissions. The important pollution control points in an automobile are shown in Fig. 20-17, and are:

- evaporation of hydrocarbons (HC) from the fuel tank,
- evaporation of HC from the carburetor,
- emission of unburned gasoline and partly oxidized HC from the crankcase, and
- CO, HC and NO/NO_2 from the exhaust.

Evaporative losses from the gas tank and carburetor often occur when the engine has been turned off and hot gasoline in the carburetor evaporates. These vapors may be

Air Pollution Control   405

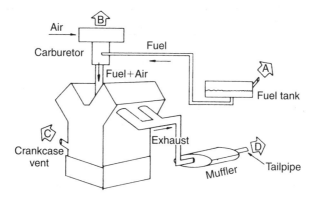

Figure 20-17. Diagram of the internal combustion engine showing four major emission points.

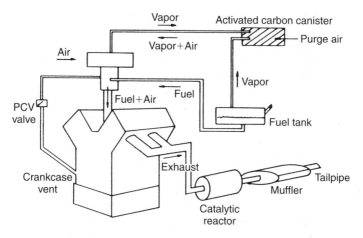

Figure 20-18. Internal combustion engine, showing methods of controlling emissions.

trapped in an activated-carbon canister, and can be purged periodically with air, and then burned in the engine, as shown schematically in Fig. 20-18. The crankcase vent can be closed off from the atmosphere, and the blowby gases recycled into the intake manifold. The positive crankcase ventilation (PCV) valve is a small check valve that prevents buildup of pressure in the crankcase.

The exhaust accounts for about 60% of the emitted hydrocarbons and almost all of the NO, CO, and lead, and poses the most difficult control problem of mobile sources. Exhaust emissions depend on the engine operation, as is shown in Table 20-2. During acceleration, the combustion is efficient, CO and HC are low, and high compression produces a lot of NO/NO_2. On the other hand, deceleration results in low NO/NO_2 and high HC because of the presence of unburned fuel in the exhaust. This variation in emissions has prompted EPA to institute a standard acceleration–deceleration cycle for

Table 20-2. Effect of Engine Operation on Exhaust Emissions, Shown as Fraction of Emissions at Idle

	CO	HC	NO/NO_2
Idling	1.0	1.0	1.0
Accelerating	0.6	0.4	100
Cruising	0.6	0.3	66
Decelerating	0.6	11.4	1.0

measuring emissions. Testing proceeds from a cold start through acceleration, cruising at constant speeds (on a dynamometer in order to load the engine), deceleration, and a hot start.

Emission control techniques include engine tune-ups, engine modifications, exhaust gas recirculation, and catalytic reactors. A well-tuned engine is the first line of defense for emission control.

A wide range of acceptable engine modifications is possible. Injection of water can reduce NO emissions, and fuel injection (bypassing or eliminating the carburetor) can reduce CO and HC emissions. Fuel injection is not compatible with water injection, however, since water may clog the fuel injectors. The stratified charge engine operates on a very lean air/fuel mixture, thus reducing CO and HC, but does not increase NO appreciably. The two compartments of the engine (the "stratification") accomplish this result: the first compartment receives and ignites the air/fuel mixture, and the second compartment provides a broad flame for an efficient burn. Better than 90% CO reduction can be achieved by this engine.

Recirculating the exhaust gas through the engine can achieve about 60% reduction of CO and hydrocarbons. The only major modification to an ordinary engine required by exhaust gas recirculation (EGR), in addition to the necessary fittings, is a system for cooling the exhaust gas before recirculating it, to avoid heat deformation of the piston surfaces. Exhaust gas recirculation, although it increased the rate of engine wear, was a popular and acceptable emission control method until 1980, but present-day exhaust emission standards require 90% CO control, which cannot be realized by this method.

New cars sold in the United States since 1983 have required the use of a catalytic reactor ("catalytic converter") to meet exhaust emission standards, and the device is now standard equipment on new cars. The modern three-stage catalytic converter performs two functions: *oxidation* of CO and hydrocarbons to CO_2 and water, and *reduction* of NO to N_2. A platinum–rhodium catalyst is used, and reduction of NO is accomplished in the first stage by burning a fuel-rich mixture, thereby depleting the oxygen at the catalyst. Air is then introduced in the second stage, and CO and hydrocarbons are oxidized at a lower temperature. Catalytic converters are rendered inoperable by inorganic lead compounds, so that cars using catalytic converters require the use of unleaded gasoline. Catalytic converters require periodic maintenance.

Diesel engines produce the same three major pollutants as gasoline engines, although in somewhat different proportions. In addition, diesel-powered heavy duty vehicles produce annoying black soot, essentially unburned carbon. Control of diesel exhaust was not required in the United States until passage of the 1990 Clean Air Act (nor is it required anywhere else in the world), and therefore little research on diesel exhaust emission control has reached the stage of operational devices.

Drastic lowering of emissions to produce a virtually pollution-free engine would require an external combustion engine, which can achieve better than 99% control of all three major exhaust pollutants. However, although work began in 1968 on a mobile external combustion engine, a working model has yet to be built. A working model requires a working fuel "steam" that is not flammable but has a lower heat capacity than water, and a linkage between the working fuel and the burning fuel that would allow the rapid acceleration characteristic of an external combustion engine-powered car.

Natural gas may be used to fuel cars, but the limited supply of natural gas serves a number of competing uses. In addition to a steady supply, a changeover would require a refueling system different from that used for gasoline. Electric cars are clean, but can store only limited power and have limited range. Generation of the electricity to power such cars also generates pollution, and the world's supply of battery materials would be strained to provide for a changeover to electric cars. The "hybrid" gasoline/electric automobiles introduced in 2001 use the engine to charge the electric motor. The 1990 Clean Air Act requires that cities in violation of the National Ambient Air Quality Standards sell *oxygenated fuel* during the winter months. Oxygenated fuel is gasoline containing 10% ethanol (CH_3CH_2OH) and is intended to bring about somewhat more efficient conversion of CO to CO_2. Its efficacy in cleaning up urban air remains to be seen.

CONTROL OF GLOBAL CLIMATE CHANGE

The two types of compounds involved in global climate change are those that produce free halogen atoms by photochemical reaction, and thus deplete the stratospheric ozone layer, and those that absorb energy in the near-infrared spectral region, which may ultimately produce global temperature change. The first group is comprised mostly of chlorofluorocarbons. Control of chlorofluorocarbon emission involves control of leaks, as from refrigeration systems, and eliminating use of the substances. While chlorofluorocarbon aerosol propellants are useful and convenient, they are not necessary for most applications. Aerosol deodorant, cleaners, paint, hairspray, and so on can be replaced by roll-on deodorant, wipe-on cleaners, rolled-on paint, hair mousse, etc. In many applications, atomized liquids appear to work as well as aerosolized liquids.

CONCLUSION

This chapter is devoted mainly to the description of air pollution control by "bolt-on" devices, but these are usually the most expensive means of control. A general environmental engineering truism is that the least expensive and most effective control point is always the farthest up the process line. Most effective control is achieved at the beginning of the process or, better yet, by finding a less polluting alternative to the process: mass transit as a substitute for cars, for example, and energy conservation instead of infinite expansion of generating capacity. Such considerations are good engineering and economics practice, as well as sensible and enlightened.

PROBLEMS

20.1 Taking into account cost, ease of operation, and ultimate disposal of residuals, what type of control device would you suggest for the following emissions?

a. Dust particles with diameters between 5 and 10 μm.
b. Gas containing 20% SO_2 and 80% N_2.
c. Gas containing 90% HC and 10% O_2.

20.2 An industrial emission has the following characteristics: 80% N_2, 15% O_2, 5% CO. What type of air pollution control equipment would you recommend?

20.3 A whiskey distillery has hired you as a consultant to design air pollution control equipment for a new facility, to be built upwind from a residential area. What problems would you encounter and what would be your control strategy?

20.4 Dust has specific gravity of 1.2 and particle size analysis as follows:

Mean diameter (μm)	% of particles by weight
0.05	10
0.1	20
0.5	25
1.0	35
5.0	10

The dust is trapped by a cyclone with diameter 0.51 m, inlet width of 15 cm, and an inlet length of 25 cm, operating at five effective turns and 2.0 m^3/s air flow.

a. What is the removal efficiency?
b. What is the pressure drop at air temperature of 20°C and 1 atm?
c. What is the separation factor? Is this a high-efficiency cyclone?

20.5 A plate-type electrostatic precipitator is to remove particles having a diameter of 0.5 μm at a flow rate of 2.0 m^3/s. The 40 plates are 5 cm apart and 3 m high. Ideally, how deep must the plates be to achieve removal?

20.6 Dust has particles with a drift velocity of 0.15 m/s. For a total air flow of 60 m^3/s, what must be the number of 10 × 10 m^2 collecting plates needed to achieve 90% removal?

20.7 Calculate the removal efficiencies for the dust described in Problem 20.4 for a standard cyclone, an electrostatic precipitator, and a fabric filter, using the curves of Fig. 20-12.

20.8 Determine the efficiency of a cyclone of diameter 0.5 m, at a flow rate of 0.4 m^3/s, and a gas temperature of 25°C. The inlet width is 0.13 m^2 and the area of the entrance is 0.04 m^2. The particles have a diameter of 10 μm and a density of 2 g/cm^3.

20.9 An electrostatic precipitator has the following specifications: height = 7.5 m, length = 5 m, number of passages = 3, plate spacing = 0.3 m. The flow rate is 18 m^3/s and the particle size is 0.35 μm, and the particles have a drift velocity of 0.162 m/s. Calculate the expected efficiency.

20.10 For Problem 20.9 above, what would be the increase in efficiency achieved by doubling the length? by doubling the number of plates and reducing the flow rate to 9 m^3/s?

20.11 A copper smelter produces 500 tons of copper per day from ore that is essentially CuS$_2$. The sulfur dioxide produced in this process is trapped in a sulfuric acid plant that produces 98% by weight sulfuric acid, which has a specific gravity of 2.3. If 75% of the SO$_2$ produced is trapped by the acid plant, how many liters of 98% H$_2$SO$_4$ are produced each day?

20.12 There are 168 h in a week, so that one 24-h day is 14.3% of a week. Therefore, you could decrease your personal electricity consumption by 14.3% by doing without electricity for one 24-h period each week. Try this for a week. What activities did you have to curtail or change? You can do this in three 8-h periods or six 4-h periods, but they should be spread out during the day to encompass heavy-use hours as well as light-use hours (do not just take three 8-h periods when you are asleep). If a fossil fuel power plant is 42% thermally efficient (that is, 42% of the heat generated is converted to electricity), what is your personal percent decrease in CO$_2$ emission?

Chapter 21

Air Pollution Law

As with water pollution, a complex system of laws and regulations governs the use of air pollution abatement technologies. In this chapter, the evolution of air pollution law is described, from its roots in common law through to the passage of federal statutory and administrative initiatives. Problems encountered by regulatory agencies and polluters are addressed with particular emphasis on the impacts the system may or may not have on future economic development. Figure 21-1 offers a road map to be followed through this maze.

AIR QUALITY AND COMMON LAW

When relying on common law, an individual or groups of individuals injured by a source of air pollution may cite general principles in two branches of that law. These branches have developed over the years and may apply to their particular cases:

- tort law and
- property law.

The harmed party, the plaintiff, could enter a courtroom and seek remedies from the defendant for damaged personal well-being or damaged property.

Tort Law

A tort is an injury incurred by one or more individuals. Careless accidents and exposure to harmful airborne chemicals are the types of wrong included under this branch of common law. A polluter could be held responsible for the damage to human health under the three broad categories of tort liability, negligence, and strict liability.

Intentional liability requires proof that somebody did a wrong to another party *on purpose*. This proof is especially complicated in the case of damages from air pollution. The fact that a "wrong" actually occurred must first be established, a process that may rely on direct statistical evidence or strong inference, such as the results of laboratory tests on rats. Additionally, intent to do the "wrong" must be established, which involves producing evidence in the form of written documents or direct testimony from the accused individual or group of individuals. Such evidence is not easily obtained.

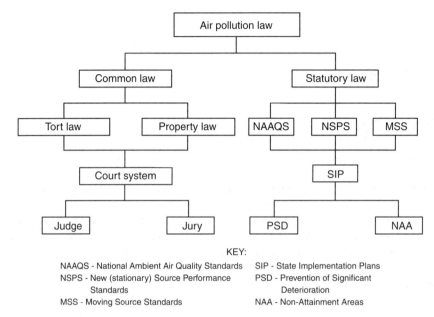

Figure 21-1. Diagram of the U.S. Federal Clean Air Act.

If intentional liability can be proven to the satisfaction of the courts, actual damages as well as punitive damages can be awarded to the injured plaintiff.

Negligence may involve mere inattention by the air polluter who allowed the injury to occur. Proof in the courtroom focuses on the lack of reasonable care taken on the defendant's part. Examples of such neglect in air pollution include failure to inspect the operation and maintenance of electrostatic precipitators, or the failure to design and size an adequate abatement technology. Again, damages can be awarded to the plaintiff.

Strict liability does not consider the *fault* or state of mind of the defendant. Under certain extreme cases, a court of common law has held that some acts are *abnormally dangerous* and that individuals conducting those acts are strictly liable if injury occurs. The court does tend to balance the danger of an act against the public utility associated with the act. An example could be the emission of a radioactive or highly toxic gas from an industrial smoke stack.

Again, if personal damage is caused by air pollution from a known source, the damaged party may enter a court of common law and argue for monetary damages to be paid by the defendant or an injunction to stop the polluter from polluting or both. Sufficient proof and precedent are often difficult if not impossible to muster, and in many cases, tort law has been found to be inadequate in controlling air pollution and awarding damages.

Property Law

Property law, on the other hand, focuses on the theories of nuisance and property rights; nuisance is based on the interference with the use or enjoyment of property, and property rights is based on actual invasion of the property. Property law is founded on ancient actions between land owners and involves such considerations as property damage and trespassing. A plaintiff basing a case on property law takes chances, rolls the dice, and hopes the court will rule favorably as it balances social utility against individual property rights.

Nuisance is the most widely used form of common law action concerning the environment. Public nuisance involves unreasonable interference with a right, such as the "right to clean air," common to the general public. A public official must bring the case to the courtroom and represent the public that is harmed by the air pollution. Private nuisance, on the other hand, is based on unreasonable interference with the use and enjoyment of private property. The key to a nuisance action is how the courts define "unreasonable" interference. Based on precedents and the arguments of the parties involved, the common law court balances the equities, hardships, and injuries in the particular case, and rules in favor of either the plaintiff or the defendant.

Trespass is closely related to the theory of nuisance. The major difference is that some physical invasion, no matter how minor, is technically a trespass. Recall that nuisance theory demands an unreasonable interference with land and the outcome of a particular case depends on how a court defines "unreasonable." Trespass is relatively uncomplicated. Examples of trespass include physical walks, vibrations from nearby surface or subsurface strata, and possibly gases and microscopic particles flowing from an individual smoke stack.

In conclusion, common law has generally proven inadequate in dealing with problems of air pollution. The strict burdens of proof required in the courtroom often result in decisions that favor the defendant and lead to smoke stacks that continue to pollute the atmosphere. Additionally, the technicality and complexity of individual cases often limit the ability of a court to act; complicated tests and hard-to-find experts often leave a court and a plaintiff with their hands tied. Furthermore, the absence of standing in a common law courtroom often prevents private individuals from bringing a case before the judge and jury unless the individual actually suffered material or bodily harm from the air pollution.

One key aspect of these common law principles is their degree of variation. Each state has its own body of common law, and individuals relying on the court system are generally confined to using the common laws of the applicable state. Given the shortcomings inherent in common law, Congress adopted a federal Clean Air Act.

STATUTORY LAW

Federal statutory law controlling air pollution began with the 1963 and 1967 Clean Air Acts. Although these laws provided broad clean air goals and research money, they did not apply air pollution controls throughout the entire United States, but only in particularly dirty communities. In 1970 the Clean Air Act was amended to cover the entire United States, and the U.S. Environmental Protection Agency (EPA) was created to promulgate clean air regulations and to enforce the Act. In 1977, provisions were added to the Clean Air Act to protect very clean areas ("protection against significant deterioration"), to enforce against areas that were not in compliance, and to extend the compliance dates for automobile emission standards. In 1990, amendments focused on toxic air pollutants and control of all vehicular emissions. The 1970 amendments, however, are the basis for existing federal clean air legislation.

National Ambient Air Quality Standards (NAAQS)

The EPA is empowered to determine allowable ambient concentrations of those air contaminants that are pollutants throughout the United States; these are the National Ambient Air Quality Standards (NAAQS) and historically have been the focus of the nationwide strategy to protect air quality. The primary NAAQS are intended to protect human health, the secondary NAAQS to "protect welfare." The latter levels are actually determined as those needed to protect vegetation. These standards are listed in Table 21-1.

NAAQS are set on the basis of extensive collections of information and data on the effects of these air pollutants on human health, ecosystems, vegetation, and materials. These documents are called "criteria documents" by the EPA, and the pollutants for which NAAQS exist are sometimes referred to as "criteria pollutants." Data indicate that all criteria pollutants have some threshold below which there is no damage.

Under the Clean Air Act, most enforcement power is delegated to the states by the EPA. However, the states must show the EPA that they can clean up the air to the

Table 21-1. Selected National Ambient Air Quality Standards

Pollutant	Primary standard		Secondary standard	
	ppm	$\mu g/m^3$	ppm	$\mu g/m^3$
Particulate matter less than 10 μm in diameter: PM_{10}				
24-h average		150		150
Annual geometric mean		50		50
Sulfur dioxide				
3-h maximum	0.5	1300		
24-h average	0.14	365	0.02	60
Annual arithmetic mean	0.03	80	0.1	260
Nitrogen oxides				
Annual arithmetic mean	0.053	100	Same as primary	
Carbon monoxide				
1 h maximum	9	10,000	Same as primary	
8 h maximum	35	40,000		
Ozone				
1 h maximum	0.12	210	Same as primary	
Lead				
Quarterly arithmetic mean		1.5	Same as primary	

Air Pollution Law   415

levels of the NAAQS. This showing is made in each State Implementation Plan (SIP), a document that contains all of the state's regulations governing air pollution control, including local regulations within the state. The SIP must be approved by the EPA, but once approved, it has the force of federal law.

Regulation of Emissions

Under Section 111 of the Clean Air Act, EPA has the authority to set *emission* standards (called "performance standards" in the Clean Air Act) only for new or markedly modified sources of the criteria pollutants. The states may set performance standards for existing sources and have the authority to enforce EPA's new source performance standards (NSPS). EPA has also delegated to certain states (e.g., Alabama) the authority to develop their own NSPS. A priority list of industries for NSPS that are to be set has been in place since 1971; new technology can also motivate revision of the NSPS. The list of industries and NSPS is too long for this chapter, but Table 21-2 gives some typical NSPS.

Categories of industrial facilities that emit these listed pollutants must use removal technologies if the facility emits 10 tons per year of any single hazardous substance or 25 tons per year of any combination of hazardous substances. Categories of industrial sources, taken from the list of hundreds, include:

Industrial external combustion boilers
Printing and publishing

Table 21-2. Some Typical NSPS as Cited in 40 CFR Part 60 (2001)

Facility	Minimum size	NSPS Particulate matter $(kg/kJ)^a$	SO_2 $(kg/kJ)^a$	NO_x $(kg/kJ)^a$
Coal-burning generating plants	250 MBtu/h^a	1.3×10^{-8}	5.2×10^{-7}	8.6×10^{-8}
Oil-burning generating plants	250 MBtu/h^a		3.2×10^{-7}	1.29×10^{-7}
Natural gas-fired generating plants				1.29×10^{-7}
H_2SO_4 plants			4 lb/ton (2 kg/metric ton) of acid produced	
MSW incinerators	250 tons/day	24 mg/dscf^b	30 ppm^c	180 ppm^c

^{a} "kg/kJ" means kg per kilojoule heat input. Emission standards are in terms of kg/kJ unless otherwise indicated.
^{b} "dscf" means dry standard cubic foot, corrected to 7% oxygen.
^{c} "ppm" means parts per million by volume.

Waste oil combustion
Gasoline marketing
Glass manufacturing

Each "major source," as defined by the 10 or 25 ton/year limit, must achieve a maximum achievable control technology (MACT) for emissions. On a case-by-case basis, an SIP could require a facility to install control equipment, change an industrial or commercial process, substitute materials in the production process, change work practices, and train and certify operators and workers. The MACT requirements continue to impact air pollution sources as the Clean Air Act is implemented.

Under Section 112 of the Clean Air Act, EPA has the authority to set national emission, or performance, standards for hazardous air pollutants (NESHAPS) for all sources of those pollutants. The new Clean Air Act required that EPA develop a "list of substances" to be regulated. Substances can be added to the list and removed from the list as a result of ongoing research. The list currently contains about 170 pollutants. Some of the compounds included are shown in Table 21-3. Some of these compounds are commonly used substances, e.g., ethylene glycol (anti-freeze), styrene (a plastic constituent), and methanol ("wood alcohol").

Table 21-3. Partial List of Regulated Hazardous Substances

Dichloroethane	Phosgene
Ethylene glycol	Phosphine
Ethylene imine (Aziridine)	Phosphorus
Ethylene oxide	Phthalic anhydride
Ethylene thiourea	Polychlorinated biphenyls
Ethylidene dichloride	Quinoline
(1,1-dichloroethane)	Quinone
Formaldehyde	Styrene
Heptachlor	Tetrachloroethylene
Hexachlorobenzene	(perchloroethylene)
Hexachlorobutadiene	Toluene
Hexachlorocyclopentadiene	Toxaphene (chlorinated camphene)
Hexane	1,2,4-Trichlorobenzene
Hydrazine	Trichloroethylene
Hydrochloric acid	2,4,5-Trichlorophenol
Hydrogen fluoride	Vinyl acetate
(hydrofluoric acid)	Vinyl bromide
Hydroquinone	Vinyl chloride
Lindane (all isomers)	Vinylidene chloride
Methanol	(1,1-dichloroethylene)

(continued)

Table 21-3. *(continued)*

Methoxychlor	Titanium tetrachloride
Methyl bromide (bromomethane)	*o*-Xylenes
	m-Xylenes
Methyl chloride (chloromethane)	*p*-Xylenes
	Antimony compounds
Methyl bromide (bromomethane)	Arsenic compounds (inorganic)
	Lead
Methyl chloride (chloromethane)	
Methylene chloride (dichloromethane)	
Methylene diphenyl diisocyanate (MDI)	
Naphthalene	
Nitrobenzene	
Parathion	
Pentachlorophenol	
Phenol	

Prevention of Significant Deterioration

In 1973, the Sierra Club sued the EPA for failing to protect the cleanliness of the air in those parts of the United States where the air was cleaner than the NAAQS, and won. In response, Congress included prevention of significant deterioration (PSD) in the 1977 and 1990 Clean Air Act amendments.

For PSD purposes, the United States is divided into class I and class II areas. Class I includes the so-called "mandatory class I" areas — all national wilderness areas larger than 5000 acres and all national parks and monuments larger than 6000 acres — and any area that a state or Native American tribe wishes to designate class I. The rest of the United States is class II.

To date, the only pollutants covered by PSD are sulfur dioxide and particulate matter. PSD limits the allowed increases in these as indicated in Table 21-4. In addition, visibility is protected in class I areas.

An industry wishing to build a new facility must show, by dispersion modeling with a year's worth of weather data, that it will not exceed the allowed increment. On making such a showing, the industry receives a PSD permit from the EPA. The PSD permitting system has had considerable impact in siting new facilities. In addition, any new project in a nonattainment area must not only have the most stringent control technology, but it must also guarantee "offsets" exist for any emissions from the new

Table 21-4. Maximum Allowed Increases under PSD

	Particulate matter		SO_2		
Class	Annual mean	24-h max	Annual mean	24-h max	3-h max
I	5	10	2	5	25
II	19	37	20	91	512
III	37	75	40	182	700

Note. Values are in $\mu g/m^3$.

facility. That is, even if the new facility uses MACT, the facility may still emit 15 tons of particles per year. Some other facility must reduce its emissions of particles by 15 tons/year before the new facility can begin operation.

Where the ambient concentrations are already close to the NAAQS and the PSD limits would allow them to exceed the NAAQS, the PSD limits clearly are moot.

Nonattainment

A region in which the NAAQS for one or more criteria pollutants is exceeded more than once a year is called a "nonattainment area" for that pollutant. If there is nonattainment of the lead, CO, or ozone standard, a traffic reduction plan and an inspection and maintenance program for exhaust emission control are required. Failure to comply results in the state's loss of federal highway construction funds.

If there is nonattainment resulting from a stationary source emission, an offset program must be initiated. Such a program requires that there be a rollback in emissions from existing stationary sources such that total emissions after the new source operates will be less than before. New sources in nonattainment areas must attain the lowest achievable emission rate (LAER), without necessarily considering the cost of such emission control.

The offset program allows industrial growth in nonattainment areas, but offers a particular challenge to the air pollution control engineer. The offset represents emission reductions that would otherwise not be required. Types of action that could generate offsets include tighter controls on existing operations at the same site, a binding agreement with another facility to reduce emissions, and the purchase of another facility to reduce emissions by installation of control equipment or by closing the facility down.

MOVING SOURCES

The emissions from cars, trucks, and buses are regulated under the 1990 Clean Air Act. Standards for gasoline engines for 1996 for CO, nonmethane hydrocarbons, and

Air Pollution Law 419

NO_x are:

- 4.2 g/mile for CO,
- 0.6 g/mile for NO_x, and
- 0.31 g/mile for nonmethane hydrocarbons.

These standards were phased for model years 1994 to 1996. Gasoline-powered automobiles from 1997 to 1999 model years were required to meet the following

- 3.4 g/mile for CO,
- 0.4 g/mile for NO_x, and
- 0.25 g/mile for nonmethane hydrocarbons.

In addition, light-duty diesel powered vehicles newer than 1996 were required to meet the 1996 emission standards. EPA has proposed that, by 2004, gasoline-powered vehicles will meet emission standards that are twice as stringent as these. Standards are also set for diesel-powered trucks, oxygenated fuels, vehicles that operate at high altitudes, and so on. Detailed discussion of these is beyond the scope of this text. The reader is referred 40 CFR Part 86 for further details.

TROPOSPHERIC OZONE

The 1990 Clean Air Act requires each state to establish categories of ozone nonattainment areas. Reasonably available control technologies (RACT) are to be required for all air polluters in such areas, depending on the degree of nonattainment. Requirements for using RACT are summarized in Table 21-5. Each state must classify areas relative to nonattainment for ozone concentration. RACT includes the use of oxygenated fuels, and many states now require suppliers to sell only oxygenated fuels from October through March each winter.

ACID RAIN

Fossil fuel electricity generating facilities have been seen as the major contributors to the acid rain problem; the 1990 Clean Air Act acknowledges the importance of

Table 21-5. Size of Pollution Sources That Must Use RACT

Level of nonattainment	Size of pollutant source that must use RACT (tons/year)
Extreme	10
Severe	25
Serious	50
Moderate	100

these sources of sulfur dioxide and nitrogen oxides. The 1990 regulations require SO_2 emissions to be reduced by 9 million tons from the 1980 nationwide emission level, and NO_2 emissions to be reduced by 2 million tons from the 1980 levels.

Reductions are to be achieved at 111 key fossil fuel facilities in 21 states encompassing the Ohio River Valley, New England and the northeast seaboard, the Appalachian states, and the North Central states. The Clean Air Act (Section 404) issues a regulated facility "allowances" for sulfur dioxide emissions. These allowances may be freely used by the facility, transferred by sale to other polluters, or held for future use. Allowances for the 111 facilities are based on historical operation data for the calendar years 1985 through 1987. An annual auction of these transferable discharge permits will be held, with anyone, including national and local environmental organizations, able to bid.

Carbon Dioxide and Climate Change

Although the United States did not sign the Kyoto Accords, by which most of the developed nations agreed to curtail CO_2 emissions, the United States recognizes the need to limit such emissions and reduce atmospheric CO_2. No law or regulation has yet been proposed. The signers of the Kyoto Accords have no regulations in place either as of this writing.

PROBLEMS OF IMPLEMENTATION

Both regulatory agencies and the sources of pollution have encountered difficulties in the implementation of air pollution abatement measures. Because most major stationary sources of air pollution now have installed abatement equipment in response to federal pollution control requirements, regulatory agencies must develop programs that promote continuing compliance. However, the track records of the agencies indicate that current programs are not always effective.

Flaws in design and construction of control equipment contribute to significant noncompliance. Problems include the use of improper materials in constructing controls, undersizing of controls, inadequate instrumentation of the control equipment, and inaccessibility of control components for proper operation and maintenance. Design flaws such as these cast doubt on how effectively the air pollution control agencies evaluate the permit applications for these sources. Additionally, many permit reviewers working in government lack the necessary practical experience to fully evaluate proposed controls and tend to rely too heavily on the inadequate technical manuals available to them.

A source permit, as it is typically designed within the regulatory framework, could ensure that an emitter plans to install necessary control equipment. It could also require the emitter to perform record-keeping functions that facilitate proper operation and maintenance. However, agencies do not always use the permit program to accomplish these objectives. For example, in some states, a majority of the sources fail to keep any operating records at all that would enable an independent assessment of past compliance. In other states, although many sources tend to keep good operating records, they are not required to do so by the control agency.

An agency's ability to detect violations of emission requirements may be critical to the success of its regulatory effort. However, many agencies rely on surveillance by sight and smell as opposed to stack testing or monitoring. For example, air pollution control agencies often issue notices of violation for such problems as odor, dust, and excessive visible emissions, even though it is recognized that this approach neglects other, perhaps more detrimental, pollutants.

The emitters of air pollution also face problems in complying with regulations. Although most large stationary sources of pollution have installed control equipment, severe emission problems may be caused by design or upsets either with the control equipment or in the production process. Poor design of the control equipment is probably the primary cause of excess emissions in the greatest number of sources, followed by process upsets and routine component failure. Other causes of excess emissions include improper maintenance, lack of spare parts, improper construction materials, and lack of instrumentation. Different industries generally experience wide variations in the frequency and duration of excess emissions.

CONCLUSION

Air pollution law is a complex web of common and statutory law. Although common law has offered and continues to offer checks and balances between polluters and economic development, shortcomings do exist. Federal statutory law has attempted to fill the voids and, to a certain extent, has been successful in cleaning the air. Engineers must be aware of the requirements placed on industry by this system of laws. Particular attention must be paid to the siting of new plants in different sections of the nation.

PROBLEMS

21.1 Acid rain is a mounting problem, particularly in the northeastern states. Discuss how this problem can be controlled under the system of common law, and compare this approach with the remedy under federal statutory law.

21.2 Pollution from tobacco smoke can significantly degrade the quality of certain air masses, and many ordinances exist that limit where one can smoke. Develop sample town ordinances to improve the quality of air (a) in shopping malls, and (b) over the spectators at both indoor and outdoor sporting events. Rely on both structural and nonstructural alternatives, i.e., solutions that mechanically, chemically, or electrically clean the air, and solutions that prevent all or part of the pollution in the first place.

21.3 Would you favor an international law that permits open burning of hazardous waste on the high seas? Open burning refers to combustion without emission controls. Would you favor such a law if emissions were controlled? Discuss your answers in detail; would you require permits that specify time of day of burning, or distance from shore of burning, or banning certain wastes from incineration?

21.4 Emissions from a nuclear power generating facility pose a unique set of problems for local health officials. Radioactive materials are included in the list of hazardous air pollutants. What precautions would you take if you were charged with protecting the air quality of nearby residents? How would your set of rules consider (a) direction of prevailing winds, (b) age of residents, and (c) siting new schools? Suggest a model NESHAPS standard for nuclear power plant emissions.

21.5 Assume you live in a small town that has two stationary sources of air pollution: a laundry/dry cleaning establishment and a regional hospital. Automobiles and front porches in the town have a habit of turning black literally overnight. What air laws apply in this situation, and given that they presumably are not being enforced, how would you: (a) determine if, in fact, ambient or emissions standards or both were being violated and, if either was, (b) advise the residents to respond to see that the laws are enforced?

21.6 Emission standards for automobiles are set in units of grams per mile. Why not ppm or grams per passenger-mile?

21.7 How would a small (25,000 persons), rather poor community handle its municipal solid waste in the cheapest way without violating any law? What would be the most environmentally sound way (rather than simply the cheapest)?

21.8 Using the latest World Alamanac or similar references, can you explain why Japan is considered a "developed nation" under the Kyoto Accords, while China is not?

21.9 What are all the ways in which the United States could reduce CO_2 emissions? Give some quantitative evidence for your answers.

Chapter 22

Noise Pollution

The ability to make and detect sound provides humans with the ability to communicate with each other as well as to receive useful information from the environment. Sound can provide warning, as in the fire alarm, information, as from a whistling tea kettle, and enjoyment, as from music. In addition to such useful and pleasurable sounds there is noise, often defined as unwanted or extraneous sound. We generally classify the unwanted sound made by such products of our civilization as trucks, airplanes, industrial machinery, air conditioners, and similar sound producers as noise.

Damage to the ear is a measurable problem associated with noise; annoyance, while not necessarily easily measured, may be of equal concern. In one Danish investigation of annoyance factors in the workplace, 38% of workers interviewed identified noise as the most annoying factor in their environment.

Urban noise is not, surprisingly, a modern phenomenon. Legend has it, for example, that Julius Caesar forbade the driving of chariots on Rome's cobblestone streets during nighttime so he could sleep (Still 1970). Before the ubiquitous use of automobiles, the 1890 noise pollution levels in London were described by an anonymous contributor to the *Scientific American* (Taylor 1970) as:

> The noise surged like a mighty heart-beat in the central districts of London's life. It was a thing beyond all imaginings. The streets of workaday London were uniformly paved in "granite" sets ... and the hammering of a multitude of iron-shod hairy heels, the deafening side-drum tattoo of tyred wheels jarring from the apex of one set (of cobblestones) to the next, like sticks dragging along a fence; the creaking and groaning and chirping and rattling of vehicles, light and heavy, thus maltreated; the jangling of chain harness, augmented by the shrieking and bellowing called for from those of God's creatures who desired to impart information or proffer a request vocally — raised a din that is beyond conception. It was not any such paltry thing as noise.

Noise may adversely affect humans both physiologically as well as psychologically. It is an insidious pollutant; damage is usually long range and permanent.

THE CONCEPT OF SOUND

The average citizen has little, if any, concept of what sound is or what it can do. This is exemplified by the tale of a well-meaning industrial plant manager who decided

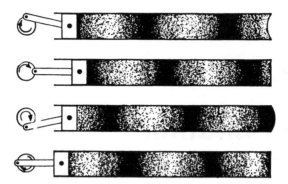

Figure 22-1. A piston creates pressure waves that are transmitted through air.

to decrease the noise level in his factory by placing microphones in the plant and channeling the noise through loudspeakers to the outside (Lipscomb 1975).

Sound is transfer of energy without transfer of mass. For example, a rock thrown at you would certainly get your attention, but this would require the transfer of the rock's mass. Alternatively, your attention may be gained by poking you with a stick, in which case the stick is not lost, but energy is transferred from the poker to the target. In the same way, sound travels through a medium such as air without a transfer of mass. Just as the stick must move back and forth, so must air molecules oscillate in waves to transfer energy.

The small displacement of air molecules that creates pressure waves in the atmosphere is illustrated in Fig. 22-1. As the piston is forced to the right in the tube, the air molecules next to it are reluctant to move and instead pile up on the face of the piston (Newton's first law). These compressed molecules now act as a spring, and release the pressure by jumping forward, creating a wave of compressed air molecules that moves through the tube. The potential energy has been converted to kinetic energy. These pressure waves move down the tube at a velocity of 344 m/s (at 20°C). If the piston oscillates at a *frequency* of, say, 10 cycles/s, there would be a series of pressure waves in the tube each 34.4 m apart. This relationship is expressed as

$$\lambda = \frac{c}{\nu}, \qquad (22.1)$$

where

λ = wavelength (m),
c = velocity of the sound in a given medium (m/s), and
ν = frequency (cycles/s).

Sound travels at different speeds in different materials, depending on such physical properties of the material as its modulus of elasticity.

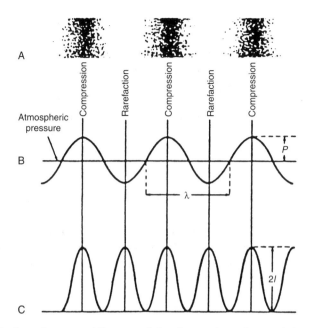

Figure 22-2. Sound waves. All parts of the figure show the spatial variation along the wave at a particular instant of time. (A) Regions of compression and rarefaction in the air. (B) Pressure wave; $P = $ pressure amplitude ($P_{rms} = P/\sqrt{2}$) and $\lambda = $ wavelength. (C) Intensity wave; the average intensity is denoted I.

EXAMPLE 22.1. In cast iron, sound waves travel at about 3440 m/s. What would be the wavelength of a sound from a train if it rumbles at 50 cycles/s and one listens to it placing an ear on the track?

$$\lambda = \frac{c}{\nu} = \frac{3440}{50} = 69 \text{ m}.$$

In acoustics, the frequency as cycles per second is denoted by the name hertz,[1] and written Hz. The range of frequency audible to the human ear is between 20 and 20,000 Hz. The middle A on the piano (concert pitch) is 440 Hz. The frequency is one of the two basic parameters that describe sound. *Amplitude*, how loud the sound is, is the other.

If the amplitude of a pressure wave of a sound with only one frequency is plotted against time, the wave is seen to produce a sinusoidal trace, as shown in Fig. 22-2.

---

[1] In honor of the German physicist Heinrich Hertz (1857–1894).

All other sounds are made up of a number of suitable sinusoidal waves, as demonstrated originally by Fourier. Although both nonrandom and random combinations of sinusoidal waves can be pleasing to the ear, noise is usually a random combination.

Although the human ear is a remarkable instrument, able to detect sound pressures over seven orders of magnitude, it is not a perfect receptor of acoustic energy. In the measurement and control of noise, it is therefore important to know not only what a sound pressure is, but also to have some notion of how loud a sound *seems* to be. Before we address that topic, however, we must review some basics of sound.

SOUND PRESSURE LEVEL, FREQUENCY, AND PROPAGATION

Figure 22-2 represents a wave of pure sound: a single frequency. A sound wave is a compression wave, and the amplitude is a pressure amplitude, measured in pressure units like N/m^2. As is the case with other wave phenomena, intensity is the square of the amplitude, or

$$I = P^2. \tag{22.2}$$

The intensity of a sound wave is measured in watts, a unit of power. When a person hears sounds of different intensities, the total intensity heard is not the sum of the intensities of the different sounds. Rather, the human ear tends to become overloaded or saturated with too much sound. Another statement of this phenomenon is that human hearing sums up sound intensities logarithmically rather than linearly. A unit called the *bel* was invented to measure sound intensity. Sound intensity level (IL) in bels is defined

$$IL_b = \log_{10}\left(\frac{I}{I_0}\right), \tag{22.3}$$

where

I = sound intensity in watts, and
I_0 = intensity of the least audible sound, usually given as $I_0 = 10^{-12}$ W.

The bel is an inconveniently large unit. The more convenient unit, which is now in common usage, is the *decibel* (*dB*). Sound intensity level in decibels is defined as

$$IL = 10\log_{10}\left(\frac{I}{I_0}\right). \tag{22.4}$$

Since intensity is the square of pressure, an analogous equation may be written for sound pressure level (SPL) in decibels

$$SPL\ (dB) = 20\log_{10}\left(\frac{P}{P_{ref}}\right), \tag{22.5}$$

where

SPL (dB) = sound pressure level (in dB),
P = pressure of sound wave, and
P_{ref} = some reference pressure, generally chosen as the threshold of hearing, 0.00002 N/m².

These relationships are also derivable from a slightly different point of view. In 1825, E. H. Weber found that people can perceive differences in small weights, but if a person is already holding a substantial weight, that same increment is not detectable. The same idea is true with sound. For example, a 2 N/m² increase from an initial sound pressure of 2 N/m² is readily perceived, whereas the same 2 N/m² difference is not noticed if it is added to a background of 200 N/m² sound pressure. Mathematically, this may be expressed as

$$ds = K \frac{dW}{W}, \qquad (22.6)$$

where

ds = is the minimum perceptible increase in a sensation (e.g., hearing),
W = the load (e.g., background SPL),
dW = the change in the load, and
K = a constant.

The integrated form of Eq. (22.6) is known as the Weber–Fechner Law

$$s = K \log_{10} W. \qquad (22.7)$$

This idea is used in the definition of a *decibel*

$$dB = 10 \log_{10} \frac{W}{W_{ref}}, \qquad (22.8)$$

where the power level W is divided by a constant W_{ref}, which is a reference value, both measured in watts.

Since the pressure waves in air are half positive pressure and half negative pressure, adding these would result in 0. Accordingly, sound pressures are measured as root-mean-square (rms) values, which are related to the energy in the wave. It may also be shown that the power associated with a sound wave is proportional to the square of the rms pressure. We can therefore write

$$dB = 10 \log_{10} \frac{W}{W_{ref}} = 10 \log_{10} \frac{P^2}{P_{ref}^2} = 20 \log_{10} \frac{P}{P_{ref}}, \qquad (22.9)$$

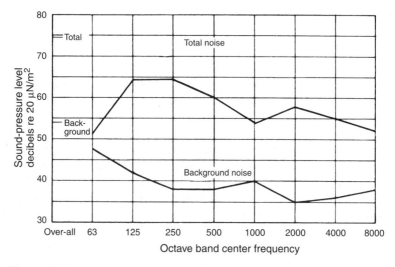

Figure 22-3. Typical analysis of a machine noise with background noise.

where P and P_{ref} are the pressure and reference pressure, respectively. If we define $P_{ref} = 0.00002 \text{ N/m}^2$, the threshold of hearing, we can define the SPL as

$$\text{SPL(dB)} = 20\log_{10}\left(\frac{P}{P_{ref}}\right). \qquad (22.5)$$

Although it is common to see the sound pressure measured over a full range of frequencies, it is sometimes necessary to describe a noise by the amount of sound pressure present at a specific range of frequencies. Such a *frequency analysis*, shown in Fig. 22-3, may be used for solving industrial problems or evaluating the danger of a certain sound to the human ear.

In addition to amplitude and frequency, sound has two more characteristics of importance. Both may be visualized by imagining the ripples created by dropping a pebble into a large, still pond. The ripples, analogous to sound pressure waves, are reflected outward from the source, and the magnitude of the ripples is dissipated as the ripples get farther away from the source. Similarly, sound levels decrease as the distance between the receptor and the source is increased.

In summary, then, the four important characteristics of sound waves are as follows:

- Sound pressure is the magnitude or amplitude of sound.
- The pitch is determined by the frequency of the pressure fluctuations.
- Sound waves propagate away from the source.
- Sound pressure decreases with increasing distance from the source.

The mathematics of adding decibels is a bit complicated. The procedure may be simplified a great deal by using the graph shown as Fig. 22-4. As a rule of thumb,

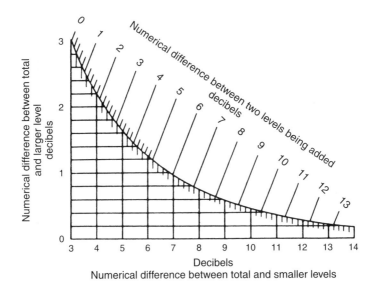

Figure 22-4. Chart for combining different sound pressure levels. For example: Combine 80 and 75 dB. The difference is 5 dB. The 5-dB line intersects the curved line at 1.2 dB; thus the total value is 81.2 dB. (Courtesy of General Radio.)

Table 22-1. Sound Pressure Levels for Speech Masking

Distance (ft)	Speech interference level (dB)	
	Normal	Shouting
3	60	78
6	54	72
12	48	66

adding two equal sounds increases the SPL by 3 dB and if one sound is more than 10 dB louder than a second sound, the contribution of the latter is negligible.

Background noise (or ambient noise) must also be subtracted from any measured noise. Using the above rule of thumb, if the SPL is more than 10 dB greater than the ambient level, the contribution may be ignored. The covering of a sound with a louder one is known as *masking*. Speech can be masked by industrial noise, for example, as shown in Table 22-1. These data show that an 80-dB SPL in a factory will effectively prevent conversation. Telephone conversations are similarly affected, with a 65-dB background making communication difficult and 80 dB making it impossible. In some cases, it has been found advisable to use *white noise*, a broad frequency hum from a fan, for example, to mask other more annoying noises.

EXAMPLE 22.2. A jet engine has a sound intensity level of 80 dB, as heard from a distance of 50 ft. A ground crew member is standing 50 ft from a four-engine jet. What SPL reaches her ear when the first engine is turned on? the second, so that two engines are running? the third? then all four?

When the first engine is turned on, the SPL is 80 dB, provided there is no other comparable noise in the vicinity. To determine, from the chart of Fig. 22-4, what the SPL is when the second engine is turned on, we note that the difference between the two engine intensity levels is

$$80 - 80 = 0.$$

From the chart, a numerical difference of 0 between the two levels being added gives a difference of 3 between the total and the larger of the two. The total SPL is thus

$$80 + 3 = 83 \text{ dB}.$$

When the third engine is turned on, the difference between the two levels is

$$83 - 80 = 3 \text{ dB},$$

yielding a difference from the total of 1.8, for a total IL of

$$83 + 1.8 = 84.8 \text{ dB}.$$

When all four engines are turned on, the difference between the sounds is

$$84.8 - 80 = 4.8 \text{ dB},$$

yielding a difference from the total of 1.2, for a total IL of 86 dB.

These characteristics ignore the human ear. We know that the ear is an amazingly sensitive receptor, but is it equally sensitive at all frequencies? Can we hear low and high sounds equally well? The answers to these questions lead us to the concept of *sound level*.

SOUND LEVEL

Suppose you are put into a very quiet room and subjected to a pure tone at 1000 Hz at a 40-dB SPL. If, in turn, this sound is turned off and a pure sound at 100 Hz is piped in and adjusted in loudness until you judge it to be "equally loud" to the 40-dB, 1000-Hz tone you had just heard a moment ago, you would, surprisingly enough, judge the 100-Hz tone to be equally loud when it is at about 55 dB. In other words, much more energy must be generated at the lower frequency in order to hear a tone at about the same perceived loudness, indicating that the human ear is rather inefficient for low-frequency tones.

Noise Pollution   431

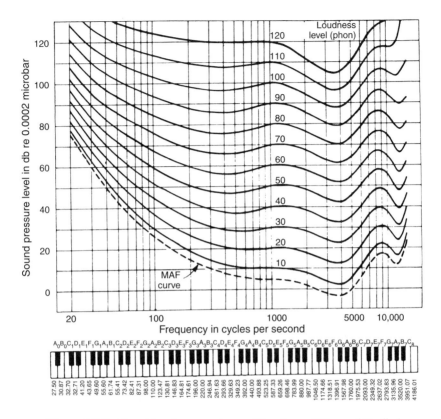

Figure 22-5. Equal loudness contours. (Courtesy of General Radio.)

It is possible to conduct such experiments for many sounds and with many people and to draw average equal loudness contours (Fig. 22-5). These contours are in terms of *phons*, which correspond to the sound pressure level in decibels of the 1000-Hz reference tone. Using Fig. 22-5, a person subjected to a 65-dB SPL tone at 50 Hz would judge this to be equally as loud as a 40-dB, 1000-Hz tone. Hence, the 50-Hz, 65-dB sound has a loudness level of 40 phons. Such measurements are commonly called *sound levels* and are not based on physical phenomena only, but have an adjustment factor, a "fudge factor," that corresponds to the inefficiency of the human ear.

Sound level (SL) is measured with a sound level meter consisting of a microphone, amplifier, a frequency-weighing circuit (filters), and an output scale, shown in Fig. 22-6. The weighing network filters out specific frequencies to make the response more characteristic of human hearing. Through use, three scales have become internationally standardized (Fig. 22-7). Figure 22-8 shows a typical hand-held sound level meter.

Note that the A scale in Fig. 22-7 corresponds closely to an inverted 40-phon contour in Fig. 22-5. Similarly, the response reflected by the B scale approximates an inverted 70-phon contour. The C scale shows an essentially flat response, giving equal

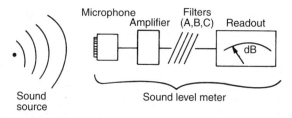

Figure 22-6. Schematic representation of a sound level meter.

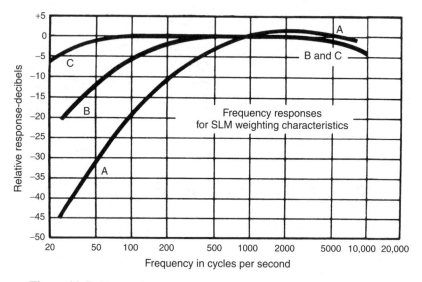

Figure 22-7. The A, B, and C filtering curves for a sound level meter.

weight to all frequencies, and approximates the response of the ear to intense sound pressure levels.

The results of noise measurement with the standard sound level meter are expressed in terms of decibels but with the scale designated. If on the A scale the meter reads 45 dB, the measurement is reported as 45 dB(A).

Most noise ordinances and regulations are in terms of dB(A), because this scale is a good approximation of human response for sounds that are not very loud. For very loud noise the C scale is a better approximation. However, because the use of multiple scales complicates matters, scales other than A are seldom used. In addition to the A, B, and C scales, a new D scale has been introduced to approximate human response to aircraft noise exclusively.

This brings up another complication involved in the measurement of noise. We do not generally respond in the same way to different types of noise, even though they may be equal in sound level. For example, a symphony orchestra playing middle C at 120 dB(A) and a jet engine at an equal 120 dB(A) will draw very different reactions

Noise Pollution 433

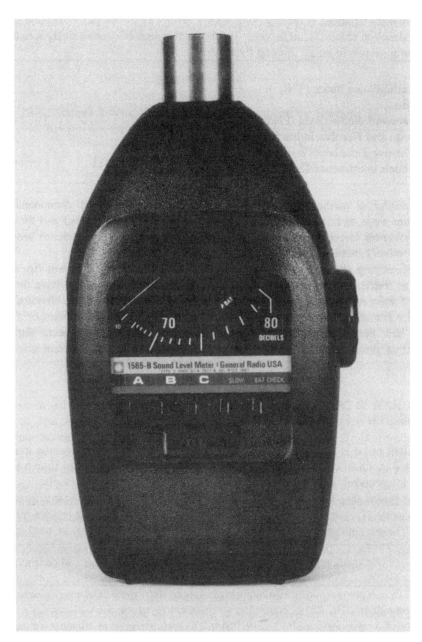

Figure 22-8. A typical sound level meter. (Courtesy of General Radio.)

434   ENVIRONMENTAL ENGINEERING

from people. A plethora of parameters have been devised to respond to this discrepancy in reactions, all supposedly "the best" means of quantitatively measuring human response to noise. Among these are:

- Traffic noise index (TNI),
- Sones,
- Perceived noise level (PNdB),
- Noise and number index (NNI),
- Effective perceived noise level (EPNdB), and
- Speech interference level (SIL).

Measuring methods have proliferated because the physical phenomenon, a pressure wave, is related to a physiological human response (hearing) and then to a psychological response (pleasure or irritation). Along the way, the science becomes progressively more subjective.

Measuring some noises, particularly those commonly called *community noise*, such as traffic and loud parties, is complicated further. Although we have thus far treated noise as a constant in intensity and frequency with time, this is obviously not true for *transient noises* such as trucks moving past a sound level meter or a loud party near the measurement location. An unusual noise that may occur during a period of general quiet is an *intermittent noise* and has a still different effect on people.

MEASURING TRANSIENT NOISE

Transient noise is still measured with a sound level meter, but the results must be reported in statistical terms. The common parameter is the percent of time a sound level is exceeded, denoted by the letter L with a subscript. For example, $L_{10} = 70\,\text{dB(A)}$ means that 10% of the time the noise is louder than 70 dB as measured on the A scale. Transient noise data are gathered by reading the SL at regular intervals. These numbers are then ranked and plotted, and the L values are read off the graph.

EXAMPLE 22.3.   Suppose the traffic noise data in Table 22-2 are gathered at 10-s intervals. These numbers are then ranked as indicated in the table and plotted as in Fig. 22-5. Note that since 10 readings are taken, the lowest reading (Rank no. 1) corresponds to a SL that is equaled or exceeded 90% of the time. Hence, 70 dB(A) is plotted versus 90% in Fig. 22-9.[2] Similarly, 71 dB(A) is exceeded 80% of the time.

---

[2] One might argue that since 70 dB(A) is the lowest value, it is the SL exceeded 100% of the time and one should plot 70 dB(A) vs 100%. The second value, 71 dB(A), is exceeded 90% of the time, etc. Either method is correct, and the error diminishes as the data increase.

Table 22-2. Sample Traffic Noise Data and Calculations for Example 22.3

Time (s)	dB(A)	Rank no.	% of time equal to or exceeded	dB(A)
10	71	1	90	70
20	75	2	80	71
30	70	3	70	74
40	78	4	60	74
50	80	5	50	75
60	84	6	40	75
70	76	7	30	76
80	74	8	20	78
90	75	9	10	80
100	74	10	0	84

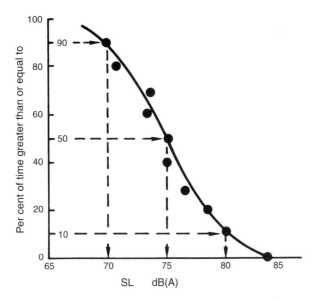

Figure 22-9. Results from a survey of transient noise. The data (from Table 22-2) are plotted as percent of time the SL is exceeded vs the SL, in dB(A).

One widely used parameter for gauging the perceived level of noise from transient sources is the noise pollution level (NPL), which takes into account the irritation caused by impulse noises. The NPL is defined as

$$\text{NPL(dB(A))} = L_{50} + (L_{10} - L_{90}) + \frac{(L_{10} - L_{90})^2}{60}. \qquad (22.10)$$

As defined earlier, the symbol L refers to the percentage of time the noise is equal to or greater than some value. The percentage is indicated by the subscript. L_{10} is thus the dB(A) level exceeded 10% of the time.

With reference to Fig. 22-9, L_{10}, L_{50}, and L_{90} are thus 80, 75, and 70 dB(A), respectively. These may then be substituted into the equation, and the NPL may be calculated. It is always advisable to take as much data as possible. The 10 readings in Table 22-2 are seldom sufficient for a thorough analysis. The percentages must be calculated on the basis of the total number of readings taken: if 20 readings are recorded, the lowest (Rank no. 1) corresponds to 95%, the second lowest, to 90%, etc.

THE ACOUSTIC ENVIRONMENT

The types of sound around us vary from a Beethoven concerto to the roar of a jet plane. Noise, the subject of this chapter, is generally considered to be unwanted sound, or the sound incidental to our civilization that we would just rather not have to endure. The intensities of some typical environmental noises are shown in Fig. 22-10. Laws against

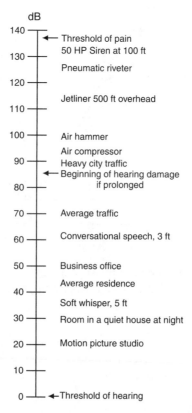

Figure 22-10. Environmental noise.

Table 22-3. OSHA Maximum Permissible Industrial Noise Levels

Sound level dB(A)	Maximum duration during any working day(h)
90	8
92	6
95	4
100	2
105	1
110	0.5
115	0.25

noise abound. Most local jurisdictions have ordinances against "loud and unnecessary noise." The problem is that most of these ordinances are difficult to enforce.

In the industrial environment, noise is regulated by the federal Occupational Safety and Health Act (OSHA), which sets limits for noise in the workplace. Table 22-3 lists these limits. There is some disagreement as to the level of noise that should be allowed for an 8-h working day. Some researchers and health agencies insist that 85 dB(A) should be the limit. This is not, as might seem on the surface, a minor quibble, since the jump from 85 dB to 90 dB is actually an increase of about four times the sound pressure, remembering the logarithmic nature of the decibel scale.

HEALTH EFFECTS OF NOISE

The human ear is an incredible instrument. Imagine having to design and construct a scale for weighing just as accurately a flea or an elephant. This is the range of performance to which we are accustomed from our ears. Yet only recently have we become aware of the devastating psychological effects of noise. The effect of excessive noise on our ability to hear, on the other hand, has been known for a long time (Suter 1992). The human auditory system is shown in Fig. 22-11.

Sound pressure waves caused by vibrations set the eardrum (*tympanic membrane*) in motion. This activates the three bones in the middle ear. The *hammer, anvil*, and *stirrup* physically amplify the motion received from the eardrum and transmit it to the inner ear. This fluid-filled cavity contains the *cochlea*, a snail-like structure in that the physical motion is transmitted to tiny hair cells. These hair cells deflect, much like seaweed swaying in the current, and certain cells are responsive only to certain frequencies. The mechanical motion of these hair cells is transformed to bioelectrical signals and transmitted to the brain by the auditory nerves. Acute damage may occur to the eardrum, but this occurs only with very loud sudden noises. More serious is the chronic damage to the tiny hair cells in the inner ear. Prolonged exposure to noise of a certain frequency pattern may cause either temporary hearing loss that disappears in a few hours or days or permanent loss. The former is called *temporary threshold shift*,

438   ENVIRONMENTAL ENGINEERING

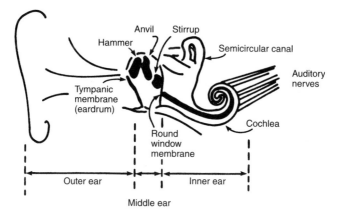

Figure 22-11. Cut-away drawing of the human ear.

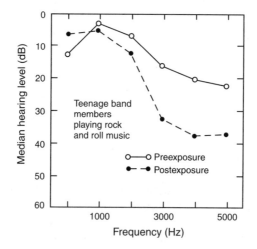

Figure 22-12. Temporary threshold shift for rock band performers. (Source: Data by the U.S. Public Health Service. See Still (1990).)

and the latter is known as *permanent threshold shift*. Literally, your threshold of hearing changes, so you are not able to hear some sounds.

Temporary threshold shift is generally not damaging to your ear unless exposure to the sound is prolonged. People who work in noisy environments commonly find that they hear worse at the end of the day. Performers in rock bands are subjected to very loud noises (substantially above the allowable OSHA levels) and commonly are victims of temporary threshold shift. In one study, the results of which are shown in Fig. 22-12, the players suffered as much as 15 dB temporary threshold shift after a concert.

Repeated noise over a long time leads to permanent threshold shift. This is especially true in industrial applications in which people are subjected to noises of a certain

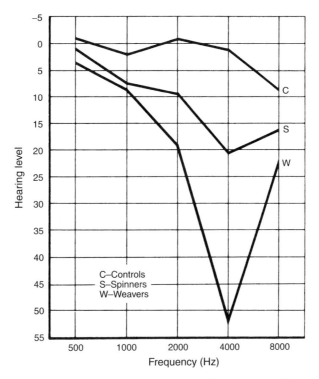

Figure 22-13. Permanent threshold shift for textile workers. (Source: W. Burns *et al.* "An Exploratory Study of Hearing and Noise Exposure in Textile Workers," *Ann. Occup. Hyg.* **7**, 322 (1958).)

frequency. Figure 22-13 shows data from a study performed on workers at a textile mill. Note that the people who worked in the spinning and weaving parts of the mill, where noise levels are highest, suffered the most severe loss in hearing, especially at around 4000 Hz, the frequency of noise emitted by the machines.

As people get older, hearing becomes less acute simply as one of the effects of aging. This loss of hearing, called *presbycusis*, is illustrated in Fig. 22-14. Note that the greatest loss occurs at the higher frequencies. Speech frequency is about 1000 to 2000 Hz, where hearing loss is noticeable.

In addition to presbycusis, however, there is a serious loss of hearing owing to environmental noise. In one study (Taylor 1970), 11% of ninth graders, 13% of twelfth graders, and 35% of college freshmen had a greater than 15 dB loss of hearing at 2000 Hz. The study concluded that this severe loss resulted from exposure to loud noises such as motorcycles and rock music and that as a result "the hearing of many of these students had already deteriorated to a level of the average 65-year-old person."

Noise also affects other bodily functions including those of the cardiovascular system. Noise alters the rhythm of the heartbeat, makes the blood thicker, dilates blood vessels, and makes focusing the eyes difficult. It is no wonder that excessive

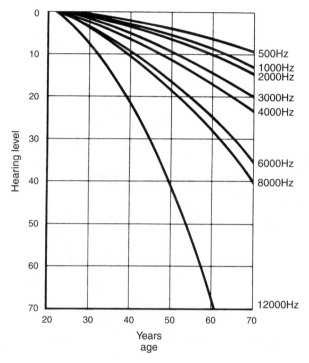

Figure 22-14. Hearing loss with age. (Source: R. Hinchcliffe, "The Pattern of the Threshold of Perception of Hearing and Other Special Senses as a Function of Age," *Gerontologica* **2**, 311 (1958).)

noise has been blamed for headaches and irritability. All of these reactions are those that ancestral cave dwellers also experienced. Noise meant danger, and senses and nerves were "up," ready to repel the danger. In the modern noise-filled world, we are always "up," and it is unknown how much if any of our physical ills are due to our response to noise. We also know that we cannot adapt to noise in the sense that our body functions no longer react a certain way to excessive noise. People do not, therefore, "get used to" noise in the physiological sense.

In addition to the noise-we-can-hear problem, it is appropriate to mention the potential problems of very-high- or very-low-frequency sound, out of our usual 20 to 20,000 Hz hearing range. The health effects of these, if any, remain to be completely documented.

THE DOLLAR COST OF NOISE

Many case histories comparing patients in noisy and quiet hospitals point to increased convalescence time when the hospital was noisy, either from internal hospital activity or from external noise. This may be translated directly to a dollar figure.

Recent court cases have been won by workers seeking damages for hearing loss suffered on the job. The Veterans Administration spends millions of dollars every year for care of patients with hearing disorders. Other costs, such as lost sleep or sleeping pills, lost time in industry, and apartment soundproofing, are difficult to quantify, but examples abound. The John F. Kennedy Cultural Center in Washington spent $5 million for soundproofing, necessitated by the jets using the nearby National Airport.

NOISE CONTROL

The control of noise is possible at three different stages of its transmission:

1. Reducing the sound produced,
2. Interrupting the path of the sound, and
3. Protecting the recipient.

When we consider noise control in industry, in the community, or in the home, we should keep in mind that all problems have these three possible solutions.

Industrial Noise Control

Industrial noise control generally involves the replacement of noise-producing machinery or equipment with quieter alternatives. For example, the noise from an air fan may be reduced by increasing the number of blades or the pitch of the blades and decreasing the rotational speed, thus obtaining the same air flow. Industrial noise may also be decreased by interrupting the path of the noise; for example, a noisy motor may be covered with insulating material.

A method of noise control often used in industry is protection of the recipient by distributing hearing protection devices (HPDs). These HPDs must be selected to have enough noise attenuation to protect against the anticipated exposures to noise. Great care must be taken to avoid interference with the ability to hear human speech and warning signals in the workplace.

Community Noise Control

The three major sources of community noise are aircraft, highway traffic, and construction. Construction noise must be controlled by local ordinances (unless federal funds are involved). Control usually involves the muffling of air compressors, jackhammers, hand compactors, etc. Since mufflers cost money, contractors will not take it upon themselves to control noise, and outside pressures must be exerted.

Regulating aircraft noise in the United States is the responsibility of the Federal Aviation Administration, which has instituted a two-pronged attack on this problem. First, it has set limits on aircraft engine noise and will not allow aircraft exceeding these limits to use the airports, forcing manufacturers to design engines for quiet operation as well as for thrust. The second effort has been to divert flight paths away from populated areas and, whenever necessary, to have pilots use less than maximum power when the takeoff carries them over a noise-sensitive area. Often this approach is not enough to prevent significant noise-induced damage or annoyance, and aircraft noise remains a real problem in urban areas.

Supersonic aircraft present a special problem. Not only are their engines noisy, but the sonic boom may produce considerable property damage. Damage from supersonic military flights over the United States has led to a ban on such flights by commercial supersonic aircraft.

The third major source of community noise is traffic. The car or truck creates noise in a number of ways. The exhaust system, tires, engine, gears, and transmission all contribute to a noise level, while the very act of moving through the atmosphere creates wind noise. Elevated highways and bridges resonate with the traffic motion and amplify traffic noise. The worst offender on the highways is the heavy truck, which generates noise in all of these ways. The total noise generated by vehicles may be correlated directly to the truck traffic volume. Figure 22-15 is a typical plot showing sound level as a function of traffic volume (measured in number of trucks per hour). Clearly, truck volume is of great importance. It is important to note that this graph is plotted as "sound level exceeded 10% of the time." Peak sound levels could be a great deal higher.

A number of alternatives are available for reducing highway noise. First, the source could be controlled by making quieter vehicles; second, highways could be routed away from populated areas; and third, noise could be baffled with walls or other types of barriers.

Vegetation, surprisingly, makes a very poor noise screen, unless the screen is 50 yards or more deep. The opposing lanes of the Baltimore–Washington Parkway are separated in many places by 100–200 yards of fairly dense vegetation, which provides an excellent noise and light screen. Newer highways rarely have the luxury of so much right-of-way. The most effective buffers have been to lower the highway, or to build physical wood or concrete barriers beside the road and thus screen the noise. All of these have limitations: noise will bounce off the walls and create little or no noise

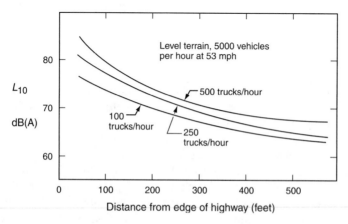

Figure 22-15. The effect of truck density and distance from a highway on noise.

Table 22-4. Design Noise Levels Set by the Federal Highway Administration

Land category	Design noise level, L_{10}	Description of land use
A	60 dB(A) exterior	Activities requiring special qualities of serenity and quiet, like amphitheaters
B	70 dB(A) exterior	Residences, motels, hospitals, schools, parks, libraries
	55 dB(A) interior	Residences, motels, hospitals, schools, parks, libraries
C	75 dB(A) exterior	Developed land not included in categories A and B
D	No limit	Undeveloped land

shadow, and walls hinder highway ventilation, thus contributing to buildup of CO and other pollutants from car and truck exhaust. The Department of Transportation has established design noise levels for various land uses, as shown in Table 22-4.

Noise in the Home

Private dwellings are getting noisier because of internally produced noise as well as external community noise. The list of gadgets in a modern American home reads like a list of New Year's Eve noisemakers. Some examples of domestic noise are listed in Table 22-5. Otherwise, similar products of different brands often will vary significantly in noise levels. When shopping for an appliance, it is just as important therefore to ask the clerk "How noisy is it?" as it is to ask "How much does it cost?"

The Eraring Power Station near Dove Creek, Australia, is an interesting example of the complexity of environmental pollution and control: all aspects of a problem must be considered before a solution is possible. During calm early mornings, residents complained of a loud "rumbling noise" coming from the power station 1.5 km away. The noise was measured in Dove Creek and for a single morning with no changes in the operation of the plant, the sound level increased from 43 dB at 5:30 AM to 62 dB at 6:00 AM. The problem: an atmospheric inversion "capturing" the noise and permitting it to reach the sleepy inhabitants of Dove Creek. The problem was solved when the duct work within the power station was attached to reactive dissipative silencers. The silencers, a series of chambers within the duct work system, were tuned to the predominant frequencies of the noise source. After a lengthy research project, the problem in Dove Creek was solved.

CONCLUSION

While noise was considered just another annoyance in a polluted world, not much attention was given to it. We now have enough data to show that noise is a definite

Table 22-5. Some Domestic Noisemakers

Item (distance from noise source)	Sound level, dB(A)
Vacuum cleaner (10 ft)	75
Quiet car at 50 mph (inside)	65
Sports car at 50 mph (inside)	80
Flushing toilet (5 ft)	85
Garbage disposal (3 ft)	80
Window air conditioner (10 ft)	55
Ringing alarm clock (2 ft)	80
Powered lawn mower (operator's position)	105
Snowmobile (driver's position)	120
Rock band (10 ft)	115

health hazard and should be numbered among our more serious pollutants. It is possible, using available technology, to lessen this form of pollution. However, the solution costs money, and private enterprise cannot afford to give noise a great deal of consideration until forced to by either the government or the public.

PROBLEMS

22.1 If an office has a noise level of 70 dB and a new machine emitting 68 dB is added to the din, what is the combined sound level?

22.2 Animals like dogs can hear sounds that have SPL *less than* 0. Show how this is possible.

22.3 Dogs can hear sounds at pressures close to 2×10^{-6} N/m². What is this in decibels?

22.4 An air compressor emits pressure waves at 0.01 N/m². What is the SPL in decibels?

22.5 Why was P_{ref} in Eq. (22.9) chosen at 0.00002 N/m²? What if a mistake was made and P_{ref} should really have been 0.00004 N/m²? This is a 100% error. How could this affect the numbers in Table 22-1?

22.6 Given the following data, calculate the L_{10} and the L_{50}.

Time (s)	dB(A)	Time (s)	dB(A)
10	70	60	65
20	50	70	60
30	65	80	55
40	60	90	70
50	55	100	50

22.7 Suppose your dormitory is 200 yards from a highway. What truck traffic volume would be "allowable" in order to stay within the Federal Highway Administration guidelines?

22.8 If the SL were 80 dB(C) and 60 dB(A), would you suspect that most of the noise was of high, medium, or low frequency? Why?

22.9 In addition to the data listed in Example 22.3 (Table 22-2), the following SL measurements were taken:

Time (s)	dB(A)	Time (s)	dB(A)
110	80	160	95
120	82	170	98
130	78	180	82
140	87	190	88
150	92	200	75

Calculate the L_{50}, L_{10}, and NPL by using all 20 data points.

22.10 If the pressure wave was 0.3 N/m², what is the SPL in decibels? If this is all the information about the sound you have, what can you say about the SL in dB(A)? What data would you need to make a more accurate estimate of the SL in dB(A)?

22.11 If you sing at a level 10,000 times greater than the power of the faintest audible sound, at which decibel level are you singing?

22.12 How many times more powerful is a 120-dB sound than a 0-dB sound?

22.13 On a graph of decibels vs hertz (10 to 50,000), show a possible frequency analysis for: (a) a passing freight train, (b) a dog whistle, and (c) "white noise."

22.14 The OSHA standard for an 8-h exposure to noise is 90 dB(A). The Environmental Protection Agency (EPA) suggests that this should be 85 dB(A). Show that the OSHA level is almost 400% louder in terms of sound level than the EPA suggestion.

22.15 A machine with an overall noise level of 90 dB is placed in a room with another machine putting out 95 dB.

  a. What will be the sound level in the room?
  b. Based on OSHA criteria, how long should workers be in the room during one working day?

22.16 What sound pressure level results from the combination of three sources: 68, 78, and 72 dB?

22.17 If an occupational noise standard were set at 80 dB for an 8-h day, 5-day-per-week working exposure, what standard would be appropriate for a 4-h day, 5-day-per-week working exposure?

22.18 Carry a sound level meter with you for one entire day. Measure and record the sound levels as dB(A) in your classes, in your room, during sports events, in the dining hall, or wherever you go during the day.

22.19 Seek out and measure the three most obnoxious noises you can think of. Compare these with the noises in Table 22-5.

22.20 In your room measure and plot the sound level in dB(A) of an alarm clock vs distance. At what distance will it still wake you if it requires 70 dB(A) to get you up? Draw the same curve outside. What is the effect of your room on the sound level?

22.21 Construct a sound level frequency curve for a basketball game. Calculate the noise pollution level.

22.22 A noise is found to give the following responses on a sound level meter: 82 dB(A), 83 dB(B), and 84 dB(C). Is the noise of a high or low frequency?

22.23 A machine produces 80 dB(A) at 100 Hz (almost pure sound).

a. Would a person who has suffered a noise-induced threshold shift of 40 dB at that frequency be able to hear this sound? Explain.
b. What would this noise measure on the dB(C) scale?

22.24 What reduction in sound intensity would have been necessary to reduce the takeoff noise of the American SST from 120 to 105 dB?

Appendix A

Conversion Factors

Multiply	By	To obtain
acre	0.404	ha
acre ft	1233	m^3
atmospheres	14.7	$lb/in.^2$
British thermal units	252	cal
Btu/ft^3	1.054×10^3	J
Btu/lb	8905	cal/m^3
Btu/lb	2.32	J/g
Btu/s	0.555	cal/g
Btu/ton	1.05	kW
Btu/ton	278	cal/tonne
calories	4.18	joules
calories	3.9×10^{-3}	Btu
cal/g	1.80	Btu/lb
cal/m^3	1.12×10^{-4}	Btu/ft^3
cal/tonne	3.60×10^{-3}	Btu/ton
centimeters	0.393	in.
feet	0.305	m
ft/min	0.00508	m/s
ft/s	0.305	m/s
ft^2	0.0929	m^2
ft^3	0.0283	m^3
ft^3	28.3	liters
ft^3/s	0.0283	m^3/s
ft^3/s	449	gal/min
ft lb (force)	1.357	joules
ft lb (force)	1.357	newton meters

(continued)

Multiply	By	To obtain
gallons	3.78×10^{-3}	m^3
gallons	3.78	liters
gal/day/ft^2	0.0407	m^3/day/m^2
gal/min	2.23×10^{-3}	ft^3/s
gal/min	0.0631	liter/s
gal/min	0.227	m^3/h
gal/min	6.31×10^{-5}	m^3/s
gal/min/ft^2	2.42	m^3/h/m^2
million gal/day	43.8	liters/s
million gal/day	3785	m^3/day
million gal/day	0.0438	m^3/s
grams	2.2×10^{-3}	lb
hectares	2.47	acre
horsepower	0.745	kW
inches	2.54	cm
inches of mercury	0.49	lb/in.^{2}
inches of mercury	3.38×10^3	newton/m^2
inches of water	249	newton/m^2
joule	0.239	calorie
joule	9.48×10^{-4}	Btu
joule	0.738	ft lb
joule	2.78×10^{-7}	kWh
joule	1	newton meter
J/g	0.430	Btu/lb
J/s	1	watt
kilograms	2.2	lb (mass)
kg	1.1×10^{-3}	tons
kg/ha	0.893	lb/acre
kg/h	2.2	lb/h
kg/m^3	0.0624	lb/ft^3
kg/m^3	1.68	lb/yd^3
kilometers	0.622	mile
km/h	0.622	mph
kilowatts	1.341	horsepower
kWh	3600	kilojoule

(continued)

Multiply	By	To obtain
liters	0.0353	ft^3
liters	0.264	gal
liters/s	15.8	gal/min
liters/s	0.0288	mgd
meters	3.28	ft
meters	1.094	yd
m/s	3.28	ft/s
m/s	196.8	ft/min
m^2	10.74	ft^2
m^2	1.196	yd^2
m^3	35.3	ft^3
m^3	264	ga
m^3	1.31	yd^3
m^3/day	264	gal/day
m^3/h	4.4	gpm
m^3/h	6.38×10^{-3}	gpm
m^3/s	35.31	ft^3/s
m^3/s	15,850	gpm
m^3/s	22.8	mgd
miles	1.61	km
miles^2	2.59	km^2
mph	0.447	m/sec
milligrams/liter	0.001	kg/m^3
million gallons	3785	m^3
mgd	43.8	liter/s
mgd	157	m^3/h
mgd	0.0438	m^3/s
newton	0.225	lb (force)
newton/m^2	2.94×10^{-4}	inches of mercury
newton/m^2	1.4×10^{-4}	lb/in.^{2}
newton meters	1	joule
newton sec/m^2	10	poise
pounds (force)	4.45	newton
pounds (force)/in.^{2}	6895	N/m^2
pounds (mass)	454	g
pounds (mass)	0.454	kg
pounds (mass)/ft^2/yr	4.89	kg/m^2/year

(continued)

Multiply	By	To obtain
pounds (mass)/yr/ft^3	16.0	kg/yr/m^3
pounds/acre	1.12	kg/ha
pounds/ft^3	16.04	kg/m^3
pounds/in.^{2}	0.068	atmospheres
pounds/in.^{2}	2.04	inches of mercury
pounds/in.^{2}	7140	newton/m^2
rad	0.01	gray
rem	0.01	sievert
tons (2000 lb)	0.907	tonne (1000 kg)
tons	907	kg
ton/acre	2.24	tonnes/ha
tonne (1000 kg)	1.10	ton (2000 lb)
tonne/ha	0.446	tons/acre
yd	0.914	m
yd^3	0.765	m^3
watt	1	J/s

Appendix B
Elements of the Periodic Table

	Symbol	Atomic number	Atomic weight
Actinium	Ac	89	227[a]
Aluminum	Al	13	26.98
Americium	Am	95	242[a]
Antimony	Sb	51	121.75
Argon	Ar	18	39.95
Arsenic	As	33	74.92
Astatine	At	85	215[a]
Barium	Ba	56	137.33
Berkelium	Bk	97	247[a]
Beryllium	Be	4	9.01
Bismuth	Bi	83	208.98
Boron	B	5	10.81
Bromine	Br	35	79.91
Cadmium	Cd	48	112.4
Calcium	Ca	20	40.08
Californium	Cf	98	251[a]
Carbon	C	6	12.01
Cerium	Ce	58	140.12
Cesium	Cs	55	132.91
Chlorine	Cl	17	35.45
Chromium	Cr	24	52
Cobalt	Co	27	58.93
Copper	Cu	29	63.55
Curium	Cm	96	244[a]
Dysprosium	Dy	66	162.5
Einsteinium	Es	99	252[a]
Erbium	Er	68	167.26

(continued)

	Symbol	Atomic number	Atomic weight
Europium	Eu	63	151.97
Fermium	Fm	100	b
Fluorine	F	9	19.0
Francium	Fr	87	223^a
Gadolinium	Gd	64	157.25
Gallium	Ga	31	69.72
Germanium	Ge	32	72.59
Gold	Au	79	196.97
Hafnium	Hf	72	178.49
Hahnium	Ha	105	b
Hassium	Hs	108	b
Helium	He	2	4.0
Holmium	Ho	67	164.93
Hydrogen	H	1	1.01
Indium	In	49	126.9
Iodine	I	53	138.91
Iridium	Ir	77	192.2
Iron	Fe	26	55.85
Krypton	Kr	36	83.8
Lanthanum	La	57	138.91
Lawrencium	Lw	103	b
Lead	Pb	82	207.19
Lithium	Li	3	6.94
Lutetium	Lu	71	174.97
Magnesium	Mg	12	24.31
Manganese	Mn	25	54.94
Meitnerium	Mt	109	b
Mendelevium	Md	101	b
Mercury	Hg	80	200.59
Molybdenum	Mo	42	95.94
Neilsborhium	Ns	107	b
Neodymium	Nd	60	144.24
Neon	Ne	10	20.18
Neptunium	Np	93	237^a
Nickel	Ni	28	58.71
Niobium	Nb	41	92.91
Nitrogen	N	7	14.01
Nobelium	No	102	b
Osmium	Os	76	190.2
Oxygen	O	8	16.00
Palladium	Pd	46	106.4
Phosphorus	P	15	30.97

(continued)

	Symbol	Atomic number	Atomic weight
Platinum	Pt	78	195.08
Plutonium	Pu	94	238^a
Polonium	Po	84	209^a
Potassium	K	19	39.1
Praseodymium	Pr	59	140.91
Promethium	Pm	61	b
Protactinium	Pa	91	231.03^a
Radium	Ra	88	228^a
Radon	Rn	86	222^a
Rhenium	Re	75	186.2
Rhodium	Rh	45	102.91
Rubidium	Rb	37	85.47
Ruthenium	Ru	44	102.91
Rutherfordium	Rf	104	b
Samarium	Sm	62	150.55
Scandium	Sc	21	44.96
Seaborgium	Sg	106	b
Selenium	Se	34	78.96
Silicon	Si	14	28.09
Silver	Ag	47	107.87
Sodium	Na	11	23.0
Strontium	Sr	38	87.62
Sulfur	S	16	32.06
Tantalum	Ta	73	180.95
Technetium	Tc	43	99^a
Tellurium	Te	52	127.6
Terbium	Tb	65	158.92
Thallium	Tl	81	204.38
Thorium	Th	90	232.04
Thulium	Tm	69	168.93
Tin	Sn	50	118.69
Titanium	Ti	22	47.9
Tungsten	W	74	183.85
Uranium	U	92	238.03^a
Vanadium	V	23	50.94
Xenon	Xe	54	131.3
Ytterbium	Yb	70	173.04
Yttrium	Y	39	88.91
Zinc	Zn	30	65.37
Zirconium	Zr	40	91.22

[a] All isotopes of these elements are radioactive. Where the mass is presented as an integer, it is the closest approximation.

[b] All isotopes of these elements are so short-lived that no mass number is given.

Appendix C

Physical Constants

UNIVERSAL CONSTANTS

Avogadro's Number $= 6.02252 \times 10^{23}$
Gas constant $R = 8.315$ kJ/kg–mole–K $= 0.08315$ bar–m^3/kg–mole–K
$= 0.08205$ L–atm/mole–K $= 1.98$ cal/mole–K
Gravitational acceleration at sea level $g = 9.806$ m/s$^2 = 32.174$ ft/s^2
Properties of water:

Density $= 1.0$ kg/L
Specific heat $= 4.184$ kJ/kg–°C $= 1000$ cal/kg–°C $= 1.00$ kcal/kg–°C
Heat of fusion $= 80$ kcal/kg $= 334.4$ kJ/kg
Heat of vaporization $= 560$ kcal/kg $= 2258$ kJ/kg
Viscosity at 20°C $= 0.01$ poise $= 0.001$ kg/m–s

Appendix D

List of Symbols

A	soil loss, tons/acre–year
A or a	area, m² or ft²
A'	surface area of the sand bed, m² or ft²
A_i	acreage of subarea i, acres
A_L	limiting area in a thickener, m²
A_P	surface area of the particles, m² or ft²
AA	attainment area
amu	atomic mass unit
B	aquifer thickness, m or ft
BAT	best available technology
BHP	brake horsepower
BOD	biochemical oxygen demand, mg/L
BOD_5	five-day BOD
BOD_{ult}	ultimate BOD: carbonaceous plus nitrogenous
Bq	Becquerel; one radioactive disintegration per second
b	slope of filtrate volume vs time curve
b	cyclone inlet width in m
C	concentration of pollutant in g/m³ or kg/m³
C	cover factor (dimensionless ratio)
C	Hazen–Williams friction coefficient
C	total percolation of rain into the soil, mm (Chap. 13)
C_d	drag coefficient
C_i	solids concentration at any level i
C_p	specific heat at constant pressure in kJ/kg–K
C_0	influent solids concentration, mg/L
C_u	underflow solids concentration, mg/L
CEQ	Council on Environmental Quality
Ci	Curie; 3.7×10^{10} Bq
CSO	combined sewer overflow
CST	capillary suction time
c	Chezy coefficient
C	wave velocity, m/s
cfs	cubic feet per second

458 ENVIRONMENTAL ENGINEERING

$D(t)$	oxygen deficit at time (t), in mg/L.
D or d	diameter, in m or ft or in.
D	deficit in DO, in mg/L
D	dilution (volume of sample/total volume) (Chap. 4)
D_0	initial DO deficit, in mg/L
DOT	U.S. Department of Transportation
d	depth of flow in a pipe, in m or in. (Chap. 7)
d'	geometric mean diameter between sieve sizes, m or ft
d_c	cut diameter, in m
dB	decibel
D_s	oxygen deficit upstream from wastewater discharge, mg/L
D_p	oxygen deficit in wastewater effluent, mg/L
du/dy	rate of shear, or slope of the velocity (u)–depth (y) profile
E	rainfall energy, ft-tons/acre inch
E	efficiency of materials separation
E	evaporation, mm
E	symbol for exponent sometimes used in place of 10.
EIA	environmental impact assessment
EIS	environmental impact statement
EIU	environmental impact unit
EPA	U.S. Environmental Protection Agency
EQI	environmental quality index
e	porosity fraction of open spaces in sand
esu	electrostatic unit of charge
eV	electron volt = 1.60×10^{-19} joule
F	final BOD of sample, mg/L
F	food (BOD), in mg/L
FONSI	finding of no significant (environmental) impact
F_B	buoyant force, in N
F_D	drag force, in N
F_g	gravitational force, in N
f	friction factor
F'	final BOD of seeded dilution water, mg/L
G	flow in a thickener, kg/m^2 × s (Chap. 9)
G_L	limiting flux in a thickener, kg/m^2 × s
G	velocity gradient, in s^{-1} (Chap. 6)
Gy	gray: unit of absorbed energy; 1 joule/kg
g	acceleration due to gravity, in m/s^2 or ft/s^2
H or h	height, m
H	depth of stream flow, in m (Chap. 3)
H	effective stack height, m (Chap. 19)

List of Symbols

H	total head, m or ft (Chap. 5)
H'	Shannon/Weaver diversity index
H_L	total head loss through a filter, m or ft
Hz	Hertz, cycles/s
h	geometric stack elevation, m
h	fraction of BOD not removed in the primary clarifier
h	depth of landfill, m
h_d	net discharge head, m or ft
h_L	head loss, m or ft
$(h_L)_t$	head loss in the ith layer of media at time t
h_s	net static suction head, m or ft
i	fraction of BOD not removed in the biological treatment step
I	rainfall intensity, in./h
I	initial BOD of sample, mg/L
I'	initial BOD of seeded dilution water, mg/L
j	fractions of solids not destroyed in digestion
J	Pielov's equitability index
K	soil erodibility factor, ton/acre/R unit
K	proportionality constant for minor losses, dimensionless
K_p	coefficient of permeability, m^3/day or gal/day
K_T	fraction of atoms that disintegrate per second $= 0.693/t_{1/2}$
k	fraction of influent SS removed in the primary clarifier
K_s	saturation constant, in mg/L
$k_{1/}$	deoxygenation constant, $(\log_e)$ s^{-1}
$k_{2/}$	reoxygenation constant, $(\log_e)$ s^{-1}
kW	kilowatts
kWh	kilowatt–hours
L	depth of filter, m or ft (Chap. 6)
L or l	length, m or ft
LS	topographic factor (dimensionless ratio)
L_o	ultimate carbonaceous oxygen demand, in mg/L
L_1	length of cylinder in a cyclone, in m
L_2	length of cone in a cyclone, in m
L_F	feed particle size, 80% finer than, μm
L_s	ultimate BOD upstream from wastewater discharge, mg/L
L_p	product size, 80% finer than, μm
L_p	ultimate BOD in wastewater effluent, mg/L
L_x	x percent of the time stated sound level (L) was exceeded, percentage
LAER	lowest achievable emission rate
LCF	latent cancer fatalities

LD_{50}	lethal dose, at which 50% of the subjects are killed
LDC_{50}	lethal dose concentration at which 50% of the subjects are killed
LET	linear energy transfer
M	mass of a radionuclide, in g
M	microorganisms (SS), in mg/L
MACT	maximum achievable control technology
MeV	million electron volts
MLSS	mixed liquor suspended solids, in mg/L
MSS	moving source standards
MSW	municipal solid waste
MW_e	megawatts (electrical); generating plant output
MW_t	megawatts (thermal); generating plant input
m	mass, in kg
m	rank assigned to events (e.g., low flows)
N	number of leads in a scroll centrifuge (Chap. 9)
N	effective number of turns in a cyclone
N^0	Avogadro's number, 6.02×10^{23} atoms/g–atomic weight
NAA	non-attainment areas
NAAQS	National Ambient Air Quality Standards
NEPA	National Environmental Policy Act
NESHAPS	National Emission Standards for Hazardous Air Pollutants
NPDES	National Pollution Discharge Elimination System
NPL	noise pollution level
NPSH	net positive suction head, m or ft
NRC	Nuclear Regulatory Commission
NSPS	new (stationary) source performance standards
n	number of events (e.g., years in low flow records)
n	Manning roughness coefficient
n	revolutions per minute (Chap. 6)
n	number of subareas identified in a region
n_c	critical speed of a trommel, rotations/s
n_i	number of individuals in species i (Chap. 3)
OCS	Outer Continental Shelf
OSHA	Occupational Safety and Health Administration
P	erosion control practice factor (dimensionless ratio)
P	phosphorus, in mg/L
P	power, N/s or ft–lb/s (Chap. 6)
P	precipitation, mm
P	pressure, kg/m^2 or lb/ft^2 or N/m^2 or atm
ΔP	pressure drop, in m of water
P_{ref}	reference pressure, N/m^2

List of Symbols 461

P_s	purity of a product x, %
PIU	parameter importance units
PMN	premanufacture notification
POTW	publicly owned (wastewater) treatment works
PPBS	planning, programming, and budgeting system
PSD	prevention of significant deterioration
Q or q	flow rate, in m^3/s or gal/min
Q	emission rate, in g/s or kg/s
Q	number of Ci or Bq
Q_h	heat emission rate, kJ/s
Q_o	influent flow rate, m^3/s
Q_p	pollutant flow, in mgd or m^3/s (Chap. 3)
Q_p	flow rate of wastewater effluent, m^3/s
Q_s	stream flow, in mgd or m^3/s (Chap. 3)
Q_s	flow rate upstream from wastewater discharge, m^3/s
Q_w	waste sludge flow rate, in m^3/s
q	substrate removal velocity, in s^{-1}
R	radius of influence of a gas withdrawal well, m (Chap. 13)
R	Rainfall factor
R	recovery of pollutant or collection efficiency, %
R	% of overall recovery of, SS in settling tank
R or r	hydraulic radius, in m or ft
R	runoff coefficient (Chap. 13)
R_x	recovery of a product x, %
R	Reynolds number
RACT	reasonable achievable control technology
RCRA	Resource Conservation and Recovery Act
RDF	refuse derived fuel
ROD	record of decision (Chap. 2)
r	radius, in m or ft or cm
r	hydraulic radius in Hazen–Williams equation, m or ft
r	specific resistance to filtration, m/kg
rad	unit of absorbed energy: 1 erg/g
rem	roentgen equivalent man
S	rainfall storage, mm
S	scroll pitch, m
S	substrate concentration, estimated as BOD, mg/L
S_o	influent BOD, kg/h
S_d	sediment delivery ratio (dimensionless factor)
S_0	influent substrate concentration estimated as BOD, mg/L
SIP	State Implementation Plans
SIU	significant individual user

SIW	significant individual waste
SL	sound level
SPL	sound pressure level
SS	suspended solids, in mg/L
Sv	sievert; unit of dose equivalent
SVI	sludge volume index
s	hydraulic gradient (Chap. 5)
s	slope (Chap. 7)
s	sensation (hearing, touch, etc.)
T	temperature, in °C
TOSCA	Toxic Substances Control Act
TRU	transuranic material or transuranic waste
t	time, in s or days
t_c	critical time, time when minimum DO occurs, in s
$t_{1/2}$	radiological half-life of a radionuclide
$\bar{t}$	time of flocculation, in min
$\bar{t}$	retention time, in s or days
UC	unclassifiable (inadequate information)
USDA	U.S. Department of Agriculture
USLE	universal soil loss equation
USPHS	U.S. Public Health Service
u	average wind speed, m/s
V	volume, in m^3 or ft^3
V_p	volume occupied by each particle, in m^3 or ft^3 (Chap. 7)
v	interface velocity at solids concentration C_i
v	velocity of flow, m/s or ft/s, and superficial velocity, m/day or ft/day
v	velocity of the paddle relative to the fluid, m/s or ft/s
v	velocity of water through the sand bed, m/s or ft/s
v_a	velocity of water approaching sand, m/s or ft/s
v_d	drift velocity, m/s
v_i	inlet gas velocity, m/s
v_O	settling velocity of a critical particle, in m/s
v_p	velocity in a partially full pipe, m/s or ft/s
v_R	radial velocity, m/s
v_s	settling velocity of any particle, in m/s
v'	actual water velocity in soil pores, m/day or ft/day
V	filtrate volume, m^3
w	specific weight, kg/m^3 or lb/ft^3
WHP	water horsepower
WEPA	Wisconsin Environment Policy Act

List of Symbols 463

WPDES	Wisconsin Pollutant Discharge Elimination System
W	power level, watts
w	cake deposited per volume of filtrate, kg/m^3
W	specific energy, kWh/ton
W_i	Bond work index, kWh/ton
X	seeded dilution water in sample bottle, mL
x	weight fraction of particles retained between two sieves
X_e	effluent SS, mg/L
X_o	influent SS, kg/h
X	microorganism concentration, estimated as SS, mg/L
X_e	effluent microorganism concentration, estimated as SS, mg/L
X_r	return sludge microorganism concentration, estimated as SS, mg/L
X_0	influent microorganism concentration, estimated as SS, mg/L
x	particle size, m
x_c	characteristic particle size, m
x_0, y_0	mass per time of feed to a materials separation device
x_1, y_1	mass per time of components x and y exiting from a materials separation device through exit stream 1
x_2, y_2	mass per time of components x and y exiting from a materials separation device through exit stream 2
x_1	mass of pollutant that could have been captured, kg
x_2	mass of pollutant that escaped capture, kg
x_0	mass of pollutant collected, kg
x	thickness, in m
Y	yield, $\Delta X/\Delta S$
Y	yield, kg SS produced/kg BOD used
Y	cumulative fraction of particles (by weight) less than some specific size
Y	volume of BOD bottle, mL
Y_F	filter yield, kg/m^2–s
y	oxygen used (or BOD) at time t, in mg/L (Chap. 4)
Z	elevation, m or ft
z	depth of sludge in a bowl, m
$z(t)$	oxygen required for decomposition, in mg/L (Chap. 4)
Σ	sigma factor
α	alpha radiation
β	beta factor
β	beta radiation
γ	gamma radiation
γ	kinematic viscosity, cm^2/s (Chap. 5)

ΔS	net BOD utilized in secondary treatment, kg/h
ΔX	net solids produced by the biological step, kg/h
η	plastic viscosity, N–s/m^2
$\Delta \omega$	difference between bowl and conveyor rotational speed, rad/s
Θ_c	mean cell residence time, or sludge age, days
η	pump efficiency (Chap. 5)
λ	wavelength, m
μ	viscosity, in N–s/m^2 or poise (lb–s/ft^2)
μ	growth rate constant, in s^{-1}
μ	maximum growth rate constant, in s^{-1}
v	frequency of sound wave, cycle/s
ϕ	shape factor
ρ	density, g/cm, kg/cm^3, lb–s/ft^4 or lb–s^2/ft^3
ρ_s	density of a solid, in kg/m^3
σ_y	standard deviation, y direction, m
σ_z	standard deviation, z direction, m
ι	rotational speed, rad/s
τ	shear stress, N/m^2
τ_y	yield stress, N/m^2
ω	rotational velocity, rad/s

Appendix E

Bibliography

Ailbmann, L. "Full-Time Hazwaste Disposal," *Waste Age* 22(3):147–151 (1991).

Allen, C.H., and E.H. Berger. "Development of a Unique Passive Hearing Protector with Level-10 Dependent and Flat Attenuation Characteristics," *Noise Control Engineering Journal* 34(3):97–105 (May–June 1990).

Ambler, C.M. "The Evaluation of Centrifuge Performance," *Chemical Engineering Progress* 48:3 (1952).

American Public Health Association. *Standard Methods for the Examination of Water and Wastewater*, 17th ed. Water Pollution Control Federation, American Water Works Association, 1989.

American Public Works Association. *History of Public Works in the United States, 1776–1976*. American Public Works Association, Chicago, 1976.

American Society of Civil Engineers and Water Pollution Control Federation. *Design and Construction of Sanitary and Storm Sewers*. American Society of Civil Engineers and Water Pollution Control Federation, New York, 1969.

Bachrach, J., and M.S. Baratz. "The Two Faces of Power," *American Political Science Review* 33:947–952 (1962).

Baker, M.N. *The Quest for Pure Water*. American Water Works Association, New York, 1949.

Baron, R.A. *The Tyranny of Noise*. St. Martins Press, New York, 1970.

Beeby, A. *Applying Ecology* (London: Chapman & Hall, 1993).

Bell, D., et al. *Decision Analysis*. Harvard University Press, Cambridge, MA, 1989.

Benefield, L.D., and C.W. Randall. *Biological Process Design for Wastewater Treatment*. Prentice-Hall, Englewood Cliffs, NJ, 1980.

Bond, F.C. "The Third Theory of Communication," *Transactions of the American Institute of Mining Engineers* 193:484 (1952).

Boorse, H.A., L. Motz, and J.H. Weaver. *The Atomic Scientists: A Biographical History*. Wiley, New York, 1989.

Cashwell, J.W., R.E. Luna, and K.S. Neuhauser. *The Impacts of Transportation within the United States of Spent Reactor Fuel from Domestic and Foreign Research Reactors* (SAND88-0714). Sandia National Laboratories, 1990.

Camp, T.R. "Sedimentation and the Design of Settling Tanks," *Transactions of the ASCE* 111:895 (1946).

Chanlett, E. "History of Sanitation," in *History of Environmental Sciences and Engineering* (P.A. Vesilind, Ed.). University of North Carolina Press, Chapel Hill, NC, 1975.

Chian, E.S.K., F.B. DeWalle, and E. Hammerberg. "Effect of Moisture Regime and Other Factors on Municipal Solid Waste Stabilization," in *Management of Gas and Leachate* (S.K. Banerji, Ed.). U.S. Environmental Protection Agency 600/9-77-026, Washington, DC, 1977.

Code of Federal Regulations of the United States, Volume 49, Parts 170–180, U.S. Government Printing Office, Washington, DC, 1991.

Code of Federal Regulations of the United States, Volume 10, Parts 20, 60, 61, 71, 100, U.S. Government Printing Office, Washington, DC, 1991.

Code of Federal Regulations of the United States, Volume 40, Part 191, U.S. Government Printing Office, Washington, DC, 1991.

Cohen, B.L. "Catalog of Risks Extended and Updated," *Health Physics* 61:317 (1991).

Commoner, B. *The Closing Circle.* Bantam Press, New York, 1970.

Darcy, H. *Les Fontaines Publiques de la Ville de Dijon*, Paris, France, 1956.

De Camp, L.S. *The Ancient Engineers*, Doubleday, New York, 1963.

Delbello, A. "A Wasted World," *Waste Age* 22(3):139–144 (1991).

Dubos, R. "A Theology of the Earth," in *Western Man and Environmental Ethics* (I.G. Barbour, Ed.). Addison-Wesley, Reading, MA, 1973.

Etzioni, A. "Mixed Scanning—A Third Approach to Decision Making," *Public Administration Review* (Dec. 1967).

Fallows, J.M. *The Water Lords.* Bantam Books, New York, 1971.

Fryer, J. "Case History: Stack Silencers for the Eraring Power Station," *Noise Control Engineering Journal* 32(3):89–92 (May–June 1989).

Glasstone, S., and W.H. Jordan. *Nuclear Power and Its Environmental Effects.* American Nuclear Society, La Grange Park, II, 1980.

"General Pretreatment Regulations for Existing and New Sources of Pollution," *Federal Register* Part IV:27, 736–773 (June 28, 1990).

Goldberg, M., City of Philadelphia, private communication (1988).

Grady, L., and H.C. Lim. *Biological Wastewater Treatment.* Marcel Dekker, New York, 1980.

Greenberg, A.E., Clesceri, L.S., and Eaton, A.D. *Standard Methods for the Examination of Water and Wastewater, 18th Edition* (Washington, DC: American Public Health Association, American Water Works Association, and Water Environment Federation, 1992).

Gunn, A.S., and P.A. Vesilind. *Environmental Ethics for Engineers.* Lewis, Chelsea, MI, 1986.

Hanna, S.R., G.A. Briggs, and R.P. Hosker. *Handbook on Atmospheric Diffusion, DOETIC-11223* U.S. Department of Energy, Washington, DC, 1986.

Hardin, G. "The Tragedy of the Commons," *Science* 162:1243 (1968).

Henriches, R. "Law—Part II: Resource Conservation and Recover Act: The Comprehensive Emergency Response, Compensation and Liability Act," *Research Journal of the Water Pollution Control Federation* 65(4):310–314 (1991).

Herricks, E.E. *Stormwater Runoff and Receiving Systems, Impact, Monitoring, and Assessment* (Boca Raton, FL: CRC Press, 1995).

Hickey, J.J., et al. "Concentration of DDT in Lake Michigan," *Journal of Applied Ecology* 3:141 (1966).

Hill, J. "Wild Cranes," *Sierra* 76:6 (1991).
Jordao, E., and J.R. Leitao. "Sewage and Solids Disposal: Are the Processes such as Ocean Disposal Proper? The Case of Rio de Janeiro, Brazil," *Water Science Technology* 22(12):33–43 (1990).
Keeney, R.A., and H. Raiffa. *Decisions With Multiple Objectives*. Harvard University Press, Cambridge, MA, 1974.
Kirby, R.S., S. Withington, A.B. Darling, and F.G. Kilgour. *Engineering in History*. McGraw-Hill, New York, 1956.
Koppel, J. "The LULU and the NIMBY," *Environment* (March):85–103 (1985).
Kozlovsky, D. *An Ecological and Evolutionary Ethic*. Prentice-Hall, Englewood Cliffs, NJ, 1975.
Lankton, L.D. "1842: The Old Croton Aqueduct Brings Water, Rescues Manhattan from Fire, Disease," *Civil Engineering* 47:10 (1977).
Lawson, L.R. *Animal Gods*. New York, 1978.
Leydet, F. *Time and the River Flowing*. Sierra Club Books, San Francisco, 1964.
Leopold, A. *A Sand County Almanac*. Oxford University Press, New York, 1949.
Liebman, J.C., J.W. Male, and M. Wathne. "Minimum Cost in Residential Refuse Vehicle Routes," *Journal of the Environmental Engineering Division of the American Society of Civil Engineers* 101(EE3):339 (1975).
Lipscomb, D.M., reported in White, F.A. *Our Acoustic Environment*. Wiley, New York, 1975.
MacIntyre, A. *A Short History of Ethics*. Macmillan, New York, 1966.
Martin, L., and G. Kaszynsk. "A Comparison of the CERCLA Response Program and the RCRA Corrective Action Program," *Hazardous Waste and Hazardous Materials* 8(2):161–184 (1991).
Marx, K. *Das Kapital* (1867).
Moldan, B., and J. Schnoor. "Czechoslovakia: Examining a Critically Ill Environment," *Environmental Science and Technology* 26(1):63–69 (1992).
Morris, R.D., A.M. Audet, I.F. Angiello, T.C. Chalmers, and F. Mosteller. "Chlorination, Chlorination By-products, and Cancer: a Meta-analysis," *American Journal of Public Health* 82:955 (1992).
Nash, R. *The American Environment*. Prentice-Hall, New York, 1976.
Nash, R. *Conservation in America*. Prentice-Hall, New York, 1970.
National Academy of Sciences Committee on the Biological Effects of Ionizing Radiation (Arthur C. Upton, Chair). *Health Effects of Exposure to Low Levels of Ionizing Radiation (BEIR V)*. National Academy Press, Washington, DC, 1990.
National Center for Health Statistics. U.S. Department of Health and Human Services, 1985.
Newhouse, J. *War and Peace in the Nuclear Age*. Knopf, New York, 1988.
O'Connor, D.J., and W.E. Dobbins. "Mechanisms of Reaeration of Natural Streams," *ASCE Trans.* 153:641 (1958).
Pasquill, F. "The Estimation of the Dispersion of Windborne Material," *Met. Mag.* 90:33 et seq. (1961).
Pasquill, F., and F.B. Smith. *Atmospheric Diffusion*. Wiley, New York, 1983.

Paulsen, I. *The Old Estonian Folk Religion*. Indiana University Press, Bloomington, IN, 1971.
Pedersea, O.J. "Noise and People," *Noise Control Engineering Journal*, 32(2):73–78 (March–April 1989).
Pielou, E.C. *Ecological Diversity* (New York: John Wiley & Sons, 1975).
Popper, F.J. "The Environmentalist and the LULU," *Environment* 27:7 (1985).
Puget Sound Air Pollution Control Agency. *Annual Report on Air Quality in the Puget Sound Air Quality Region* (1985).
Raverat, G. *Period Piece*, as quoted in Reyburn, W. *Flushed with Pride*. McDonald, London, 1969.
Research Journal of the Water Pollution Control Federation 63(4):302–307 (1990).
Reyburn, W. *Flushed with Pride*. McDonald, London, 1969.
Rheinheimer, G. *Aquatic Microbiology, 3rd Edition* (New York: John Wiley & Sons, 1985).
Ridgway, J. *The Politics of Ecology*. Dutton, New York, 1970.
Rosin, P., and E. Rammler. "Laws Covering the Fineness of Powdered Coal," *Journal of the Institute for Fuel* 7:29 (1933).
Rowland, F.S. "Chlorofluorocarbons and the Depletion of Stratospheric Ozone," *American Scientist* 77:37–45 (1989).
Schama, S. *Citizens: A Chronicle of the French Revolution*. Knopf, New York, 1989.
Schoenfeld, C. (Cornell University) as reported in the *Washington Post* (7 January 1981).
Schrenk, H.H. et al. "Air Pollution in Donora, Pa.," *U.S. Public Health Service Bulletin No. 306* (1949).
Shannon, C.E., and W. Weaver. *The Mathematical Theories of Communication* (Urbana, IL: University of Illinois Press, 1949).
Shell, R.L., and D.S. Shure. "A Study of the Problems Predicting Future Volumes of Wastes," *Solid Waste Management* (Mar. 1972).
Shuster, K.A., and D.A. Schur. "Heuristic Routing for Solid Waste Collection Vehicles" (U.S. Environmental Protection Agency OSWMP SW-113, Washington, DC, 1974).
Slovic, P. "Perception of Risk," *Science* 236:280–285 (1987).
Smith, G. *Plague on Us*. Oxford University Press, Oxford, 1941.
Solzhenytsin, A. *The Gulag Archipelago*. Bantam Books, New York, 1982.
Still, H. *In Quest of Quiet*. Stackpole Books, Harrisburg, PA, 1970.
Stone, C.D. *Should Trees Have Standing?* Kaufman, Los Angeles, 1972.
Stratton, F.E., and H. Alter. "Application of Bond Theory of Solid Waste Shredding," *Journal of Environmental Engineering Division of American Society of Civil Engineers* 104 (EEI) (1978).
Streeter, H.W., and E.B. Phelps. "A Study of the Pollution and Natural Purification of the Ohio River," *Public Health Bulletin* 146, USPHS, Washington, DC (1925).
Strumm, W., and J.J. Morgan. *Aquatic Chemistry, 3rd Edition* (New York: John Wiley & Sons, 1996).
Suter, A. "Noise Sources and Effects: A New Look," *Sound and Vibration* 26(1):18–31, (1992).
Suter, G.W. II "Environmental Risk Assessment/Environmental Hazards Assessment: Similarities and Differences," in *Aquatic Toxicology and Risk Assessment* (W.G. Landis, and W.H. van der Schalie, Eds.), Vol. 13, p. 5. ASTM STP 1096, 1990.

Taylor, R. *Noise*. Penguin, New York, 1970.
Thomas, H.A., Jr. "Graphical Determination of BOD Curve Constants," *Water Sewer Works* 97: 123 (1950).
Tschirley, F.H. *Scientific American* 254:29 et seq. (1986).
Udall, S. *The Quiet Crisis*. Addison-Wesley, Reading, MA, 1968.
U.S. Code 42 Sec. 7401 et seq. *The Clean Air Act* (PL101-549) as amended (1990).
U.S. Code 42.10101 et seq. *The Nuclear Waste Policy Act* (PL97-425) as amended (1987).
U.S. Code of Federal Regulations, Volume 40, Parts 50, 60, 61 (1991).
U.S. Department of Energy. *Health and Environmental Consequences of the Chernobyl Nuclear Power Plant Accident*, DOE/ER-0332. U.S. Department of Energy, 1987.
U.S. Environmental Protection Agency. *Standard Methods for Water and Wastewater Analysis*. Washington, DC, 1991.
U.S. Geologic Survey. *Geologic Disposal of Radioactive Waste (USGS Circular 779)*. Washington, DC, 1979.
USDA Soil Conservation Service, "Sedimentation," *National Engineering Handbook*, Section 3. U.S. Government Printing Office, Washington, DC, 1978.
USDA Agricultural Research Service. "Present and Prospective Technology for Predicting Sediment Yield and Sources." *Proceedings of the Sediment Yield Workshop*, U.S. Department of Agriculture Sedimentation Laboratory, Oxford, MS, 1972.
USEPA Office of Research and Development, *Loading Functions for Assessment of Water Pollution from Nonpoint Sources*. National Technical Information Service, Springfield, VA, 1976.
USEPA. *Results of the Nationwide Urban Runoff Program* (Washington, DC: U.S. Environmental Protection Agency, 1983).
USEPA Environmental Research Laboratory. *Silvicultural Activities and Nonpoint Pollution Abatement: A Cost-Effectiveness Analysis Procedure*. U.S. Environmental Protection Agency, Athens, GA, 1990.
USEPA. *Guidance Specifying Management Measures for Sources of Nonpoint Pollution in Coastal Waters, EPA-840-B-93-001c* (Washington, DC: U.S. Environmental Protection Agency, 1983).
Vesilind, P.A. *Treatment and Disposal of Wastewater Sludges*. Ann Arbor Science, Ann Arbor, MI, 1979.
Vesilind, P.A., Rimer, A.E., and Worrel, W.A. "Performance Testing of a Vertical Hammermill Shredder," *Proceedings of the National Conference on Solid Waste Processing*. American Society of Mechanical Engineers, New York, 1980.
Wark, K., and C.F. Warner. *Air Pollution, Its Origin and Control*. Harper & Row, New York, 1986.
Weschmeier, W.H., and Smith, D.D. *Predicting Rainfall-Erosion Losses from Cropland East of the Rocky Mountains, USDA Handbook No. 282* (Washington, DC: U.S. Department of Agriculture, 1965).
Wetzel, R.G., and G.E. Likens. *Limnological Analysis, 2nd Edition* (New York: Springer-Verlag, 1991).
Wetzel, R.G. *Limnology, 2nd Edition* (Philadelphia, PA: Saunders College Publishing, 1983).

Wisely, W.H. *The American Civil Engineer, 1965–1974.* American Society of Civil Engineers, New York, 1974.

World Resources Institute. *World Resources 1987.* Basic Books, New York, 1987.

Worrell, W.A., and P.A. Vesiland. "Testing and Evaluation of Air Classifier Performance." *Resource Recovery and Conservation* 4:247 (1979).

Worster, D. "The intrinsic Value of Nature, "*Environmental Review* 4:1–23 (1981).

Index

2,4,5-T, 308, 416
2,4-D, 308

Absorbance, 98
Absorbed radiation dose, 319
  equivalent, 319
Acid formers, 210–211, 267
Acid mine drainage, 54, 93, 326
Acid rain, 7, 287, 419–420
Acoustic environment, 428–429
Activated carbon adsorption, 197
Activated sludge, 184–195, 199, 205, 207, 209–210, 214, 230
  biological processes, 187–195
Adiabatic lapse rate, 355–358, 360–361
Adsorption, 169–170, 197, 399
Aerated lagoon, 196, 299
Aeration
  activated sludge, 183–187, 190
  diffused, 187
  extended, 184–185
  step, 185
  tank, 183–185, 191, 193–194, 197, 199
  tapered, 185–186
  zone of, 108–109
Aerobic decomposition, 58–59, 69, 86, 97, 263, 268, 290
Aerosols, 361
Agent Orange, 308
Agricultural chemicals, 40, 233–234
Agricultural wastes, 53
Agriculture
  sediment loading function, 236–241
  water pollution, 51–52, 54, 73, 75–76, 235, 241–243
Air classifiers, 283–285
Air dispersion of pollutants, 362–366
  dispersion coefficient, 362–365
  horizontal dispersion, 352–355
  vertical dispersion, 355–362
Air pollution control, 385–409
Air pollution law, 411–422
Air quality measurement, 375–382
Aircraft noise, 432, 441–442
Algae, 55, 57, 70, 71–73, 92, 196, 206
Algal blooms, 73
Alkali scrubbing, 402
Alkalinity of water, 94–95
Alpha radiation, 313–318, 320, 326, 329–330, 332, 337
Aluminum
  cans, 251–252, 261, 273, 279
  phosphorus removal, 198, 206, 208
  recovery, 273–275, 284–286
Aluminum sulfate (alum), 2, 135, 137, 140, 219, 222
Ambient air quality measurement, 375
Ambient air quality standards, 39, 373, 407, 412, 414–415
Ambient lapse rate, 355–360, 364
Ammonia, 58–60, 151, 347
  measurement of, 90, 97–99, 379
  wastewater, 169, 197, 266
  water pollution, 69–70, 75, 235
Anaerobic decomposition, 59–60, 69, 72, 262, 267–268, 290
Antagonism, 39
Anthropogenic pollutants, 21, 52, 321
Anticyclones, 351–352
Aquacludes, 111, 115
Aquatic ecology. *See* Ecology, aquatic
Aquifers, 109, 111–115, 242, 341–342, 344
  confined, 111
  drinking water, 75, 247, 297–298
  unconfined, 112–114

471

472 ENVIRONMENTAL ENGINEERING

Arsenic, 417
  cancer, 35
  measurement, 102–103
  water pollution, 76
Atmospheric cleansing, 368–371
Atmospheric dispersion, 333, 361–368
Atomic absorption spectrophotometry, 103
Automobiles
  air pollution, 54, 245, 333, 375, 379, 381, 385–386, 404, 414, 419
  risk, 31, 35–37
  water pollution, 54, 233, 245
Autotrophs, 55–56, 73
Axial Flow, 129

Background radiation. *See* Radiation, background
Backwashing, 141, 143
Bacteria, 59–60, 71–73
  measurement of, 85
  water and wastewater treatment. *See also* Activated sludge, 135, 143, 150–151
  water pollution, 51–53, 68, 75–76, 100
Becquerel, 318–320
Beer-Lambert Law, 97–98
Bel, 426
Belt filter, 218–219, 221
Belt magnet, 286
Bernoulli theorem, 122
Beta equation, 226
Beta radiation, 313–318, 329–330, 332, 337
Beverage container law, 261
Binary separation, 280–281
Bioaccumulation, 39
Biochemical oxygen demand (BOD)
  landfills, 266
  measurement of, 62–64, 84–90
  nonpoint source pollution, 234
  sludge, 187–189, 192–196, 205–207
  ultimate, 64, 90
  wastewater treatment, 154, 168–169, 171–172, 182, 184–185, 197, 199
Bioconcentration, 39, 297
Biodegradation, 57–59
Biosorption. *See also* Contact stabilization, 185–186

Black lung disease, 39
BOD. *See* Biochemical oxygen demand
Body burden, 39
Boilers, 287, 306, 415
Bond work index, 278
Bonds, 25–26
Brake horsepower (BHP), 130
Breeder reactor, 331
British thermal units/pound, 288
Bulking sludge, 185–186

Cadmium, 76
  sludge concentration, 209, 344
Calcium bicarbonate, 94
Calcium hydroxyapatite, 198, 206
Calcium sulfate
  air pollution, 402
Cancer, 34–35, 38, 40–46, 151, 323, 337
Capillary suction time (CST) apparatus, 223–224
Capillary water, 108
Carbon dioxide, 72, 90–91, 94
  biodegradation, 57–60, 184, 187, 205, 228
  global atmospheric change, 338, 404, 420
  incineration, 198, 287–288, 304, 307, 386, 399
  landfills, 267–268
Carbon monoxide, 38, 414
Carbon. *See also* Total organic carbon, 21, 197, 247, 254, 309, 369, 406
  activated, 197, 399, 405
  organic, 52–53, 59–60, 71–73
Carbon-14., 315
Carcinogens. *See also* Cancer, 7, 9, 38, 40–41, 45, 76, 91, 103, 151, 296, 308, 323–324
Carman-Kozeny equation, 143
Cascade impactor, 377
Cascading, 282
Catalytic combustion, 399–401, 403
Catalytic converters, 385, 406
Cataracting, 282
Catch basins, 243, 246, 248
Cation exchange capacity (CEC), 344
Cells
  activated sludge, 188, 190–193
  landfills, 265

Cellulose, 91, 271
Centipoise, 120–121
Centrifugal pump, 129
Centrifugation
　air pollution control, 389–393
　resource recovery, 282–283
　sludge, 218–219, 224–228
Checklists, 14, 16–19, 21, 31, 349
Chemical oxygen demand (COD), 72, 91, 169, 188, 266
Chernobyl nuclear reactor, accident at, 9, 322, 333–334
Chezy open channel flow equation, 158
Chlor-alkali process
　mercury, 76
Chloride, 266
Chlorinated hydrocarbons, 76, 403
Chlorine. See also Disinfection
　byproducts, 347
　wastewater processing, 86, 187
　water treatment, 2, 76, 136, 150–151
Chlorofluorocarbons (CFCs), 407
Cholera, 5, 51, 100
Chronic obstructive pulmonary disease (COPD), 34, 43–44
Cigarette smoking, 34, 39, 42–43
Civil engineering, 186, 189, 195, 198–199, 201, 205–208
Clarifiers, 175
Clean Air Act, 32, 295, 341, 345, 348, 385, 407, 412, 449
Clean Water Act, 4, 295, 346, 348
Climate change, 407, 420
Coagulation, 135–138
Coal
　combustion, 333, 338
　desulfurization, 385, 400–401
　dust, 39
　mining, 19
Coal burning power plant, 49, 351, 382, 396, 415
COD. See Chemical oxygen demand
Code of Federal Regulations, 299, 336, 380
Cohort studies, 35
Coliform bacteria, 53, 61, 100–103
Collecting sewers, 155–156
Collection efficiency. See Separation efficiency

Color in water
　measurement, 92, 135
Colorimetric measurement, 97–99, 103, 377, 379
Combined sewers, 53
Combustion, heat of, 288–289
Comminutor, 4, 172, 174, 200
Committed effective dose equivalent, 320
Common law
　air pollution, 411–413, 421
　hazardous waste, 341–342, 345
Community noise, 434, 441–443
Compactors, 255, 441
Composite sampling, 81–82
Composting, 290–292
Composting toilet, 169
Conduits, 119–120, 154–155
　circular, 159
　closed, 120–128, 159
Cone of depression, 112–114
Coning plume, 359–360
Consumers, 55, 57
Contact stabilization, 185
Control of airborne particles, 387–398
　baghouses, 393–394
　cyclones, 389–393, 398
　electrostatic precipitators, 389, 396–397, 412
　fabric filters, 389, 393–394, 398
　gravity settling, 387–388
　HEPA filters, 399
　scrubbers, 389, 394–396, 398–399
Control of gaseous pollutants, 399–404
　carbon dioxide, 404
　nitrogen oxides, 403
　odors, 403
　organic compounds, 403
　sulfur dioxide, 399–403
Control of vehicle emissions. See also Automobiles, 404–407, 418–419
Copper
　mining and smelting, 10, 35, 333
　piping, 126, 135
　water pollution, 76, 102, 135
Council of Environmental Quality, 13, 24
Counterions, 136–138

Critical mass, 328
Critical time, 65
Cropland, sedimentation loading.
  See Agriculture, sediment loading
Curie, 318–320
Cut diameter, 392
Cyclones (air pollutant devices).
  See Control of airborne particulates, cyclones
Cyclones (storms), 351–352

Darcy-Weisback head loss equation, 110, 143, 224
dB, 427
dbA, 433–434
dbN, 433–434
DDT, 39, 103
Deadheading, 256
Decay, radioactive, 313–316
Decibels, 426
Decomposers, 55, 57, 205
Decomposition. See Biodegradation
Denitrification, 197
Density of fluid, 121, 139
Deposition of air pollutants, 54, 233, 245, 333, 369–371, 385
Detention basins, 243, 246
Detergents, 53, 73, 103, 233–234, 245, 247, 401
Diaphragm pumps, 129
Diesel engines, 333, 369, 404, 406–407, 419
Diffusion
  filtration process, 141–143
Dioxin. See TCDD
Disease vectors, 247, 265, 345
Disinfection byproducts, 76, 91, 103
Disinfection. See also Chlorine, 2, 93, 102, 150–151, 230
Disintegration constant, 314, 319
Dispersion of pollutants
  atmospheric, 326, 333, 351–352, 361–371, 417
  horizontal, 352–355
  vertical, 334, 355–357
Displacement pump, 129
Dissolved organic carbon. See Carbon, organic

Dissolved oxygen (DO). See also Biochemical oxygen demand.
  deoxygenation, 62–63
  lakes, 71–73
  measurement of, 81–84
  reoxygenation, 62–63
  sag curve, 61–66, 85
  water pollution, 21, 52, 57, 68–69
Dissolved solids, 94–95, 233, 242
Diversity of species index, 68–69
Domestic sewage, 88, 155, 167–169
Dose-response relationship, 38–40, 45
Drawdown, 112–115
Drinking water
  disinfection, 2, 91, 100–101, 151
  hazardous waste, 297, 336
  quality measurement, 92
  solid waste disposal, 251
Drinking water. See also Water treatment, 10, 51, 75, 135, 235, 247, 298
Drought, 117–119
Dry weather flow, 53, 154
Dumps, 115, 259–260, 263, 265, 345
Dust, 39, 326, 360, 389, 393–394, 397, 421

Ecology, 1, 4, 5–7, 20, 303
  aquatic, 54–57, 72
Economic impact, 15–16, 24–28
Ecosystems, 5, 54–56, 75, 414
  risk assessment, 47
Eddy current devices, 286
Effective dose equivalent, 320–321
Effective Perceived Noise Level (EPNdB), 434
Effective stack height, 362, 366–369
Efficiency curve, 131, 143
Effluent standards, 171–172
Electric cars, 407
Electrostatic precipitators. See Control of airborne particulates, electrostatic precipitators
Emissions, atmospheric. See also Radiation
  incinerators, 229, 288, 298, 308, 345–346
  measurement of, 375
  regulation, 415–421
Endangered species, 303, 342

Index   475

Endpoints, assessment, 47
Energy
  energy from waste combustion, 269, 287–290
Environmental ethics, 7, 10, 29
Environmental impact assessment, 13, 23–25, 29, 324
Environmental impact statement, 13–21
Environmental impact unit, 21
Environmental Protection Agency (EPA), 10
  air quality standards, 371, 380, 382, 413–417
  carcinogens, 41
  hazardous waste, 295, 347–349
  radiation, 320
  risk analysis, 9, 23–24, 45–46
  water pollution law, 244, 343
Environmental quality index (EQI), 21
Epidemics, 2, 5, 51
Epilimnion, 70–71
Erosion. *See* Sedimentation
Escherichia coli, 100–101
Estuaries, 52, 73, 75, 233, 236
Ethics, 7–10
Eutrophic. *See also* Eutrophication, 73
Eutrophication, 72–73, 197
Evaporation, 107–108, 218–219, 235
Evapotranspiration, 108
Event tree, 37–38
Exhaust gas recirculation (EGR), 406
Exposure time, 39, 45
Exxon Valdez spill, 54, 347

Fabric filters. *See* Control of airborne particulates, fabric filters
Fall turnover, 71
Fanning plume, 359–360
Federal Aviation Administration, 441
Fertile isotopes, 330
Fertilizer
  sludge, 209, 228, 230
  water pollution, 52–53, 73, 75–76, 97, 233, 241–243, 245–247
Filter beds, 141, 143–150, 183
Filter run, 143, 148–149
Filter yield, 95, 222–223

Filtration
  head loss, 143–145
  sludge, 140
  water treatment, 2, 141–150
Financing of capital expenditures, 25–27
Finding of no significant impact (FONSI), 13–14, 17–18, 29, 31
Fish
  mercury poisoning, 76–77
  oil spills, 54
  oxygen depletion, 21, 68, 82
  sedimentation, 52
Fisherman's Contingency Fund, 346
Fission chain reaction, 328
Fission products, 329–331
Fixed solids, 95–96
Flocculation, 135–140, 180, 309
Flotation thickener, 214, 217–218
Flow weighted composite, 81–82
Flue-gas desulfurization, 401–402
Fluid bed incinerator, 228–229
Flux plot, 216
Fog, 361, 371
Food and Drug Administration, 76, 345
Food chain, 7, 39, 55, 76, 102, 209, 333–334, 342, 344
Food-to-microorganisms ratio (F/M), 184–185, 187
Formerly Used Sites Remedial Action Program (FUSRAP), 326
Free available chlorine, 150
Frequency analysis, 119, 428
Friction loss, 122–123, 131–132
Froth flotation, 286
Fuel injection engines, 406
Fumigation, 359–360

Gamma radiation, 313–318, 320, 329–330, 332, 337
Garbage, 251, 274
Garbage collection, 255–259
Garbage compactors, 255
Garbage grinder, 254, 259
Gas bubbler, 378
Gas chromatography, 103–104, 379
Gases, 68, 325, 333
  control, 208, 269, 310, 369–370, 399–407

476 ENVIRONMENTAL ENGINEERING

Gases (*continued*)
  measurement, 377–382
  stack, 289, 362, 386–387
Gaussian dispersion. *See* Air dispersion of pollutants
General obligation bonds (GO), 25–27
Germ theory, 5
Grab samples, 81–82, 381
Gravity settling tank, 140
Gravity thickener, 214–215, 217–218
Gravity water, 108
Gray, 318–320
Grit chamber, 172–173
Groundshine, 334
Groundwater, 108–109, 247
  draw down, 112–115
  hazardous waste, 297, 301, 303, 310, 349
  pollution, 53, 73–75, 97, 235, 241
  radioactive materials, 334
  solid waste disposal, 251, 265–266, 342, 344

Half life, physiological, 39
Half-life, 314–316, 319, 328–330, 332–334
Hazardous waste, 295–311
  accidents, 301
  chemical treatment, 298
  classification, 297
  clearinghouses, 302
  detoxification, 298
  groundwater, 115, 297–298
  hazardous waste exchanges, 301–303
  incinerators, 304–308
  land fills, 308–310
  law, 75, 345–350
  manifests, 299–300
  materials exchange, 302–303
  recovery alternatives, 301
  siting, 303–304
  transportation, 299–301
  volume reduction, 298–299
Haze hood, 353
Hazen-Williams formula, 122–126
Head capacity curve, 131–132
Health effects. *See also* Bacteria; Carcinogens
  air pollution, 375

  chlorine, 151
  dose-response, 40–41
  noise pollution, 437–440
  radioactivity, 321–325
  risk assessment, 34, 41–43
  thresholds, 9, 38–40, 323–325
Heat island, 353–354, 364
Heat value, 288
Heavy metals
  measurement, 102–103
  sludge, 207, 209–210, 230
  toxic substances, 76, 298
  wastewater treatment, 168, 187
HEPA filters. *See* Control of airborne particulates, HEPA filters
Hertz, 425
Heuristic routing, 256–260
Homeostasis, 6
Hydraulic elements graph, 159, 162
Hydraulic ram, 254
Hydraulics
  conduit flow, 120–128
  pumps, 128–131
  sewer, 157–164
Hydrocarbons, 54, 76, 103, 236, 244–245, 247
  automobile emissions, 233, 403–406, 418–419
Hydrogen
  acceptor, 60, 68
  donor, 57
  fuel cells, 211
  ions, 91–95
  solid waste energy recovery, 288–289
Hydrogen peroxide, 187
Hydrogen sulfide, 60, 68, 399, 401
Hydrologic cycle, 107–108, 132
Hydrolysis, 150, 299
Hygroscopic moisture, 108
Hypolimnion, 71–72

Impeller, 129–130
Incinerators
  air emissions, 396
  fluidized bed, 228–229
  gas pollutant controls, 385, 400, 403
  hazardous waste, 298, 303–308
  multiple hearth, 228–229
  radioactive waste, 332

siting, 9, 26
sludge, 228–229
solid waste, 259
Indicator organisms, 100
Industrial waste
 biological oxygen demand, 88
 clearinghouse, 302
 laws, 341
 wastewater, 153, 167
Infiltration, 107, 153–154, 167, 243, 245–248, 251, 295, 309, 310, 344
Inflow, 153–154
Intensity
 light, 92, 97–98, 103, 377, 379
 precipitation, 107, 235, 237
 sound, 425–426, 434
Interaction matrix technique, 16, 18
Intercepting sewers, 156
Interception, 141–143
Intermittent noise, 434
Inversions, 303, 356–361, 364, 371, 443
Iodometric titration, 82
Ion exchange, 99, 298–299, 331
Ionization chamber, 320–321
Ionizing radiation, 297, 313–314, 316–318, 332
 somatic and genetic effects of, 321–326
 units for measuring, 318–321
Irrigation water, 241–242

Jet pump, 129
Jigs, 286

Kinematic viscosity, 121

Lakes
 effect of pollution on. *See also* Water pollution, 4, 52–54, 70–73, 76, 197, 236, 297, 342
 oxygen, 57, 82
 temperature, 70–71
Land application, 97, 229–230
Land cover
 runoff, 237, 240
Landfarming, 299
Landfills
 closure and ultimate use of, 268–270
 design, 265–266
 gas, 266–268

hazardous waste, 308–310
operation, 266–268
subsidence, 264–265
volume reduction, 269–270
Lapse rate. *See also* Adiabatic lapse rate, Ambient lapse rate, 357–362
Latency period, 40–42
Latent cancer fatalities, 42, 44
Law
 air pollution, 411–422
 common, 341, 411–413, 421
 hazardous waste, 75, 341–350
 implementation problems, 420–421
 noise, 436–437
 property, 411–413
 solid waste, 341–350
 statutory, 411, 413–418, 421
 tort, 411–412
LC50, 40, 297
LD50, 40, 297, 308
Leachates, 251, 265–268, 309–310, 325, 334, 342, 344
Lead, 39, 266, 316–317, 331
 air pollution, 385, 405–406, 414, 417–418
 water pollution, 76, 135
Lime, 94, 198, 206, 210–211, 213–214, 222, 402
Limiting nutrient, 72
Limnology, 70
Linear energy transfer (LET), 319, 322, 324
Linear non-threshold theory, 323–325
Liquid scintillation counters, 304, 332
Litter control, 245, 251–252, 261
Loading functions
 sedimentation, 236–241
Loading, and activated sludge systems, 184–195
Looping plume, 359–360
Low Level Radioactive Waste Policy Act of 1980, 336–337
Lowest achievable emission rate (LAER), 418
LULUs (locally undesirable land uses), 9, 303

Magnetic field, 313, 316
Magnetic separation, 273, 286–287, 290

478 ENVIRONMENTAL ENGINEERING

Manholes, 4, 153–154, 156–157, 162–167
Manifest system for transport of solid waste, 299–300, 348
Manning formula, 122, 158–162
Marsh gas. *See* Methane
Masking noise, 429
Mass curve, and reservoir calculation, 116–117
Material recovery, 280–281, 298–299
Materials Recovery Facility (MRF), 273, 286
Maximum achievable control technology (MACT), 416, 418
Mean cell residence time, 190–194
Measurement of airborne particles, 375–377
Measurement of gaseous pollutants, 377–380
Measurement techniques
  air pollution, 375–384
  alkalinity, 94
  ambient air quality, 375
  biochemical oxygen demand (BOD), 84–90
  carbon monoxide in air, 379–380
  coliforms, 99–102
  color, taste, odors in water, 92
  dissolved oxygen, 82–84
  dust, 375–377, 382
  gases, 377–380
  heavy metals in water, 102–103
  ionizing radiation, 320–321
  nitrogen in water, 97–99
  noise pollution, 434–436
  organic compounds in water, 91, 103–104
  particles in air, 375–377, 382
  pathogens in water, 99–102
  pH, 92–93
  phosphorous in water, 97–99
  solids in water, 94–96
  stack gases, 381
  sulfur dioxide in air, 378–379
  turbidity of water, 92
  water quality, 81–106
Measuring radiation. *See* Measurement techniques, ionizing radiation
Membrane filtration test (MF), 101–102
Mercury, 76–77, 102–103

Mesophilic digestion, 213
Mesotrophic stage, 73
Metalimnion, 71–72
Meteorology, 351–355
Methane formers, 210–213
Methane. *See also* Natural gas, 59–60, 91
  landfills, 267–268, 310
  resource recovery, 243, 288
  sludge digestion, 210–213
  solid waste disposal, 342
Methyl mercury, 76
Milorganite, 230
Minamata disease, 76
Mining
  water pollution, 52, 54, 233, 240
Mixed flow, 129
Mixed liquor suspended solids (MLSS), 184–185, 193–195, 199
Mixed waste, 304–305
Mobile sources. *See* Control of vehicle emissions
Monod model, 189, 192
Moody's Investor Service, 26
Most probably number (MPN) test, 101
Mud valve, 140
Multi-attribute utility analysis, 23
Multiple hearth incinerator, 228–229, 307
Municipal solid waste (MSW). *See also* Solid waste, 25, 251–261, 269, 276, 278–279, 283–284, 286–292, 415

National Ambient Air Quality Standards (NAAQS), 407, 414–418
  nonattainment, 407, 415
National emission standards for hazardous air pollutants (NESHAPs), 416
National Environmental Policy Act (NEPA), 9, 13–14, 24–25, 29
National Pollutant Discharge Elimination System (NPDES), 243–244, 343
Natural gas. *See also* Methane, 59, 211, 306, 328, 400, 403, 407
Net discharge head, 130
Net positive suction head (NPSH), 130
Neutral stability, of atmosphere, 356, 358, 364, 367
Neutralization, 135–136, 298, 331, 349
New source performance standards (NSPS), 412, 415–417

Newtonian fluids, 120, 177, 208
Newton's First Law, 424
NIMBY ("not in my back yard"), 9–10
Nitrate, 58–60, 69, 90, 97–99, 197
  water pollution, 55, 70, 75, 235
Nitrification, 90, 197
Nitrites, 58–60, 70, 90, 97–99, 235
Nitrobacter, 197
Nitrogen oxides, 287
  acid rain, 418
  automobile emissions, 404–407, 415
  control of, 399, 403, 415, 418–420
Nitrogen. *See also* Nutrients
  biodegradation, 57–60, 291
  incineration, 287, 304
  measurement of, 89–90, 97–99
  wastewater treatment, 195, 197–199, 209
  water pollution, 52–53, 69, 71–73, 75, 235, 241–242, 244
Nitrosomas, 197
Nodes, 256
Noise and Number Index (NNI), 434
Noise pollution, 423–446
  aircraft, 432, 441–442
  background, 427–429
  community, 434, 441–443
  control, 441–443
  economic effects, 440–441
  health effects, 20, 437–440
  in home, 443
  industrial, 429, 437
  laws, 436–437
  levels, 430–434
  measurement, 426–436
  traffic, 434–435, 441–442
  transient, 434–436
Non-attainment of clean air standards, 417–419
Non-dispersive infrared spectrophotometry, 379–380
Non-Newtonian fluids, 120, 208
Nonpoint source pollution, 51, 77, 233–249
Nuclear fission products. *See* Fission products
Nuclear fuel cycle, 321, 325–331
Nuclear reactions, 317–318, 330

Nuclear Regulatory Commission, 9, 297, 325, 335, 346, 368
Nuclear Waste Policy Act of, 1987, 336–337
Nuisance. *See* Law, property
Nutrients. *See also* Nitrogen, Phosphorus, 52–53, 55, 57, 72–73, 97, 101, 172, 187, 196, 200, 233–235, 241–242, 244–246, 291

Occupational Safety and Health Act (OSHA), 437–438
Oceans
  ecosystem, 6–7, 55–56, 404
  effect of pollution on, 75–77
  hazardous waste dumping, 301, 346–347
  sludge disposal, 228
  solid waste disposal, 263
Odors
  air pollution, 385, 403, 421
  environmental impact, 17–18, 21–23
  landfills, 265, 270
  sludge, 175, 208, 210, 219, 230
  water treatment, 135, 172
  water, measurement of, 92
Offshore Oil Spill Pollution Fund, 346
Oil
  pollution, 9–10, 53–54, 233, 244–245, 346–347
  sulfur, 385, 400–401
  residues, measurement of, 103–104, 168
Oligotrophic state, 72
Opacity, 382
Operating point, 132–133
Organ dose, 39
Organic carbon. *See* Carbon, organic
Orthophosphates. *See* Phosphorus
Outer Continental Shelf (OCS) Lands Act, 346
Overflow rate, 178–179, 181–182, 195
Overflow. *See* Product
Ownership rights, 8
Oxidation ponds, 196, 206
Oxygen demand. *See* Biochemical oxygen demand
Oxygen demanding substances, 52, 244
Oxygen probe, 82–84
Oxygen. *See* Biochemical oxygen demand (BOD); Dissolved oxygen (DO)

Oxygenated fuel, 407
Ozone, 91, 151, 378, 407, 418–419

Packed scrubbers, 399
Parameter importance units (PIU), 21
Parshall flume, 153
Particle counter, 320
Particles. *See also* Control of airborne particulates
 measurement of radioactive particles, 313, 316–318, 320, 321–322, 329
 nonpoint pollution, 235–241, 246
 resource recovery, 275–279, 283–284, 286
 sludge, 214
 soil, 109
 solid waste, 252–253, 268
 wastewater treatment, 169, 176–182, 190, 197
 water treatment, 135–150
Pathogens, 5, 51–52, 92, 99–102, 135, 150, 208, 210, 213, 233–236, 241, 244, 247, 345
Perceived Noise Level (PNdB), 434
Percolation test, 170, 251, 266, 267
Permanent hearing threshold shift, 438
Permeability, 110–111, 115, 394
Permeameters, 110–111
Pesticides, 40, 52–54, 57, 76, 103, 210, 233–236, 241–242, 244–247, 299, 347
pH
 measurement of, 81, 92–94, 99
 sludge, 207, 210–213
Phons, 431
Phosphates, 58–60, 73, 97, 99, 333
Phosphorus. *See also* Nutrients, 57–60, 71–72
 measurement of, 97–99
 removal, 53, 195–196, 200, 206
 sediment erosion, 236, 247
 sludge, 209, 212
 wastewater, 52, 169, 172
 water pollution, 72–73, 75–76
Photographic film
 radiation, 320–321
Photometer, 98
Photosynthesis, 55, 57–59, 70, 396
Phreatic surface. *See* Water table
Physiological half-life, 39

Phytoplankton, 71
Phytotoxicity, 297
Piping. *See also* Conduits
 friction loss, 122–123, 131–132
 hydraulics of, 119–132
Plasticity. *See* Rheological characteristics
Plumes. *See* Stack gases
PM10
 particulates, 375, 414
Pneumatic pipes, and garbage collection, 254–255
Point source pollution, 52, 77
Polishing pond, 196
Pollution rose, 354–356
Polychlorinated biphenyls (PCBs), 266, 344, 416
Polychlorinated dibenzodioxins (PCDDs), 308
Polychlorinated dibenzofurans (PCDFs), 308
Polynary separation, 280
Population dose equivalent, 320
Population equivalent, 52–53
Porosity of soil, 109–110, 144–147
Positive crankcase ventilation (PCV), 405
Power curve, 131
Precedents, 412–413
Precipitation, 107–108, 235–237, 243, 245, 247, 267, 351, 368, 370–371
President's Council on Environmental Quality (CEQ), 13, 24
Pressure filter, 218–220
Pressure head of water, 110, 121
Pressure sewers, 151–152
Prevailing lapse rate. *See* Ambient lapse rate
Prevention of significant deterioration (PSD), in air quality, 417–418
Price-Anderson Act, 346
Primary clarifier, 175, 179, 181–185, 206–208
Probability. *See also* Risk analysis, 35–38
Process loading factor, 192
Producers, 55, 57
Product
 material recovery, 280–281, 283–287
 risk, 37–38, 41, 57
Propeller pump, 129
Property law, 411–413

Index   481

Public health, 4–5, 251–252
Pulp and paper mills, 65, 196–197
Pumps, 128–134
　characteristic curves, 131
　operation head, 132
　sewage systems, 156–164
　system head curve, 131–132
Pyrolysis, 269–270

Quality factor, 319–320
Quicklime. *See* Lime

Rad, 318–320
Radial flow, 112, 115, 129
Radiation
　background, 321
　decay, 313–316
　health effects, 321–325
　measurement, 318–321
Radiation inversion, 358
Radiation. *See also* Ionizing radiation, 313–321
Radioactive waste, 313–339
　by-product, 326
　contaminative cycle, 326–331
　health effects, 321–325
　high-level (HLW), 20, 31, 326, 335–336
　low-level (LLW), 326, 336–337
　management, 334–337
　measurement units, 318–321
　nuclear power, 404
　reprocessing, 331
　sources, 325–332
　transportation, 337
　waste management, 20, 23–24, 304
Radioactivity-to-mass conversion, 319
Radioisotopes. *See also* Radionuclides, 332, 334
Radiological half-life, 315, 319, 328–330, 332–334
Radionuclides, 315–316, 318–320, 325–326, 329–333
Radionuclides
　movement of, through environment, 333–334
Radium, 313, 319, 323, 326
Radon, 320–321, 332–333, 338
Rain energy
　erosion, 235–237

Rain gauge, 107
Rainout of pollutants, 370
Raw primary sludge, 205
Raw sludge, 175, 182, 209, 219
Reasonable Available Control Technologies (RACT), 419
Reciprocating pump, 129
Recycling, 191, 273–293, 386
Refuse, 115, 246, 251–262, 263–270, 273–292, 341, 343
Regulations. *See* Law
Reject, 280, 284
Relative biological effectiveness (RBE). *See* Quality factor
Relative risk, 42
Relaxation length, 317
Rem, 318–320, 325
Reservoirs, 2, 116–121, 132
Resource extraction, 18, 29
Resource Conservation and Recovery Act (RCRA), 260, 295, 297, 303, 326, 332, 342, 348–350
Resource recovery, 253, 260, 273–293, 295, 298, 310
Respirable particles, 377
Respiratory system, 44
Return activated sludge, 184, 187, 199
Reuse, 273–293
Revenue bonds, 25–26
Reverse osmosis, 197, 309
Revised Universal Soil Loss Equation (RUSLE). *See also* Universal Soil Loss Equation, 240–241
Reynolds number, 145, 147, 177
Rheological characteristics, 208–209
Risk analysis, 23, 33–47
Risk assessment, 33–49
　ecosystem risk assessment, 47
Risk factors, 33–35, 41
Risk perception, 46
Rosin-Rammler distribution function, 276, 279
Rotary pump, 129
Rotodynamic pump, 129
Roughness coefficient, 158
Route optimization, 255
Rubbish, 251
Runoff process, 233

Salmonella typhosa, 100
Salt, and radioactive waste disposal, 336
Sand beds, 218–219
Sand filters, 140–146, 246, 248, 394
Sand, removal of, from wastewater, 172
Sanitary landfills. *See* Landfills
Sanitary sewers, 4, 153–154, 167
Saturation, zone of, 108–109
Screens, 4, 172, 282–283
Sedimentation tanks. *See* Settling tanks
Sedimentation, 141–143, 174, 245, 247, 309
Sediments. *See also* Suspended Sediments; Turbidity
  erosion, 8, 16, 235–245
Seeding, 86, 244
Separate sewerage, 4
Separation efficiency, 388
Separation factor, 391
Separation of waste materials, 274–280, 282–286, 298
  hammermills, 275
  shredding, 268, 275–280
Septic tanks, 169–171, 233, 243
Settling tanks, 137, 140–141, 173–180, 184, 191, 195
Settling velocity, 176–177, 179, 216, 284, 369–370, 397
Sewage systems. *See* Wastewater
Shredding, 268, 275–280
Sievert, 318–320
Sigma equation, 225
Significant deterioration, prevention of, 412–413, 417–418
Silt. *See also* Sediments, 243–244, 246
  water treatment, 135
Silviculture, 230, 345
Siting
  hazardous waste facilities, 10, 25, 303–305, 308
Siting
  landfills, 264–265
Size reduction. *See* Shredding
Slope factors
  runoff, 236, 239
Sludge, 205–232
  as fertilizer, 209, 228, 230
  characteristics, 207–210
  dewatering, 218–228

  sources, 205–207
  stabilization, 210–213
  thickening, 214–218
  ultimate disposal, 228–230
  viscosity, 208–209
  yield stress, 208–209
Sludge volume index (SVI), 186, 195, 214–216
Smog, 287, 361
Smoke, 34, 39, 306, 330, 361, 377, 382
Soil characteristics
  runoff, 235–241
Soil Conservation Service, 237
Soil erodibility factor, 224–226
Solid waste, 251–263
  air classifiers, 283–287
  characteristics, 252–254
  collection, 254–259
  composting, 290–291
  disposal, 259–261, 263–271
  energy recovery, 286–290
  law, 341–350
  leachate, 266
  moisture concentration, 253–254
  municipal, 252–259
  plastics, 251, 253, 261, 284–285
  resource recovery, 280–281
  screens, 281–283
  separation process, 282–286
  volume reduction, 269–270
Solid waste Disposal Act of 1965, 342
Sones, 434
Sound, 423–426
  frequency, 423–426
  level, 430–434
  level meter, 431–433
  pressure, 427–430
  wavelength, 423–426
Special nuclear material, 331
Specific gravity of liquid, 81, 120
Specific resistance to filtration test, 219–223
Specific weight, 120–121, 130
Specific yield, 109–110
Speech Interference Level (SIL), 434
Spray tower, 387, 394–395, 398–399
Stack gases
  atmospheric dispersion, 362, 367
Stack height. *See* Effective stack height

Index

Stack sampling, 375, 381–382
Standard mortality ratio (SMR), 42, 44–45
Standing, legal, 413
Static discharge head, 130
Static head, 121, 130–132
Static suction head, 129–130
Statutory law
  air pollution, 413–414, 421
  hazardous waste, 341, 345, 350
Step aeration, 185–186
Stokes equation, 177, 369
Storm sewers, 3, 153
Stormwater runoff, 244–245
Straining, 141–143
Streams. *See also* Water pollution
  BOD, 84–90
  effect of pollution on, 60–70, 76, 81–82, 233
Subadiabatic conditions, 356–358, 360, 364, 367
Subatomic particles, 313, 318
Subsidence inversion, 358
Suction lift, 130
Sulfur, 58–60, 287–288, 304
Sulfur dioxide
  control of, 355–356, 385, 399–403, 417–420
  measurement, 378–379
Sulfuric acid
  air pollution, 379, 382, 401–402
  mining, 54
Superadiabatic conditions, 356–357, 360, 364, 367
Superfund Act, 75, 347–348, 350
Surcharges, 168, 347
Surface sink absorption, 370
Surface water
  reservoir capacity, 115–119
Suspended solids
  environmental assessment, 22–23
  measurement, 94–96
  sludge, 206–207, 214–216
  wastewater treatment, 172–182
  water pollution, 52, 75, 233, 241, 244, 247
  water treatment, 135–150
Synergism, 39
System head curves, 131–132

Tapered aeration, 185–186
TCDD (2,3,7,8-tetrachlorodibenzo-*p*-dioxin), 308
TCM (Tetrachloromercurate), 379
Temperature inversion, 356, 364
Temporary hearing threshold shift, 437–440
Thermal stratification, 73
Thermocline, 71
Thermoluminescent detectors, 321
Thermophilic digestion, 213
Thickeners, 195, 216–218
Three Mile Island, 9
Threshold limit value (TLV), 38–39
Thresholds of health effects, 38–39
Tile field, 169–170
Time
  incineration of hazardous waste, 306
  vs. dosage, 39
Tip, 263
Tipping fees, 252, 268
Topographic factor, 237
Torrey Canyon disaster, 63
Tort law, 411–412
Total body burden, 39
Total organic carbon. *See also* Carbon, organic, 91, 103, 188
Total suspended solids, 95, 212, 375
Toxic substances
  heavy metals, 76, 209, 246
Toxic Substances Control Act, 349
Traffic Noise Index (TNI), 434
Transfer stations, and solid waste, 255
Transient noise, 434–436
Transpiration, 107–108, 267
Transportation pollution, 15, 24, 31
Transuranic waste (TRU), 325, 331, 336–337
Trash, 251, 259, 308
Trespass, 413
Trichlorophenol herbicides, 308
Trickling filter, 183, 206–207, 218
Tritium, 315, 318, 329–333
Trivalent cations, 91, 136–138
Trommel screens, 282–283
Trophic levels, 55–56
Troposphere, 351
Turbidimeters, 92

Turbidity of water. *See also* Suspended sediments, 68, 92, 135, 148
Turbulence, 140, 173, 306, 369
Tympanic membrane, 437–438
Typhoid, 100
Typhus, 251

U.S. Coast Guard, 346
U.S. Congress, 4, 14, 341–342, 413, 417
U.S. Department of Agriculture, 240
U.S. Department of Commerce, 346
U.S. Department of Engergy, 16, 332, 335, 368
U.S. Department of Interior, 343
U.S. Department of Transportation, 297, 299, 346, 443
U.S. Geological Survey, 335
U.S. Public Health Service, 170
Ultimate oxygen demand, 62–64, 90
Universal Soil Loss Equation (USLE). *See also* Revised Universal Soil Loss Equation, 236, 240
Uranium, 313, 322, 325–328, 330–335
  mining waste, 325–326
User charges, 25, 27–28, 304

Vacuum filter, 95, 218
van der Waals force, 136
Vehicle emissions. *See* Control of vehicle emissions
Velocity head of water, 121
Venturi scrubbers, 394
Vinyl chloride, 41, 309
Viruses, 52, 101, 143
Viscosity of fluid, 120–121
V-notch weir, 174
Volatile organic compounds control, 403
Volatile solids, 95, 208, 210, 212–213

Washout of pollutants, 370
Waste activated sludge, 184, 191, 199, 205, 210
Waste exchanges, 302
Waste Isolation Pilot Project, 336

Waste stabilization ponds, 299
Wastewater, 2
  central treatment, 171–200
  characteristics, 167–169
  collection of, 153–166
  onsite treatment, 169–171
  primary treatment, 172–182
  quantities, estimation of, 153–154
  secondary treatment, 182–195
  tertiary treatment, 195–200
  treatment costs, 25–28
Water closet, 153, 170
Water horsepower (WHP), 130
Water pollution
  effects of, on groundwater, 73–75
  effects of, on lakes, 70–73
  effects of, on oceans, 75–76
  effects of, on streams, 60–70
  sources, 51–54
  water quality measurement, 81–106
Water quality. *See also* Water pollution, 2, 10, 22, 53–54, 64, 73, 81–106, 243–248, 342–345, 346–347
Water supply, 2
  sources, 107–119
  transmission, 119–132
Water table, 109
Water treatment, 135–152
Weber-Fechner Law, 427
Wells, drawdown of, 112–115
Wet collectors (scrubbers), 394
White noise, 429
Wind motion, 353–354, 362
Wind rose, 353–355
Wind speed, 108, 354–356, 366–367
Windrows, 290–291
Wood pulping waste, 91

Yellowcake, 326
Yield, 109, 188–189, 192, 206, 209, 222–223

Zeta potential, 136
Zinc, 135, 266, 275
Zoning, 295, 341
Zooplankton, 71–72